Visions

Readings for a Changing World

Myron C. Tuman
University of Alabama

Allyn and Bacon
Boston
London
Toronto
Sydney
Tokyo
Singapore

Vice President, Humanities: Joseph Opiela
Editorial Assistant: Mary Beth Varney
Executive Marketing Manager: Lisa Kimball
Editorial-Production Service: Omegatype Typography, Inc.
Manufacturing Buyer: Suzanne Lareau
Cover Administrator: Linda Knowles
Text Design: Carol Somberg/Omegatype Typography, Inc.
Electronic Composition: Omegatype Typography, Inc.

A Pearson Education Company
160 Gould Street
Needham Heights, MA 02494

Internet: www.abacon.com

Between the time Website information is gathered and published, some sites may have closed. Also, the transcription of URLs can result in typographical errors. The publisher would appreciate notification where these occur so that they may be corrected in subsequent editions.

Library of Congress Cataloging-in-Publication Data
Tuman, Myron C.
Visions: readings for a changing world / Myron C. Tuman.
p. cm.
Includes bibliographical references and index.
ISBN 0-205-29122-8 (alk. paper)
1. Readers—Technological innovations. 2. English language—Rhetoric Problems, exercises, etc. 3. Technological innovations Problems, exercises, etc. 4. Information technology Problems, exercises, etc. 5. Report writing Problems, exercises, etc. 6. Readers—Information technology. 7. College readers. I. Title.
PE1127.T37T84 2000
808'.0427—dc21 99-33112
CIP

Credits appear on pages 508–510, which constitute an extension of the copyright page.

Printed in the United States of America
10 9 8 7 6 5 4 3 2 1 04 03 02 01 00 99

The radio, and the telephone, and the movies that we know
May just be passing fancies, and in time may go.
—George and Ira Gershwin, 1938

For George and Ira, and Sarah and Ella—
for showing us that some things really are here to stay.

CONTENTS

part 5 At Work 327

13 Working More 329

14 Working Less 350

15 Women, Work, and Technology 372

PREFACE

"We must find topics for our writing courses," writes Robert Scholes in *The Rise and Fall of English,* "that enable students to focus on their culture at the points where it most clearly impinges upon them, where they already have tacit knowledge that needs only to be cultivated to become more explicit" (p. 102). *Visions* attempts to be a college reader providing just such topics—a college reader for students who find themselves in the midst of a new information age. While there are many other thematic readers that raise large questions about human beings and their place in the world, as well as an increasing number of more specialized thematic readers that ask students to read and write about computers and the Internet, *Visions* is the first reader that brings a broad humanistic focus (what some might call a "non-techie" focus) to important questions relating to technology.

The readings are organized in six parts, each of which asks students to consider important questions not just about how technology is changing all our lives for better or worse, but also about what it is we really want from technology—or, stated differently, what is our vision for the future? Why is it, for instance, that at some moments most of us seem attracted to simple, elegant solutions to problems (and hence seek the kinds of things that technology seems so good at providing), whereas at other times we seem more interested in creating obstacles for ourselves: in climbing a mountain instead of driving around it, in trying to catch fish with a lure instead of bait? What vision do we have for our own lives? What problems emerge when our lives become too simple or too complex, too easy or too difficult? And what role does technology play in helping us make the right choices?

This is a reader for teachers who want their students to think deeply, not only about the future and about the best uses of technology but also about the puzzling relationship between technology and human nature—and about how the interplay of work and play remains a crucial factor in our leading happy, productive lives and building solid communities. The one issue throughout most of the readings and Explorations in *Visions* is the conflicting, often competing roles that simple and complex technologies play in our lives.

The book consists of eighteen chapters grouped into six parts. The three chapters that make up the first part (Objects of Desire) deal with everyday objects (basically tools, toys, and gadgets); the three chapters that form the last part (Future Thoughts) deal most broadly and most directly with issues related to technology and the future. In between, there are clusters of three chapters under parts entitled At Home, At Play, At School, and At Work. Although technology is a common thread in the eighteen chapters, the intent throughout is to offer an extremely wide range of readings and, hence, discussion and writing topics—on the different ways (high tech versus traditional) that we make new friends; decide where we live; and choose where we shop, what we eat, how we entertain ourselves, how we keep in shape, and how we earn a living, to list just some of the most basic issues in our lives shaped in part by technology.

The readings themselves are clustered on the assumption that the best discussions will involve student responses not to individual readings but to the series of broad topics raised directly or indirectly in each chapter's set of four or five readings. In the case of the first chapter (Simple Things), for example, class discussion and writing would normally follow the students' reading all five of the selections: Thoreau's praise for his simple homemade furniture; George Carlin's monologue about his stuff; Henry Petroski's tribute to the lowly pencil; Eric Brende's analysis of Amish technology; and Alan Thein Durning's list of seven ultraefficient, albeit thoroughly unglamorous, tools.

What broadens the assignments even more is that each group of readings is followed by a set of six Explorations related to the readings. Each Exploration is itself based on a new but related topic and contains *(a)* a group activity, *(b)* Web work, or links for further research, and *(c)* a suggested writing assignment. Embedded in this reader, therefore, are 108 directed assignments that take students out from the cluster of readings in many exciting, unexpected ways.

Just what students discuss and write about will vary greatly from class to class. Each individual reading has a suggested reading prompt that can be the basis of a writing assignment, just as each of the eighteen chapters has an initial set of questions for all the readings. Yet each of the eighteen chapters concludes with Explorations that, while thematically related to the readings, can take the class as a whole or students working individually or in small groups in countless directions as they respond to the suggested writing topics or to other writing assignments generated within the class.

Following the five readings in Chapter 1 listed above, for example, are six related Explorations, on (1) hammers and other low-tech tools, (2) what gives some objects "glamour," (3) packaging, (4) bicycles, (5) clutter, and (6) organizers. When discussing simple tools, for instance, students are directed to the Best Buys page at Ace Hardware's Web site; when discussing clutter, they are asked to visit the Dallas-based Container Store on the Web; when discussing organizers, the Dayrunner Corporation. The goal here is to use the Web as a quick and easy resource through which students can do field research; this is comparable to asking students to visit a local hardware or office supply store when brainstorming for their papers. Links to all the sites mentioned in the text, plus some more up-to-date links, are included in the Web site that accompanies this book (www.abacon.com/tuman).

While *Visions* can be used strictly as a traditional reader, teachers with limited experience using online resources with their students will find the

Web work section that accompanies each Exploration an engaging, non-threatening way to extend the boundaries of their classes. The dual goal in having students visit the suggested Web sites is to raise students' interest in the topic and to enrich their base of knowledge—all in preparation for a writing assignment that will ask them to reflect on their own vision of the future and to ponder just what kind of simple and complex structures they want in their own lives. The Web work sections of each Exploration, therefore, are part of the larger aim of the book to help students become more familiar with, and, hence, less in awe of, many aspects of the new information age. The Web is a strange and wonderful place where students can do research on Wal-Mart, for instance, in part by visiting the Wal-Mart corporate site—likely not a place for an objective assessment of the company and yet a site rich with information for anyone trying to understand problems confronting small towns and small businesses. What students learn on the Web, in other words, is designed to complement, not to replace, the reading and writing that has long formed the core of college writing classes.

All of us—the most critical of students and most critical of teachers alike—are already awash in expertly produced advertising designed to flatter and cajole us: to play effectively to our longings and insecurities even when we know better, even when we are intellectually aware of just how we are being manipulated. As skillful flatterers know, we can all be seduced by our desires. "I can resist anything but temptation," quipped Oscar Wilde—and perhaps the defining experience of modern life is the dangerous, alluring interplay of desire and technology. That interplay forms the backdrop of *Visions:* the way our craving for new solutions—for progress—is sometimes a rationalization for new pleasures.

One of the goals of this text is to help students attain a critical awareness of this rationalization process. Such awareness can be difficult, given that our attachments to technology are often much deeper, much harder to pin down than we may care to admit. Analysis has its limits, in other words, even for the most intelligent among us. This is why the pedagogy at the core of this book constantly asks students for personal reflection as well as research and analysis. True understanding of how we as individuals fit into the complex interplay of new technology and ordinary experience (what might be called *deep* or *authentic understanding* of the interplay of technology and desire) is likely to come only when students move back and forth between detailed narrative accounts of their own experience and the ideas of others as presented in the readings, class discussions, and Web work. The critical moment for students using this book will come in their writing, when they test for themselves what they encounter in readings here and on the Web (for example, about the value of technology in the schools) against their own real-life experiences. Central to *Visions* is the premise that students cannot achieve critical understanding of the Web or of the world at large without thinking and writing deeply about their own lives.

THE *VISIONS* WEB SITE

References to Web materials appear in two recurring places in the text—in the headnote for each of the readings and in the Web work section of each of the 108 Explorations. This printed text ordinarily contains a brief reference to these Web materials, but more complete references as well as the actual links to these places (as well as links to sites mentioned in the readings) can be found at the accompanying Web site: www.abacon.com/tuman.

ACKNOWLEDGMENTS

I can still remember a fellow teacher raving about a student of his who had won an art contest with a drawing she had executed by looking only at the model, not the picture. Why bother? was my first response then. How wonderful! my second. Here some twenty years ago was the core idea of this book: just how, when, and why we shun the easy way and instead cultivate difficulty. How can we explain the perseverance of low technology in a high-tech world?

Now, all these years later, I still owe a real debt of gratitude to Tim Trapolin and the entire faculty at the McGehee School for helping to teach me that much of the joy of teaching and of learning comes from doing things the hard way.

I also owe a different debt to my father, Walter Tuman, who, at age 86, provided serendipitous help with the title of this book, when in a phone conversation of March 1999, he referred to his own interests and current readings in new, visionary technologies. Once again, thanks, Dad.

At a more immediate level, I have benefited from the meticulous attention of Genie Davis in the preparation of the manuscript, from the general editorial assistance of Kathryn Tuman, and from the computer expertise of Genie and Warren Eckstein in the creation of the accompanying Web site. I would also like to thank the following helpful readers of the manuscript: Joe Opiela at Allyn and Bacon; Nick Carbone, Marlboro College; John Clark, Bowling Green State University; Ray Dumont, University of Massachusetts, Amherst; Douglas Eyman, Cape Fear Community College; Paul Heilker, Virginia Tech; Will Hochman, University of Southern Colorado; Joseph Janangelo, Loyola University; Bill Lalicker, West Chester University; James McDonald, University of Southwestern Louisiana; Clyde Moneyhun, Youngstown State; David Roberts, Samford University; Peter Sands, University of Wisconsin, Milwaukee; and Margaret Syverson, University of Texas at Austin.

introduction

Envisioning the Future

The great American writer Henry David Thoreau (a person of intense desires and strong opinions about technology and progress) referred to most human inventions as "improved means to an unimproved end." This textbook seeks to encourage students and their teachers to look closely at what things are changing in our world and what things are staying more or less the same—and, in particular, to consider the often competing effects of increasing simplicity and complexity in our lives. What kinds of changes really do make our lives better (lead to progress), and in what areas of our lives are we content with older and possibly simpler processes? And finally, can we use the readings and explorations offered here to separate what is truly important (improving our goals or ends) from the constant distraction and hype related to new, "improved" ways of doing things?

The goal of *Visions,* then, is to get people to think about the interplay between technology and four other key concepts that continue to reshape our world: progress, community, desire, and sustainability.

TECHNOLOGY

How does technology make my life better—easier, safer, more convenient, even simpler? Does it work much the same for other people of my generation? For people generally? And how does technology make my life (and the lives of other people) worse—more expensive, less convenient, more complex? What new gadgets do I absolutely depend on each day—and how did others ever live without them? What new gadgets am I convinced I could live happily without the rest of my life? What would the world be like if everyone had some wonderful new technology? Or if suddenly some seemingly indispensable technology such as electricity disappeared, and no one had it?

PROGRESS

Will our lives be better than those of our parents? Will our children's lives be better than ours? And their children's lives better than theirs? And just what do we mean by *better:* more material prosperity (more cars, TVs), more leisure time (to do what?), better health, more stable families and communities? Is life today getting simpler or more complex, better or worse, for me personally? For my friends and family? For people around me? And if things are not getting better, then why not? What happened; when did things all of a sudden start getting worse?

COMMUNITY

What activities do you enjoy doing with others rather than alone? At what times and in what places do you willingly get together with your friends, with your family, or with people you really don't know very well—and for what reasons? How does technology help you in these common efforts; for example, in planning them and in notifying and then bringing the other people together? How does technology add difficulties to these activities—separating people, giving them less in common? What changes in the world (new technologies) might make coming together easier?

DESIRE

Under Technology, above, we asked, What new gadgets am I convinced I could live happily without the rest of my life? Here we ask, What new gadgets am I convinced I could live happily without the rest of my life—and why, if I know better, do I still want them? We tend to think of technology as the mainstay of the rational part of our world: technology as the sum of the processes we use to make the world a safe and orderly place. Yet how much of the better world we want to build for ourselves individually or for the society collectively is merely an expression (a rationalization) of other, deeper, not well expressed desires we may have? At the core of any discussion of progress, community, and technology has to be a heightened awareness of the crucial role that our desires—perhaps a bit more safely defined as our dreams—play in determining the future we finally want for ourselves and for others. When futurists and Luddites clash (as they do in Chapter 16), they may seem to be discussing technology and progress on the surface; but invariably their real text is the dreams each side has for a better future—what each desires personally, although usually expressed in terms of the common good.

The interplay of technology and desire—all our various reasons for wanting to build a better world—goes to the heart of this anthology: namely, the concept that all of us constantly measure our personal needs against global concerns and, vice versa, evaluate the soundness of global solutions in terms of our personal experiences of the world. All of us are faced with personal problems relating to how we will earn our own living and help support our family and friends. Yet, as we begin the twenty-first century, it is hard to ignore comparable global problems: questions about how other people are going to earn their living and support their families and friends. Just two countries in

Asia—China and India—have more than a third of the world's population. So far neither country has embarked fully on the technological expansion and the comparable belief in the value of "progress" that has defined life in the United States and Western Europe for the past two centuries, but both of these countries, and maybe Asia generally, seem poised for some great leap forward into modern consumerism in the new millennium. What sort of world awaits us all if huge numbers of people in China and India abandon their traditional, often agricultural lifestyles for our fast-changing, high-tech way of life? And can we expect Chinese and Indians to do anything else but become Western-style consumers of the latest products if we are unwilling to reconsider altering how we ourselves live? Key questions we should ask are just what we mean by *progress* and *community* and what *technology* truly supports a better life—that is, what is it we truly *desire*. What is our vision for the future?

SUSTAINABILITY

Related to these questions about progress is the notion of *sustainability*, the idea that it is possible (and, who knows, it may not be) for us in the United States to have a high standard of living that leaves enough resources for the other people of the world, including our children and their children. Although *Visions* has this serious subject at its core, the text's goal is not to preach conservation or anything else but to provoke and to inform in equal amounts, mainly by bringing together readings that as a group raise questions more than they provide clear and simple answers. Indeed, one problem with sustainability addressed in the opening unit and throughout the text is that, for all its high moral standing, the "better" world sustainability offers us often seems a plainer, less attractive, hence less desirable one. Will we have to give up, or at least redefine, pleasure in order to produce a sustainable world?

As we discuss the many issues related to technology in a changing time, it will soon become clear that, at least for most of us who are interested both in improving our own lives and in helping the world at large, there are no simple solutions or clear-cut answers. Nor are there any omniscient experts to tell us just what to do. The issues are really too large for any one field, much less one person, to control. This book, then, presents series of clustered readings on topics related to the complex (good and bad) ways that technology continues to affect our lives.

Visions is not just *about* technology; it also engages its readers directly with one revolutionary technology of our age: the World Wide Web. In asking students to expand all their discussions by considering material relevant to their topic that is available on the Web, this text tries to exemplify some of the best aspects of a technological change that promises, or threatens, to transform education itself. That this is still a book—and not purely a Web site—reflects an equal belief in the value of tradition.

Writing about Change

Is a new Wal-Mart outside of a small town a sign of progress or a sign of decay, something to celebrate or lament? Do gadgets serve primarily to frustrate us, or do some of them actually make our lives more efficient? Will students read and write better or worse because of computers, or just differently? Is it fair to consider race in awarding college scholarships? Should we allow all-male or all-female schools? Should executions be televised? Ideas, opinions, insights, and put-downs fly everywhere, seeming to invite us to join in. "Minority scholarships are an outrage, an attack on the notion of merit," we hear on one side. "Minority scholarships are just a continuation of colleges' long-standing policy of helping students in need," we hear on the other. With little effort we can find ourselves drawn into a debate, quickly taking sides and, in many cases, feeling our blood pressure rise as we listen to arguments we oppose.

It is not that difficult to generate intense classroom debate. What do you think? What do your classmates think? Why are you right, and they wrong? Presumably any writing class will provide a suitable format for giving your opinions and listening and responding to the opinions of your classmates. The readings or Web materials in any writing class can hardly fail to generate some heated feeling and discussion, and such debate is certainly an important step in producing powerful writing. But it is just that—a "step" to powerful writing, and not the writing itself. The connection between opinions and writing, moreover, is not nearly as simple and clear as we might like. Strong opinions may be an important first step in the genesis of a piece of writing, but a string of opinions ("I think this, that, and this other . . .") is rarely more than that—two or three sentences thrust into a conversation, often interrupting someone else and in turn to be interrupted by someone else's comments.

Debate and argument, regardless of how heated, normally produce sentences that are largely unplanned and strung together: You say what you want, usually in a sentence or two, and then someone else has his or her say. Writing, especially in college courses, involves an entirely new dimension of difficulty. Here what you say (or "write") has to be both planned and sustained. Instead of commenting and responding orally in units of two or three sentences, you will likely be required by your instructor to comment and respond in writing in larger units (called papers) of at least two or three typed pages—perhaps a minimum of twenty to thirty sentences, a unit ten times longer than your oral response. Try organizing your writing as you do your oral response—your most important point first (before someone interrupts you) followed perhaps by a sentence or two of explanation—and what happens? You have exhausted much of what you have to say on the topic, usually in 50 words or less, a fraction of the 500-word minimum requirement of even short college essays. When you write a paper it is as if, instead of taking part in a lively class discussion, you are all of a sudden alone in front of an audience, a group of people who are gathered to listen to you but who do not know much about you and (unlike your friends or classmates) are not predisposed to like you. "What are you doing standing before us?" this audience seems to ask. "What is the point of this discourse, this five-minute speech, this short paper you are bothering us with?"

In such a situation all the fire and spontaneity of a debate is likely to flow completely out of the speaker. Often, when placed in such a cool, even hostile situation, the inexperienced writer feels like mumbling, "Nothing—I really don't have anything to say; you see it was my teacher [or "my parents" or "my future employer"] who made me get up here, made me make this speech, write this paper, take this course." While we all may feel sympathy with this person's plight, the task still remains: to give him or her the skills necessary to generate and organize writing that has within itself the power to command and keep the attention of an audience of reluctant readers. "Pay attention!" your writing must say. "Here is something you need to consider!"

And here the fire of classroom debates may not be especially helpful. Here what we need is something a bit more old-fashioned: technique. We need specific skills in organizing, developing, and presenting a formal essay. We need to know how to begin an essay, then how to organize it, develop its main points, and finally conclude it.

UNIFYING AN ESSAY: THE THESIS

The first key in writing an organized essay is to recognize a fundamental difference between writing and class discussion. In class discussion the topic—the subject of inquiry—is publicly determined, often by the teacher or, in this class, by the Web experiences themselves. All you have to do is add your opinion. But in writing—even in the same class in which you have the class discussions—the burden of establishing the topic is entirely the writer's responsibility. As the writer, you must specify exactly what it is you are talking about; and you must do it for a general audience, one that has not necessarily taken part in class discussions or explored the Web with you. Indeed, establishing your topic—part of the task of defining your thesis—is one of your main jobs in your introduction.

A thesis statement, the governing idea of a piece of writing, is the clearest signal that what you are producing, unlike random conversation, does have a clear and specific purpose. As a practical matter, you may find it helpful to think of a thesis as consisting of two parts: your topic (executions, for instance) and the assertion of an attitude about that topic ("executions should be televised"). The key to a thesis is this attitude about a subject, this assertion of a belief on the writer's part that readers or listeners would not likely accept or understand completely without further explanation. A thesis, in other words, can be thought of as the statement of an opinion about a specific topic, an opinion that you plan on developing and supporting over the length of the essay.

Strong theses involve controversial assertions; for example, a statement that women should never be allowed to serve in military combat. Weak theses make less controversial assertions; for example, that women are playing an increasingly important role in the military. Strong theses have the advantage of commanding attention and quickly establishing your sense of purpose: "Women have no place in combat, and let me tell you why." Such strong theses help keep your focus and your readers' attention on one central concern. Yet strong theses at times also have the disadvantage of encouraging oversimplification, reducing complex problems to black-or-white options—an especially troublesome aspect if you are able to see merit on both sides of an issue. The lesson here is that one should not feel compelled always to use strong thesis statements. Don't be afraid to use a weak thesis if it more accurately reflects your position.

THESIS AND PURPOSE

Many writing teachers will insist that you define your thesis early in the writing process so as to force you to think about and openly confront the question of your purpose and thus to ensure unity. Some instructors may even require you to state your thesis in a single sentence and to place it in a specific location, usually the last sentence of the first paragraph. There is a danger, however, of mistaking a readily produced thesis ("high schools should be able to distribute condoms," for example) for a genuine interest in the material. Just as not clearly stating a thesis may lead to disorganized writing, overemphasizing it can lead to unified writing that has not very much to say; for example, an essay with three simple reasons why schools should have the right to distribute condoms, each reason repeated two or three times. The ideal approach to this dilemma would seem to be to combine the best of both approaches: to formulate a thesis early on in the writing process but then to revise it on a regular basis as you get a clearer sense of your actual purpose, your real interest in the essay you are writing.

The thesis is a formal statement of your attitude or opinion about your topic and is often formulated in response to a teacher's assignment. A purpose, in contrast, has to do with why you are interested in that topic in the first place; it is the real reason you are expressing your opinion, and as such it really cannot be made up. Your purpose in writing an essay for school can be, and unfortunately too often is, to please the teacher, to do the work required to get a grade—but in such a situation there is obviously little psychological connection between what you are doing (articulating and supporting an opinion) and why you are doing it (following someone else's instructions). As a rule we seldom do things well without a high level of motivation. When we really are stating our opinions we have a deep personal stake in what and how we are communicating—after all, it is our belief that is on the line. When we are following someone else's directions, however, we usually have other things on our minds—things such as doing as little as possible, avoiding mistakes, pleasing someone else, and the like. We may try, but without true conviction something is often lacking, just as our facial expression may lack a certain something when out of politeness we eat a dish we don't really like. Professional actors spend hundreds of hours studying techniques (such as "method" acting) that enable them to become more convincing in playing parts precisely by generating a real inner purpose for the lines they have to read.

What then are students to do when asked by teachers to write papers for which they have no motives? Excellent teachers will, of course, expend considerable energy on presenting Web and other materials to arouse your genuine interest in topics. Excellent teachers also offer students a range of topics and suggested approaches for their responses. But, as will be obvious to anyone who has ever taught or taken a writing course, there are no surefire solutions provided entirely by the teacher or the materials. The only surefire solution is the student's own determination to find a personal connection to the material, a way of uncovering a genuine interest in the material at hand.

The discussion and writing suggestions that accompany each round of readings in this text are organized with this precise point in mind—to help you make this personal connection. Finally, however, you must push yourself to link the public issues of Web and classroom debates with your own personal experiences.

THE INSIDE–OUTSIDE APPROACH

One technique for generating a genuine purpose and thus for linking the public and private worlds is to use the inside–outside approach.

Above we talked about the thesis as containing the one subject of your essay—what your paper is about. Often, however, college essays, even short ones, are really about two things, not one: the "outside" material that explains why you are interested in a subject, and the "inside" material that develops your subject. This inside material includes your thesis statement and what is traditionally referred to as the body of your essay—the series of paragraphs that systematically develop your thesis. Depending on the specific instructions you receive from your teacher for different papers you may be writing, this inside material can focus either on the public issues discussed in the readings or personal issues that you draw from your own experiences. (In this book the suggestions for writing at the end of each chapter ask you to focus on both public and personal matters.)

The outside material, on the other hand, is a short explanation or presentation to readers that shows why you are interested in this topic in the first place and, by extension, why they also should be interested. We call this "outside" material because it is normally placed at the outer edges of your paper: in your first paragraph before your statement of thesis (where it thus forms your introduction) and in your last paragraph after you restate your thesis (where it forms your conclusion). This outside or framing material, directly connected to your purpose and forming the introduction and conclusion, can often make up as much as a third of the total length of a short paper, and thus can rightly be considered a second, related subject. Furthermore, as with inside material, outside material can be drawn from either public or personal experience, usually according to the following simple formula: public inside, personal outside; personal inside, public outside.

Beginning writers can use this natural inside–outside tension of college essays to connect the public issues of the text, Web material, and other writings with their own private experiences. Thus, whenever they want to concentrate on a public issue (e.g., Web censorship, the spread of casino gambling) in the center or inside of their paper, they should draw upon their own experience in the outside framework, talking about something they have personally experienced in the introduction and then again in the conclusion. Conversely, when they want to focus on their own experiences, they might use material from the public debate in their introduction and conclusion. The goal here is both technical—to help students achieve a sense of craft to their writing with carefully worked-out introductions and conclusions—and ideological—to help students see the many connections (and perhaps at times contradictions) between their own practical experiences and the ideas of experts.

The following sample student essay utilizes this inside–outside technique, placing brief references to public material in the first and last paragraphs and extended personal material in the body of the paper.

History in the Un-making

Americana these days seems to be pretty equally divided into corporate-sponsored homogeneity and the unique kind of mom-and-pop shops that used to line the roads. Today, as corporate America pounds us with the message that McDonald's is the "normal" place to eat and Nikes are the "right" shoes to wear, it's hard not to fall for all the hype and start to feel a need to trudge in line with everyone else. In spite of our strong desire for originality, we're still sensitive to those who say we're simply weird and not with it. Some comfort may be drawn from the rapidly expanding World Wide Web, which brings into our homes images of Roadside America (www.roadsideamerica.com) and other off-the-main-road unique attractions. The proliferation of sites like these makes us feel more comfortable, less alone, in our passion for the one-of-a-kind experience, the kind of story each person has to tell. Absorbing the life of a real person, not just the corporate-fed, corporate-dressed, enriches our own lives like no other experience.

About three miles off U.S. Highway 82, near Gordo, Alabama, this kind of enrichment permeates the atmosphere. Be forewarned. This is not the Smithsonian. There's no shiny Spirit of St. Louis here, no immaculately glass-encased Dolly Madison inaugural ball gown, no perfectly taxidermied woolly mammoth towering over the rotunda. Instead your fascination will be piqued by the cracked porcelain-faced Pretty Patty with the dirty lacy dress that probably used to be pink, or maybe red; by the occasional glimmer of mostly dusty shards of emerald and amber glass, no longer recognizable as 19th-century cough syrup bottles; and, most hauntingly of all, by the mangy-coated brown Hereford bull, now buckled at the knees. This is Ma'Cille's Museum.

Around Pickens County, the natives call this local landmark "Marcellie's," although it's hard to figure out why. Not only does the hand-painted sign clearly spell out Ma'Cille's name, but everyone knows the story of how Lucille House began collecting and "preserving" the cows, horses, and family dogs around the farm. The locals know the proper name, but, over the years, the modified version, bearing the mark of out-in-the-sticks Alabama slang, has been adopted as their personal name for the beloved museum; it shows their affection, their close kinship, their ownership. Marcellie's has become a part of the identities of all the residents of the county now, nearly a hundred years after its birth.

As a young girl, early in the century, Lucille tried to—but just couldn't—quell her intense interest in preserving animals. She was a girl, and she wasn't supposed to be interested in the pigs and chickens, in the squirrels and deer, beyond the best recipe for making the meat tender and tasty. On the sly, Lucille began visiting Old Man Elmore's taxidermy in town whenever she could get away, at first just silently observing his careful work, later asking infinite questions, which he laughingly answered. When she began to experiment on the dog that died of old age and then the squirrel she shot in the woods, her family intensified their objections. She was actually pretty good at taxidermy, they had to admit, but she should concentrate on dolls if she wanted to collect something.

So she did. Her doll collection soon filled up her taxidermy workshop, and she asked her father for more space. The old barn wasn't being used for anything except storing old tractors and hay-raking discs that they should get rid of anyway, so he agreed to her expansion. The barn inspired her to display her animal trophies in realistic positions, turning her creative energy back to her original obsession, taxidermy. Within a few years, cows, mules, and horses filled the hay-lined stalls, squirrels scampered motionless along the rafters, dogs and cats lay sleeping in the sun that streamed in the windows. Lucille placed in the sunlight the colorful old bottles she found during her safaris around the farm, illuminating the barn with a festive kaleidoscope of light. The mesmerizing carnival effect impressed her family and neighbors. Soon people were coming from all over the county to see her collections of dolls, bottles, and, of course, animals. In the 1940s, Lucille began charging admission and made enough money to support herself and her young sons. Ma'Cille's Museum by then encompassed three buildings: the original taxidermy shop, where she displayed her dolls; the old barn, filled with animals; and a long, low building, erected just for the purpose of housing all the leftover bottles and memorabilia she collected.

The museum doesn't open very often now. In the 1970s, Ma'Cille, as the old woman was then affectionately called because everyone claimed her as aunt or grandmother, began leaving the doors locked during the week, only opening up for weekend tourists, whose numbers were dwindling. She and her son Glenn continued well into the 1980s to hold their annual barbecue and open house, but when Ma'Cille died in 1986, motivation to prop up the crumbling animals and wash the dolls' faces seemed to die too. Glenn Jr. still lives on the homeplace and welcomes the opportunity to lead a tour through his grandmother's lifelong collection, but today all the windows are dirty, almost opaque. The dusty, fading dolls stare forward lifelessly, and the decimated cows just want to lie down and rest. The smell of history surrounds you as you wander through the dark buildings—and haunts you.

Compared to Ma'Cille's Museum, the sterile, perfectly preserved exhibits of the Smithsonian contain little history. Sure, you'll find unique and famous artifacts there, but you won't feel anyone's life's work entering your pores, taking possession of your soul. Ma'Cille's fascination with cows and squirrels would today be called eccentric by polite people, loony by others. Her work was certainly considered out of the mainstream, even by her country contemporaries, and in these times of corporate definition of the norm, we, the consumers, risk being labeled wacky just for our interest; but it's worth the risk.

Ma'Cille's Museum could have been preserved in its original illuminated, kaleidoscopic glory if some corporation had bought it. They would probably have moved all three buildings 25 miles east to allow easy access from the interstate where they could proudly display their Exxon emblem as owner and preserver of history, or they might have built new concrete and steel housing. But the dusty history of Ma'Cille would have been lost—along with part of the community's basic identity. To get the full effect of the haunting journey, it's necessary to take the turn off the interstate, then off the U.S. highway, then off the state road, and onto the potholed county road that leads to the family farm. That's the only place to experience Ma'Cille's life.

NOTE: This description is reproduced from fading memories of a tour 12 years ago. Many facts and dates are fictionalized.

As this student paper illustrates, it is not enough for students using the inside–outside approach to write two separate essays—a public discussion of a global problem (how small towns generally are changing, losing much of what made each one distinctive) and a second, private account of a special place in the author's own experience. Both essays may be quite acceptable in many situations, but the key to the inside–outside approach is to combine the two in a single essay: to introduce and conclude the formal argument with reference to one's own life, or, conversely and as this student has done here, to introduce and conclude the recounting of one's own life with references to a relevant public argument.

In systematically working to redefine and sharpen their own sense of purpose—what they really have invested at a personal level in a variety of widely circulated public debates—students will have the chance to experience the true power of writing as reflection. They will also discover that the most important, the most compelling texts in the course—like the most important voices in society—are not necessarily those included in this reader (as good and as interesting as many of them are), but may be their own essays. Public debates today, in these readings and in society generally, are often loud and strident, at times seeming to be the personal property of professionals on both sides of issues. One response is to shout back and talk only about issues; another is to withdraw and talk only about ourselves. The better solution, and the one advocated here, is to combine the two: to talk about the connections between issues and our own lives. It is in our own stories—the writings of ourselves and our classmates, the public tales of individuals—that we are fully capable of expressing our connection to the world: not in our thesis, but in our purpose.

GETTING STARTED II

An Introduction to the World Wide Web

Essay writing is a decidely low-tech craft, normally taking hours of careful, concentrated effort. Surfing the World Wide Web, on the other hand, is about as high-tech as one can get, and is often done for fun, without plan or purpose. Now let's see if we can start the process of connecting the two.

BRIEF HISTORY

The World Wide Web began with the efforts of researchers at major universities in the 1960s to share their work by using phone lines to connect the big mainframe computers on their campuses. The resulting network of mainframe computers came to be called the Internet—the Net, for short—and consisted of what can be described as a series of standards (called protocols, or IPs for Internet protocols) for allowing people to share information while working on many different kinds of computers scattered throughout the world. One of these early sets of protocols that has had tremendous impact allows for the sharing of electronic messages (or email).

But as good as the Internet was, and continues to be, for sharing messages directly between individuals or between those who had subscribed to a mailing list in order to receive all postings to that list, there was no comparably easy way to share information with strangers—specifically, to put information on your computer (text files, pictures, or software) that you wanted others to be able first to locate and then to access on their own (without having first to contact you for permission). In other words, what was missing in the original Internet was a way to use this global computer network as a vast publishing system, a way to post something on one computer so that it could be easily found and viewed by anyone, anywhere.

In the mid-1990s, Internet users in Europe developed a revolutionary and spectacularly easy way to overcome this weakness, as well as to exploit the capacity of new personal computers to display color pictures and eventually sounds and movies as well. At the heart of this technique (maybe the greatest discovery since sliced bread) is *hypertext linking,* which works by using a new kind of software program (called a *browser*) to retrieve and display ordinary text and pictures (written according to a new standard, sort of like email, called *hypertext markup language,* or *html*). What one sees through a browser is easily understood text or pictures; however, if the text is highlighted or if the cursor changes shape when over a picture, there is a corresponding link associated with that material. This then is the only command structure one needs to learn in order to navigate around the entire World Wide Web, jumping from site to site all over the world. If something looks interesting and is highlighted (or, if it is a picture, if the cursor changes shape when placed over it), then all you need to do to learn more is to click on it. The browser then goes out on the Net to the specific location indicated (the location itself having the woeful acronym *URL,* for "Uniform Resource Locator") and retrieves and displays this new page. The technical process of clicking-and-jumping from place to place was given the uninviting name of hypertext transfer protocol (or http), while the whole system of globally linking texts and images on the Internet was given the more inviting name of the World Wide Web, or WWW, or just the Web. Any questions?

THE WEB AS AN INFORMATION SOURCE

Historically most writers, students and professionals alike, have relied largely upon library resources in documenting their topics but have paid little attention to one key function of the library that is no longer performed by the Web: the filtering out of unreliable and highly partisan information. For example, over the years logging companies have prepared volumes of materials about their environmental efforts; but although these companies have often tried to give these materials to K–12 schools as classroom resources for units on forest conservation, for the most part such materials never made it into research libraries. Similarly, practically every industry and every special interest or political cause has a professional association designed both to lobby Congress and, through its publications, to affect public opinion. In the past, one had to write these companies or organizations to get their publications—librarians regularly excluded much of this material from limited shelf space. Today, however, all of this highly partisan, one-sided material is immediately available via the Web. Meanwhile, more judicious journalistic or scholarly assessments are often not available on the Web but appear in books and magazines that depend for their existence on sales through traditional outlets.

Remember: What is easiest to find on the Web may be highly partisan material, put there by an advocate. Conversely, the most thoughtful analysis of any problem may not even be available on the Web, or may be available to students only via a library service that offers restricted access to online versions of some, certainly not all, magazines.

Another word of caution: Organizations that publish information on key issues via their Web sites are not always open in telling you who they are and who is paying for their site. If there were a Cheeseburger.Com on the Web, a site designed to tell you everything you wanted to know about this wonderfully unhealthy sandwich, it might well be sponsored by an independent-sounding front organization called something like Americans for Good Eating, even if this organization received all its funding from the cattle and dairy industries. Traditional library sources may help you identify biases in Web sites, but often the only reliable resource is your own knowledge of the issue itself: a knowledge, for example, that nutritionists, cattle ranchers and dairy farmers, fast-food operators, and vegetarians are all liable to have different assumptions and different things to say about our eating habits.

The Web is truly a wonderful and a strange place. If you are taking a creative writing class, your short stories and those of your classmates can be globally published on the Web via the click of a switch and thus made available for students all over the world to read and respond to—while the short stories of the most praised or most widely sold contemporary writers (such as Stephen King, for example) are for the most part available only through traditional booksellers or libraries. Surprised?

Figure 1 presents a three-step process for evaluating material you find on the Web. For information on citing Web resources in your writing, see the Appendix: Guidelines for Using Web Resources in Your Writing.

FINDING IT ON THE WEB

One incredible feature of the Web is the power and ease of search engines for locating information—maybe too much ease, if you find too much information. There are now Web-crawling programs that systematically visit sites on the Web to compile a huge index of all key terms in every single Web document—an electronic version of having a vast army of underpaid workers meticulously reading through and indexing every published page in the world. AltaVista is one of the most powerful of these search services, containing (as of September 1998) a huge index to more than 125 million different Web pages. AltaVista is amazingly fast, maybe because it was started by a computer manufacturer in order to promote the sale of its fast Web servers.

As fast and as comprehensive as AltaVista may be, it can also be a little dumb: It tries to find everything related to your search item. Entering *gadgets,* for instance, produces some 30,000 hits, or nearly 3,000 pages of sites, listed 10 sites per page. Searching under *wacky* produces even more, some 50,000 hits; and combining the two (searching for *wacky gadgets*) still produces 31,000. Although AltaVista may be trying to create a more hierarchical interface through what it calls "Content Zones," it remains most helpful when you are looking for something very specific—the proverbial needle in the haystack—such as pages related to a particular person; say our Chapter 1

A Three-Step Process for Evaluating Web Resources

Step 1: Investigating the Author

Does the entry you are using have an author? If so, what can you find out about that person at the site, and elsewhere on the Web? (Place the person's name in quotation marks when using a search engine.) Check the card catalog of your campus library, or any periodical database your library may provide, for that person.

Step 2: Investigating the Text

Identify the range of positions one might take on a given position and attempt to locate your source within that range. In general, what other groups take a similar position? An opposing one? Are there other texts at this same site on this same topic? If so, do they seem to reflect a similar attitude? Can you tell when this text was written? Is the text carefully edited for writing as well as factual content?

Step 3: Investigating the Site

What can you learn about the party responsible for providing the content of the Web site (as opposed to the party responsible for designing and maintaining the site)? How does the content provider identify itself and its interests? What else is on this particular site or on other sites hosted by this group that would indicate its feelings about particular issues? How recently has the Web site been updated? What interest does this party seem to have that leads it to undertake the cost and effort to maintain this Web site? Do the links offered by this site tell us anything about its interests? Do a Web and a library search of this party to learn more.

Figure 1

author Henry Petroski. Typing *"Henry Petroski"* (with the quotation marks) forces AltaVista to look only for pages with both names and produces a large, barely manageable number of hits (514 as of September 1998), with the "best" hits (as computed by software developed by AltaVista) listed first.

With research related to this course and with college research generally, you are more likely to need to work in the opposite direction: starting not with a specific name but with a broad topic, such as immigration reform. For such a search, a directory service like Yahoo! will often be a better jumping-off point.

To understand how a directory service like Yahoo! works, think of a great transit station, maybe a cross between a hub airport and huge subway terminal, located in the middle of nowhere. People from all over the world start out each morning in places A, B, and C and want to get to destinations X, Y, and Z; yet at one point or another they all wind up, at least for a limited time, at Q, the transit hub. Most people arrive with a single goal in mind: to find their way inside the huge terminal to the gate or track that will take them directly to their final (or next) destination. In other words, they arrive knowing where they are going and leave again to complete their journey as quickly as possible. Other people, though, may just be out for the fun of it—especially because travel on this system is free. They arrive at the transit station and spend hours right there, dining in its restaurants, shopping at it stores, maybe making short trips to other locations before quickly returning to the station.

Yahoo! can be seen as the largest virtual transit station on the World Wide Web—the Web's Grand Central Station. People normally go to Yahoo! to get directions to another, final destination: the information they are searching for on the Web. Yet Yahoo!, like other Web portal services, is a for-profit company that hopes to make money (just as malls try to do) by encouraging you to linger there as long as possible, and hence by providing its own set of tempting attractions. Specifically, Yahoo!, along with the handful of other Web portal services with which it is in fierce competition, hopes to make money both by selling you products themselves (such as a subscription to its own magazine, *Yahoo! Internet Life*) and, more importantly, by selling ad space on its screens to companies whose services are only a mouse-click away.

Yahoo! and its competitors are thus in a battle to offer users an ever increasing array of services—although, as often happens with mass-market competition, with each vendor trying to match any new feature offered by any of its competitors. The three basic services Yahoo! and its competitors offer are as follows:

1. Comprehensiveness: The ability to search everything on the Web and sometimes everything that is readily accessible in the older discussion groups (called Usenet groups) that flourished on the Internet a decade or more before the emergence of the Web after 1994. A comprehensive transit service is one that provides a user with a direct transfer to any spot on the Web.
2. Speed: The ability to allow thousands of individuals to search this massive amount of information quickly to find exactly what they are looking for (by means of excellent searching software, and with the latest algorithms), all using the latest and fastest computers serviced by the fastest and most expensive telecommunications lines.
3. Guidance: The help offered by the search service to those people who don't know precisely where they are going, as when a traveler at the transit station knows only that she wants to get to Ohio or, just as common, a traveler wants advice on finding a quiet cabin in the mountains.

Historically, pure search engines like AltaVista have emphasized comprehensiveness; in contrast, directory structures like Yahoo! and now its recent competitor, E-Blast (from the *Encyclopaedia Britannica*) have stressed guidance—creating a kind of electronic department store or place where people can move through the aisles or floors on their own with some reasonable assurance of being able to find what they are looking for and, as with shopping in such stores, with the added advantage of seeing lots of fun and possibly interesting things along the way.

Researching Wacky Gadgets through Yahoo!

Yahoo!, unlike AltaVista, lists its own categories (if available) at the top of any search. Although there are, surprisingly, no Yahoo! categories for *gadgets* (a key topic in Chapter 3), there are three categories under *novelties,* including Business and Economy > Companies > Novelties. Clicking on Novelties leads to a page of thirteen related categories (see Figure 2) plus a listing of numerous retailers and other collectors of odd products. Clicking on Wacky Products brings one to a list of such useless things as a blenderphone ("another combinational appliance, like a clock radio, except less useful") and a remote finder (makes finding your lost TV remote control "as easy as clapping your hands").

The numbers after each category indicates the number of sites under that new category; the @ symbol indicates that this category is crosslisted under another category.

Entering *novelties* in a Yahoo! search, then, leads one to an organized path of offbeat items on the Web. And with Web searching often all one needs is one good lead.

Searching the Web for In-Depth Material

Now let's look at a real school project: trying to find interesting, relevant material on a potentially large topic such as U.S. immigration policy. What one wants from a Web search of so broad a topic is not the track number of every train on the information railroad heading to an immigration-related destination (AltaVista, a search service that provides such information, lists nearly a quarter of a million such sites) but directions to one or more staging areas where various experts discuss and otherwise make recommendations regarding immigration sites. The key to using Yahoo! to access Web information is learning how to find the way to such staging areas, which Yahoo! refers to as *categories.*

Home > Business and Economy > Companies > Novelties

- 3D Glasses@
- Apparel *(8)*
- Cheeseheads@
- Computer and Internet *(27)*
- Glow-in-the-Dark *(11)*
- Magnets *(18)*
- Novelty Music *(8)*
- Political *(24)*
- Prank Phone Calls *(5)*
- Rubber Band Guns *(4)*
- Tasteless *(25)*
- Teeth *(6)*
- Wacky Products *(57)*

Figure 2

Like any good computer product, Yahoo! does much of the work for us. Hence, at a basic level, it conveniently organizes the world of the Web into fourteen major categories, all of which fit on a single screen: Arts and Humanities / Business and Economy / Computers and Internet / Education / Entertainment / Government / Health / News and Media / Recreation and Sports / Reference / Regional / Science / Social Science / Society and Culture. But Yahoo!'s real power resides in the manner in which it moves directly and transparently to these high-level and many more lower-level categories in response to any search request we make—that is, without any of the complex numbering sequences of a typical library catalog.

When we search for *immigration,* Yahoo! returns, not a list of the 200,000 possible Web sites out there, but only *twenty-three matching categories,* along with nearly 600 end-sites classified under Immigration by Yahoo! along with the Yahoo! categories where those end-sites are contained. Furthermore, it attempts to match those twenty-three categories on the basis of relevance, starting with business and government sites and then moving to sites dealing with regional issues.

Immigration: Track 1

About halfway down this list is one that looks particularly promising for students working on a research assignment in an English class, who may be interested in personal narratives and colorful details. Working downward from the top category Arts and Humanities, it is under Humanities, then under History, then under U.S. History, and then finally under Immigration; or, as listed compactly in Yahoo!, as Arts > Humanities > History > U.S. History > Immigration.

Clicking on this link brings us to three interesting-looking sites, all still in Yahoo!:

- Ellis Island@
- Ethnic Mosaic of the Quad Cities—this exhibit of photographs and texts celebrates the different immigrant groups that form the heritage of these midwestern towns.
- Migration to America—discover how immigration and diversity have defined our nation in the past and present.

If we do not select Ellis Island, then the other two links are to end-sites outside of Yahoo!. The first is to a fascinating study of immigration in the Quad Cities of Rock Island and Moline, Illinois, and Davenport and Bettendorf, Iowa. The second (as of April 18, 1997) is indicative of an ongoing problem with all Web transit systems: It is a dead link, likely because the site has changed location without putting in a forwarding pointer or has folded altogether. At some point this particular link will get removed, but there is as yet no design feature in the Web itself that automates the never ending task of keeping links current.

The @ after Ellis Island indicates that it is a Yahoo! category appearing in multiple places within the hierarchy, and that clicking on it will take one to what Yahoo! considers to be its primary location (see Figure 3).

In the case of Ellis Island this primary location appears five levels below the top Immigration category, Regional > U.S. States > New York > Cities > New York City > Ellis Island.

The second Ellis Island listing is to www.ellisisland.org—an excellent site run by the Statue of Liberty–Ellis Island Foundation with links to the Ellis Island Museum; the American Family Immigration History Center; and, best of all for research purposes, another transit area—that is, another set of immigration-related links, although here offered by the Statue of Liberty–Ellis Island Foundation. These links provide access to separate Web sites devoted to two different television documentaries on Ellis Island, one by the History Channel site and one by the i-channel. The i-channel, among other links, has one to the *Ellis Island Cookbook,* from which five wonderful recipes are given by children of the immigrants along with profiles (some spoken) of their relatives. The recipes are for Croatian nut roll, potatoes and dandelions, cabbage rolls, lemon pudding with gingersnaps, and beet and carrot salad. (There's also information for ordering the entire cookbook for $16.95 plus shipping and handling.)

The hotlinks at Statue of Liberty–Ellis Island Foundation also contain a link to reproductions of stereoscopic photographs of Ellis Island that were widely circulated in the early twentieth century. These photos are part of the Keystone–Mast Collection at the California Museum of Photography (Riverside).

Home > Arts > Humanities > History > U.S. History > Immigration > Ellis Island

- American Immigrant Wall of Honor - search on-line to see if your family name is already inscribed at Ellis Island.
- Ellis Island - explore this fascinating immigration museum.
- Ellis Island [fieldtrip.com]
- Ellis Island Station - history of immigrant processing, photos, and information for genealogical researchers.
- Ellis Island Virtual Tour - follow students from Queensbury Middle School as they learn about immigration during an Ellis Island reenactment. A site designed for teachers and students to study immigration.
- Ellis Island, New Jersey - contains information on the recent Supreme Court battle over the island, as well as historical and genealogical research information about the immigration station, 1892-1954.
- Ellis Island: Sacrifice of a Shrine - analysis and critique of plans for Ellis Island.
- Immigration: 1900-1920 - photographs documenting the immigration process taken by publishers of stereoscopic photographs.

Figure 3

Immigration: Track 2

Our original Yahoo! search of *immigration* produced twenty-three categories; and three spots above the link to Arts > Humanities > History > U.S. History > Immigration is another interesting link for student purposes—to Immigration Reform, which appears under Government > U.S. Government > Politics > Political Issues along with twenty-two other issues, from abortion and affirmative action to the war on drugs and welfare reform (in other words, a veritable table of contents for student research papers). The Immigration Reform link leads to nine rich end-sites, as shown in Figure 4.

While there is a wealth of information in each of these nine sites, including links to other sites, the link to reprints of articles on immigration from the *Atlantic Monthly* magazine provides a gold mine of information and points of view, including James Fallows's provocative mock–presidential memo, "Immigration: Lock and Load."

In addition to this mock action memo, there are links to a 1997 article on English-only laws, two 1996 articles on the impact of immigration on low-wage earners, and a half dozen earlier articles on different aspects of the problem. There is also a fascinating section reprinting much older *Atlantic Monthly*

Home > Government > U.S. Government > Politics > Political Issues > Immigration Reform

- Citizens and Immigrants for Equal Justice - group of immigrants and family fighting harsh new immigration laws regarding deportation. They feel deportation of longtime legal permanent residents is inhumane.
- Federation for American Immigration Reform - F.A.I.R.
- Immigration - articles from The Atlantic Monthly's archive and related links.
- Immigration Forum
- Norman Matloff's Immigration Forum - analyses of the issue and components thereof by this University of California at Davis professor.
- NumbersUSA - supporting immigration reform. Discusses the effects of legal immigration on the US population.
- Stop Immigration Reform - A reference to inform the reader of the real facts surrounding immigration and attempts at immigration reform.
- U.S. Border Control - supports legislation to stop illegal immigration and drug smuggling; restore national security and develop fair and effective immigration policy.
- United States: Locked Away - report on immigration detainees in jails in the United States from Human Rights Watch.

Figure 4

pieces on immigration. Among these are two 1906 essays, one advocating restrictions on immigration by southern Europeans (to protect what is best in American life from "vast throngs of ignorant and brutalized peasantry") and the other protesting policies discriminatory to Chinese immigrants.

With a little forethought and planning, as well as an eye for good leads, the Web can be a wonderful source to expand practically any topic in a college writing course. Meanwhile, all the topics in *Visions,* beginning with those in Part One, were selected because they offer especially rich opportunities for Web exploration.

part I

Objects of Desire

The first three chapters explore our attachment to, and occasional frustration with, objects—especially the tools, toys, and gadgets that we keep around us to work with or to play with, or some combination of the two. The first chapter (Simple Things) deals with the attraction of low-tech, unglamorous things such as tools that we keep, and often prize, because they work so well. The chapter opens with a startling contrast between two oddly kindred spirits: the great nineteenth-century thinker Henry David Thoreau (who sings the praises of traveling light) and the contemporary stand-up comedian George Carlin (who muses on our collective inability to follow Thoreau's advice, suggesting in his comic way that maybe we would be better off if we could).

The second chapter (Toys, Simple and Not) deals with our attachment to objects that by definition are not immediately connected to the world of work and that hence, at least at some level, we do not absolutely need. A common thread here (perhaps most apparent in the last reading, "Dip into the Future") is what our toys tell us about ourselves. The third of these chapters (Gadgets and the Flight from Simplicity) deals with objects that do some of the work of tools while often having some of the playful elements of toys. Gadgets that are all-work are really clever tools, yet without some element of fun associated with these items we might not refer to them as *gadgets*.

One central question running throughout these first three chapters, and an important aspect of this text generally, is the notion of *sufficiency*—having enough, but no more than enough, of anything to get a specific job done or, more philosophically, to lead a good life. The key notion here is that one great advantage of a simple life is that it may leave enough resources for others (including future generations) also to live well. Hence the important juxtaposition between the chapter on simple things and the two chapters on toys and gadgets: By definition toys and gadgets are objects that we do not absolutely need—except, that is, for the pleasure they provide us. Our continual fascination with toys and gadgets, in other words, may point to the possibility that excess or at least uselessness is a basic component of pleasure and thus at odds with the notion of sufficiency.

This struggle with pleasure, and with excess feeling generally, is at the core of many of the readings and topics in these chapters. In Exploration 1b, for instance, you will be asked to consider something that may seem a little odd: What makes a tool or any other objects *glamorous*? And for those who say that tools cannot be glamorous, we take them to the Web sites of two companies built entirely on marketing "glamorous tools": Brookstone and The Sharper Image. In Exploration 1e, you will confront one of simplicity's archenemies, clutter, and will be asked to consider why most of us surrender and just live with it; and in 1f, we will consider one of the chief contemporary tools for dealing with the clutter in our lives related to scheduling: organizers.

One possible unexpected Exploration in Chapter 2 entails a look at sweets as indisputable examples of the key emotional role that nonessential things play in our lives. Chapter 3 ends with an Exploration that touches on a serious theme running throughout all three chapters in this opening part, although perhaps most visible in "Rubes in Training" and Lloyd Sieden's piece on Buckminster Fuller: namely, the connection between madness and genius and the assumption that both (like gadgets in general) involve a level of excess far beyond what humanity may need just to get by. The interplay of tools, toys, and gadgets in these three chapters opens up insights into the complexity of human needs and hence into the complex demands we place on technology in our lives—not just to provide for our physical needs (increasingly, in environmentally nonintrusive ways) but to amuse and, perhaps, to inspire us as well.

A long-running TV commercial for the Isuzu Trooper in the late 1990s certainly brought out this connection. In the commercial a father is pushing a shopping cart, with his young, acquisitive son in it, through the crowded aisles of a toy store. The boy is constantly grabbing at the toys and the father is repeatedly telling him no; that is, until Dad comes across a store display of a life-sized Trooper wrapped up as a model toy. "Oooooo," he moans, reaching out to the package as if to put it in his check-out basket.

It is easy to condemn the excess at work in this ad—the sanctioning of the urge to overspend and hence to waste, especially on what is portrayed as an overblown toy. Toys to some extent, and gadgets even more, do seem at odds with a fundamental moral and ecological imperative of our age to conserve, to seek sufficiency and no more. Yet without some element of excess, there would be few toys, fewer gadgets, and perhaps no geniuses. Is it not possible that what some would call waste, others might call exuberance or, more simply, joy?

c h a p t e r 1

Simple Things

READING QUESTIONS

- What pleasures do we get from our possessions? How do our possessions create problems for us?
- What are the joys of things' working well, working simply; of things' being in order? The frustrations of things' not working well, being out of order?
- What are the limits, why are we sometimes bored, with things' working well, working simply, being in order?

QUOTATIONS

Machines are worshipped because they are beautiful, and valued because they confer power; they are hated because they are hideous, and loathed because they impose slavery.

— *Bertrand Russell (1928), British philosopher and social critic, 1872–1970*

If the automobile had followed the same development cycle as the computer, a Rolls-Royce would today cost $100, get a million miles per gallon, and explode once a year, killing everyone inside.

— *Robert X. Cringely, computer columnist*

When the only tool you own is a hammer, every problem begins to resemble a nail.

— *Abraham Maslow, American psychologist, 1908–1970*

Man is a tool-using animal. . . . Without tools he is nothing, with tools he is all.

— *Thomas Carlyle, British social critic, 1795–1881*

KISS—Keep it simple, stupid!

— *Motto at Lockheed's advanced deployment facility*

My Furniture *Henry David Thoreau*

Henry David Thoreau (1818–1862) remains an influential writer and thinker nearly 140 years after his death, in no small measure because of his celebration (most vividly in his classic *Walden*) of living simply, closely, and honestly with nature. Nor does Thoreau seem any less relevant today as he speaks out against the materialism that confuses possessions with progress and, as in the following passage, on the value of making do, of simplifying one's life by existing with a sufficiency, not a maximum, of goods.

The following selection is from the opening section of *Walden*, entitled "Economy."

At the *Visions* Web Site Links to more on Thoreau's life and times; to furniture manufacturers and retailers.

READING PROMPT

This is a short passage, a single paragraph; but, as is customary with Thoreau, it speaks on many levels, all tied together by the different meanings Thoreau assigns to furniture. What is Thoreau saying here specifically about furniture and more generally about life?

My furniture, part of which I made myself—and the rest cost me nothing of which I have not rendered an account—consisted of a bed, a table, a desk, three chairs, a looking-glass three inches in diameter, a pair of tongs and andirons, a kettle, a skillet, and a frying-pan, a dipper, a wash-bowl, two knives and forks, three plates, one cup, one spoon, a jug for oil, a jug for molasses, and a japanned lamp. None is so poor that he need sit on a pumpkin. That is shiftlessness. There is a plenty of such chairs as I like best in the village garrets to be had for taking them away. Furniture! Thank God, I can sit and I can stand without the aid of a furniture warehouse. What man but a philosopher would not be ashamed to see his furniture packed in a cart and going up country exposed to the light of heaven and the eyes of men, a beggarly account of empty boxes? That is Spaulding's furniture. I could never tell from inspecting such a load whether it belonged to a so-called rich man or a poor one; the owner always seemed poverty-stricken. Indeed, the more you have of such things the poorer you are. Each load looks as if it contained the contents of a dozen shanties; and if one shanty is poor, this is a dozen times as poor. Pray, for what do we *move* ever but to get rid of our furniture, our *exuviae;* at last to go from this world to another newly furnished, and leave this to be burned? It is the same as if all these traps were buckled to a man's belt, and he could not move over the rough country where our lines are cast without dragging them—dragging his trap. He was a lucky fox that left his tail in the trap. The muskrat will gnaw his third leg off to be free. No wonder man has lost his elasticity. How often he is at a dead set! "Sir, if I may be so bold, what do you mean by a dead set?" If you are a seer, whenever you meet a man you will see all that he owns, ay, and much that he pretends to disown, behind him, even to his kitchen furniture and all the trumpery which he saves and will not burn, and he will appear to be harnessed to it and making what headway he can. I think that the man is at a dead set who has got through a knot-hole or gateway where his sledge load of furniture cannot follow him. I cannot but feel compassion when I hear some trig, compact-looking man, seemingly free, all girded and ready, speak of his "furniture," as whether it is insured or not. "But what shall I do with my furniture?" My gay butterfly is entangled in a spider's web then. Even those who seem for a long while not to have any, if you inquire more narrowly you will find have some stored in somebody's barn. I look upon England today as an old gentleman who is traveling with a great deal of baggage, trumpery which has accumulated from long housekeeping, which he has not the courage to burn; great trunk, little trunk, bandbox, and bundle. Throw away the first three at least. It would surpass the powers of a well man nowadays to take up his bed and walk, and I should certainly advise a sick one to lay down his bed and run. When I have met an immigrant tottering under a bundle which contained his all—looking like an enormous wen which had grown out of the nape of his neck—I have pitied him, not because that was his all, but because he had all that to carry. If I have got to drag my trap, I will take care that it be a light one and do not nip me in a vital part. But perchance it would be wisest never to put one's paw into it.

A Place for Your Stuff

George Carlin

Comedian George Carlin (born 1937) has long delighted audiences by skewering the inconsistencies that seem to be at the heart of everyday life. Perhaps his most celebrated routine, first performed in 1972, ridicules the seven dirty words then not allowed on network television. Here, in an almost equally famous routine, he picks up on Thoreau's critique of American materialism. Carlin suggests, however, that while our attachment to possessions may not be logical or, as Thoreau might argue, healthy, it is certainly heartfelt.

At the *Visions* Web Site Links to the George Carlin fan page; to California Closets and other storage retailers.

READING PROMPT

Carlin's attitude about our possessions is in many ways as complex and maybe even as profound as Thoreau's. What attitudes do Carlin and Thoreau share? How do they differ?

Hi! How are ya? You got your stuff with you? I'll bet you do. Guys have stuff in their pockets; women have stuff in their purses. Of course, some women have pockets, and some guys have purses. That's okay. There's all different ways of carryin' your stuff.

Then there's all the stuff you have in your car. You got stuff in the trunk. Lotta different stuff: spare tire, jack, tools, old blanket, extra pair of sneakers. Just in case you wind up barefoot on the highway some night.

And you've got other stuff in your car. In the glove box. Stuff you might need in a hurry: flashlight, map, sunglasses, automatic weapon. You know. Just in case you wind up barefoot on the highway some night.

So stuff is important. You gotta take care of your stuff. You gotta have a *place* for your stuff. Everybody's gotta have a place for their stuff. That's what life is all about, tryin' to find a place for your stuff! That's all your house is: a place to keep your stuff. If you didn't have so much stuff, you wouldn't *need* a house. You could just walk around all the time.

A house is just a pile of stuff with a cover on it. You can see that when you're taking off in an airplane. You look down and see all the little piles of stuff. Everybody's got his own little pile of stuff. And they lock it up! That's right! When you leave your house, you gotta lock it up. Wouldn't want somebody to come by and *take* some of your stuff. 'Cause they always take the *good* stuff! They don't bother with that crap you're saving. Ain't nobody interested in your fourth-grade arithmetic papers. *National Geographics,* commemorative plates, your prize collection of Navajo underwear; they're not interested. They just want the good stuff; the shiny stuff, the electronic stuff.

So when you get right down to it, your house is nothing more than a place to keep your stuff . . . while you go out and get . . . *more stuff.* 'Cause that's what this country is all about. Tryin' to get more stuff. Stuff you don't want, stuff you don't need, stuff that's poorly made, stuff that's overpriced. Even stuff you can't afford! Gotta keep on gettin' more stuff. Otherwise someone else might wind up with more stuff. Can't let that happen. Gotta have the most stuff.

So you keep gettin' more and more stuff, and puttin' it in different places. In the closets, in the attic, in the basement, in the garage. And there might even be some stuff you left at your parents' house: baseball cards, comic books, photographs, souvenirs. Actually, your parents threw that stuff out long ago.

So now you got a houseful of stuff. And, even though you might like your house, you gotta move. Gotta get a bigger house. Why? Too much stuff! And that means you gotta move all your stuff. Or maybe, put some of your stuff in storage. Storage! Imagine that. There's a whole industry based on keepin' an eye on other people's stuff.

Or maybe you could sell some of your stuff. Have a yard sale, have a garage sale! Some people drive around all weekend just lookin' for garage sales. They don't even have enough of their own stuff, they wanna buy other people's stuff.

Or you could take your stuff to the swap meet, the flea market, the rummage sale or the auction. There's a lotta ways to get rid of stuff. You can even give your stuff away. The Salvation Army and Goodwill will actually come to your house and pick up your stuff and give it to people who don't have much stuff. It's part of what economists call the Redistribution of Stuff.

OK, enough about your stuff. Let's talk about other people's stuff. Have you ever noticed when you visit someone else's house, you never feel quite at home? You know why? No room for your stuff! Somebody *else's* stuff is all over the place. And what crummy stuff it is! "God! Where'd they get *this* stuff?"

And you know how sometimes when you're visiting someone, you unexpectedly have to stay overnight? It gets real late, and you decide to stay over? So they put you in a bedroom they don't use too often . . . because Grandma died in it eleven years ago! And they haven't moved any of her stuff? Not even the vaporizer?

Or whatever room they put you in, there's usually a dresser or a nightstand, and there's never any room on it for your stuff. Someone else's shit is on the dresser! Have you noticed that their stuff is shit, and your shit is stuff? "Get this shit off of here, so I can put my stuff down!" Crap is also a form of stuff. Crap is the stuff that belongs to the person you just broke up with. "When are you comin' over here to pick up the rest of your crap?"

Now, let's talk about traveling. Sometimes you go on vacation, and you gotta take some of your stuff. Mostly stuff to wear. But which stuff should you take? Can't take all your stuff. Just the stuff you really like; the stuff that fits you well that month. In effect, on vacation, you take a smaller, "second version" of your stuff.

Let's say you go to Honolulu for two weeks. You gotta take two big suitcases of stuff. Two weeks, two big suitcases. That's the stuff you check onto the plane. But you also got your carry-on stuff, plus the stuff you bought in the airport. So now you're all set to go. You got stuff in the overhead rack, stuff under the seat, stuff in the seat pocket, and stuff in your lap. And let's not forget the stuff you're gonna steal from the airline: silverware, soap, blanket, toilet paper, salt and pepper shakers. Too bad those headsets won't work at home.

And so you fly to Honolulu, and you claim your stuff—if the airline didn't drop it in the ocean—and you go to your hotel, and the first thing you do is put away your stuff. There's lots of places in a hotel to put your stuff.

"I'll put some stuff in here, you put some stuff in there. Hey, don't put your stuff in *there*! That's my stuff! Here's another place! Put some stuff in here.

And there's another place! Hey, you know what? We've got more places than we've got stuff! We're gonna hafta go out and buy . . . *more stuff!!!*"

Finally you put away all your stuff, but you don't quite feel at ease, because you're a long way from home. Still, you sense that you must be OK, because you do have some of your stuff with you. And so you relax in Honolulu on that basis. That's when your friend from Maui calls and says, "Hey, why don't you come over to Maui for the weekend and spend a couple of nights over here?"

Oh no! Now whaddya bring? Can't bring all this stuff. You gotta bring an even *smaller* version of your stuff. Just enough stuff for a weekend on Maui. The "third version" of your stuff.

And, as you're flyin' over to Maui, you realize that you're really spread out now: You've got stuff all over the world!! Stuff at home, stuff in the garage, stuff at your parents' house (maybe), stuff in storage, stuff in Honolulu, and stuff on the plane. Supply lines are getting longer and harder to maintain!

Finally you get to your friends' place on Maui, and they give you a little room to sleep in, and there's a nightstand. Not much room on it for your stuff, but it's OK because you don't have much stuff now. You got your 8 × 10 autographed picture of Drew Carey, a large can of gorgonzola-flavored Cheez Whiz, a small, unopened packet of brown confetti, a relief map of Corsica, and a family-size jar of peppermint-flavored, petrified egg whites. And you know that even though you're a long way from home, you must be OK because you do have a good supply of peppermint-flavored, petrified egg whites. And so you begin to relax in Maui on that basis. That's when your friend says, "Hey, I think tonight we'll go over to the other side of the island and visit my sister. Maybe spend the night over there."

Oh no! Now whaddya bring? Right! You gotta bring an even smaller version. The "fourth version" of your stuff. Just the stuff you *know* you're gonna need: money, keys, comb, wallet, lighter, hankie, pen, cigarettes, contraceptives, Vaseline, whips, chains, whistles, dildos, and a book. Just the stuff you *hope* you're gonna need. Actually, your friend's sister probably has her own dildos.

By the way, if you go the beach while you're visiting the sister, you're gonna have to bring—that's right—an even smaller version of your stuff: the "fifth version." Cigarettes and wallet. That's it. You can always borrow someone's suntan lotion. And then suppose, while you're there on the beach, you decide to walk over to the refreshment stand to get a hot dog? That's right, my friend! Number six! The most important version of your stuff: your wallet! Your wallet contains the only stuff you really can't do without.

Well, by the time you get home you're pretty fed up with your stuff and all the problems it creates. And so about a week later, you clean out the closet, the attic, the basement, the garage, the storage locker, and all the other places you keep your stuff, and you get things down to manageable proportions. Just the right amount of stuff to lead a simple and uncomplicated life. And that's when the phone rings. It's a lawyer. It seems your aunt has died . . . and left you all her stuff. Oh no! Now whaddya do? Right. You do the only you can do. The honorable thing. You tell the lawyer to stuff it.

Rub-Out

Henry Petroski

Henry Petroski is an engineering professor with an unusual fascination with the mechanics of simple things that work well. He's carried this fascination so far as to write an entire book on the history of the pencil: *The Pencil: A History of Design and Circumstance* (1990). Petroski's *The Evolution of Useful Things* (1994), has chapters on such helpful inventions as the paper clip, Post-it notes, and clamshell fast-food containers, without which there might not be a McDonald's. In this piece Petroski considers the fate of the pencil in an age when writing is increasingly being done on computers.

At the *Visions* Web Site Links to pencil manufacturers and collectors.

READING PROMPT

What are Petroski's feelings about the future of the pencil? In this essay Petroski does not come out and talk openly about his views on the larger issue of technology and progress, but can some of his ideas nonetheless be discerned here? How then does his interest in the pencil reflect his feelings about technology and progress generally?

The announcement by the Educational Testing Service this week that it plans to replace pencil and paper with computers on its standardized tests prompts a question. Is this:

(a) the end of No. 2 pencils in the exam room?
(b) an end to No. 2 pencils?
(c) an end to all pencils?
(d) none of the above?

The answer—despite the doom and gloom front-page stories in *The New York Times* and *The Washington Post*—is (d). After all, this is not the first time that the sturdy, low-tech pencil has been threatened. Technological advances and materials shortages have written its obituary many times before. And each time, the plucky implement, seemingly on the verge of extinction, left for dead, has bounced back to life.

The modern pencil dates from the middle of the sixteenth century, when a new mineral that made a dark but removable mark was discovered in England's Lake District. The new mineral was called black lead, because it made a blacker line than metallic lead, which had been among the most common means of making a mark on paper without the mess and bother of ink and quill. Black lead also had military applications, especially in casting cannon balls, and so the English mine from which it came was closely guarded and its output regulated. This made the smallest sliver of black lead very costly. Thus the lead, as it had come to be called, was encased in wood not only to keep the fingers clean, but to protect the pricey substance from breaking. The assemblage of lead and wood was called a pencil, after the fine brush known as a penicillum that dated back to Roman times.

With the growth of chemical knowledge, pencil lead was found in 1779 to be a form of carbon. It was then given the name graphite, after the Greek word meaning "to write." (The substance that was found so effective in rubbing out pencil marks was called rubber.) When at war with the English in the late eighteenth century, the French could not get pure graphite, and so an engineer named Nicolas-Jacques Conté was assigned the task of coming up with a way of making good pencils out of poor graphite. He found that if he recombined refined graphite with a clay binder, and if he baked the mixture into a ceramic, he could make pencils as good as the renowned English ones. He found, further, that by altering the proportions of graphite and clay, he could make pencils that made lighter and darker marks. Conté used the numbers one, two, etc., to designate different degrees of hardness in his pencils.

Although Henry David Thoreau did not need to take any standardized test to enter college, pencils were instrumental in paying his way through Harvard. Thoreau's father, John, was an early American pencil-maker, and when Henry David could not keep a teaching job after his graduation in 1837, he worked at the family business, seeking ways to improve the product. He developed the French process to such a degree that by the mid-1840s Thoreau pencils were the best made in America, and Ralph Waldo Emerson bragged that they were made right in Concord. Soon, however, the expanding German pencil firms flooded the American market with their less expensive products, and the Thoreaus were driven out of the pencil business.

By the middle of the nineteenth century, the English source of graphite had been exhausted. At about the same time an equally rich and pure source had been discovered in Asia, and the German firm A. W. Faber gained exclusive rights to it. Soon, Siberian graphite became the world standard for pencils, and other manufacturers began to color their pencils yellow and give them names like Mongol and Mikado to suggest that they too originated in Asia. This influence remains today, with about three out of every four pencils finished in yellow.

From the beginning of pencil-making, the quality of the wood has been as important as that of the graphite, and red cedar from the southeastern United

States was long the wood of choice. By the early twentieth century, however, supplies of red cedar became so scarce that pencil-makers bought old barns and fence posts to stay in business. In time, incense cedar from the western United States was found to be a substitute wood, and that is what is most likely to be in a good No. 2 today. Of late, there has been some concern that jelutong, a tropical rain forest tree, is being used increasingly for less expensive pencils. To counter concerns that pencils are wasteful of resources, one company came out with the American ECOwriter pencil, which uses recycled cardboard and newspaper fiber. Perhaps it will be more readily accepted than the plastic variety of some years ago.

The humble wood-case pencil has survived the exhaustion of its earliest and best sources of graphite and red cedar; it has survived the introduction of the mechanical pencil, the fountain pen, the ball-point; it has survived the development of the typewriter and the personal computer (which killed the typewriter); and it will survive the recasting of standardized tests into a new cyber-tech format. There are some things, simply, that technology cannot improve on. Compared to the handy, durable, effective pencil, the mouse is, well, a mouse.

Technology Amish Style

Eric Brende

In this selection Eric Brende focuses on the life of what might seem to be a simple people, the Amish—people who live in a series of church and farming-based rural communities most widely associated with the area of Pennsylvania around Lancaster (and called the Pennsylvania Dutch Country) and briefly brought into national attention with the popular 1985 film *Witness*. Although one view of the Amish is as people who reject modern technology, and thus as a kind of Luddite (see Sale's article in Chapter 13, Working More), a richer view and one put forth here by Brende is that they are people who define themselves largely in terms of the mastery, indeed the near perfection, of an earlier technology—that of pre-industrial agricultural life.

At the *Visions* Web Site Links to Amish Country Welcome Center (near Lancaster, PA); to sites on Shaker art and music.

READING PROMPT

What does Brende find most and least attractive about the Amish way of life?

We were busily working, the Amish crew and I, dismantling an old house to gather a cheap lumber supply. The wood would be used for a new dwelling now partially completed on the southeast slope of my recently purchased property. After a year and a half living amongst the Amish, my wife Mary and I have decided to stay, and although we have declined to join the Amish church, our "advisory board"—a plain-dressed widower and his four grown children—has exceeded what we thought possible in their friendliness and eagerness to assist.

"Eric . . . !" someone was calling to me. "Maybe you could come over here and do this for us." What special task have they singled out for me? I wondered. I tiptoed across the exposed floor joists and looked. On the floor below, one of my new Amish friends gazed up plaintively and pointed to an overhead light fixture near my feet. "Could you please take those lights down?" he asked. I suppressed a giggle. I no more knew how to take apart an electric light than how to tie a double half-hitch. Did he actually suppose that because I had grown up in an electrified environment I knew how electric lights were put together?

In fact, he might well have made such an assumption. In many or most Amish settlements, everyone does know how to do everything. Many women know carpentry, including the daughter in the family that was helping me tear down the old house and build the new one. The other day I happened upon an Amish wife sawing boards for a piece of furniture she was building. Cabinetry is her hobby. Simply speaking, most work demands that have been broken down into innumerable fields of specialization in modern technological culture have remained whole among the Amish. Families will fail to make ends meet unless all household members know how to do sundry interconnected tasks—among them growing, preserving, and preparing for the table most of their food; constructing and maintaining homes; making clothes; and often even medicating themselves.

Underlying the reluctance to specialize is an awareness of how well shared know-how helps keep Amish society stable. Without it, how could Amish neighbors so readily join together in undertakings such as tearing down or building a house? Or threshing wheat? This is not to say the Amish are necessarily experts in each of the tasks to which they aspire. "I don't do anything long enough to get good at it," an Amish fellow once told me with a self-effacing chuckle. Still, the extent of their mastery is considerable. A glance at their efforts reveals beautiful, well-maintained farms thriving in an age when many others cannot even survive. Inhabitants of what seems to be a provincial backwater, the Amish turn out to be surprisingly enlightened, at least with regard to the material conditions of their life. Next to them I feel sheltered from the world around me.

I feel oddly ungrounded as well. Life in a fast-paced society is next to impossible without narrowing one's focus—few of us have the time to learn the inner workings of such conveniences-cum-necessities as cars and computers—and it is axiomatic that in narrowing one's focus one can lose sight of the larger picture. But that's not the only problem. New developments continually evolve in the myriad areas that this narrowed focus excludes. The result is that people are often plagued by what seems like uncontrollable change.

The Amish, by contrast, consciously steer their cultural course in the sea of alternatives opened by technological advance, accepting only those that enhance their way of life. If the Amish plead ignorant on the subject of electrical wiring, for instance, it is because they have made an active decision not to avail themselves of the technology. Installing electricity would only permit them to plug in clever contraptions that could, at the push of a button, shift much skilled work away from them, reducing the need for shared know-how and the opportunities for community building.

Word of how successful the Amish way of life can be is spreading, and for not a few, curiosity has enkindled conversion. In the community near us, I can think of at least 10 separate parties, either individuals or families, who have initiated the process of baptism into the Amish Church. But I doubt that it is necessary to embrace the Amish religion to attain their community cohesion or their sense of connection to the material world.

A TRADITION OF INNOVATING

The full extent of the differences between Amish society and mainstream America becomes clear every time the Amish council near us meets. Members present are allowed to speak their minds on any matter before the community: since everyone has intimate knowledge of how the whole of Amish life works, everyone is considered competent to volunteer insights and ideas as well as assist in decisions. If anyone has clout, it's principally because of experience—first speaking rights go to older members. The practice of drawing lots for central Amish leadership positions underscores people's confidence in their mutual abilities.

Interestingly, one of the recurring items on the agenda is technology. Unlike mainstream technologists, who never test the social impact of their products but unscientifically presume a beneficial effect, the Amish often adopt new technologies only after a trial period in which they assess the effects. For example, when members of the local community began using a nearby pay phone to facilitate sales of produce to regional grocery-store chains, a council was called to discuss the possible ramifications. Participants pointed out that lightning-fast communication across the miles was undermining the face-to-face intimacy so important to their community. But they also observed that without the phone they would be at a competitive disadvantage among suppliers to grocery stores, which prefer up-to-the-minute information on shipments. Out of this dilemma came the decision to urge only sparing use of the phone.

At times the Amish actually promote technological development. After all, some technologies help perpetuate a tight-knit community. The Pioneer Maid wood cookstove is a case in point: conceived by two Amish brothers, Elmo and Mark Stoll of Aylmer, Ontario, and now produced by an Amish factory, it was the first wood cookstove in North America to employ an airtight combustion compartment. The only significant advance in household wood-fired stoves in 300 years, it efficiently cooks food, heats a water supply, raises bread dough, dries vegetables, and warms 2,000 square feet of living space all at the same time. The stove does require a sizable repertoire of domestic skills and knowledge to operate, but because everyone needs to operate it, vital resources and activities remain in the purview of the community and help hold it together.

Similarly, an Amish blacksmith in Indiana has refitted the motor-driven John Deere 24-T hay baler with a slick assemblage of gears, drive-chains, and a steel wheel from the 1940-vintage W-30 International Tractor. Unlike the balers used by the Amish everywhere else, which are drawn along by horses but use fuel-powered engines for the actual task of baling, this one is powered entirely by horses. And in the community next to us, Amish farmers have resurrected an old stationary threshing machine. They salvaged the transmission of a junked cement truck and fitted it with wooden shafts that fanned out from the hub. Five teams of two horses each are hitched to the shafts, and when they move forward, horsepower is redirected along a rubber belt that connects the transmission with the threshing machine. Having threshed with the innovative Amish crew two summers in a row, I can avow that the contraption leaves ample work to bond together a gang of laborers.

Indeed, the Amish were, from their beginnings in seventeenth-century Europe, leaders and innovators in agricultural techniques. While most Alsatian peasants were still using the sickle, the Amish had moved to the more efficient scythe, which allowed a greater range of motion and required less stooping. They were also forerunners in practices like rotating crops, irrigating meadows, and using clover and alfalfa to build up the soil. Hoping to improve the productivity of the land in their jurisdictions, progressive eighteenth-century noblemen enticed the Amish to settle in Poland and Russia, while more recently Mexico, Belize, and Paraguay have welcomed Amish immigrants with open arms.

Even though traditionalism may prevail among the Amish, often the tradition itself is one of tinkering, adaptation, and innovation. Ivan Glick, a farmer and writer especially interested in agricultural history, points to the large number of surprisingly competent self-taught engineers in the Amish settlement of Lancaster County, Pa. And in some cases where the Amish have resisted technology, history has vindicated them. For instance, they have been the living cornerstone for a nationwide return to horse farming, which they have proven is "much more profitable than tractor farming if done correctly," according to agricultural industry observer Baron Taylor of Lancaster County.

Though feasible only if the scale is not too large, the practice keeps capital costs low and promotes sustainable soil management, enhancing land value and use. The number of non-Amish horse farmers has tripled over the last 20 years and now exceeds that of the Amish.

More striking than Amish technical command, however, is Amish emotional health. According to a study by psychiatrist Janice Egeland, Amish people suffer non-organically based depression at a small fraction of the national rate. And then there's the strong Amish family. The Amish divorce rate is next to nil, and only about one in four Amish offspring leaves the community.

While there are doubtless a variety of reasons for such impressive statistics, my firsthand acquaintance with the Amish leads me to believe that co-possession of vital know-how is certainly one of them. It helps unite spouses by giving women a fuller sense of partnership in the overall household enterprise and a more visible, respected place in the community. I suspect that it also puts parents in an excellent position to be parents—to instill emotional stability and adult responsibility in their progeny. Amish parents are their children's primary educators, employers, and financiers—a position that would hardly be possible if they had not acquired the lion's share of the society's knowledge.

The work of child psychologist Urie Bronfenbrenner reinforces this notion. In his book *Two Worlds of Childhood* comparing child-rearing practices in the United States and the Soviet Union, Bronfenbrenner asserts that children who can meaningfully participate in the world of adult work enjoy far greater psychological well-being than children deprived of such participation and so do the adults those children become.

BRAVE NEW BOUNDARIES

Unusual as the Amish may seem, their roots are not very much different from those of mainstream society, which may help explain why some conversions to their way of life succeed. Sociologist Max Weber employed the term Protestant ethic to account for the meticulous industry widely in evidence among Calvinists and their cultural descendants, including white Anglo-Saxon Protestants. It could be argued that this industry has been duplicated by the Amish in agriculture. Historically, both Calvinism and the Amish church have sparked behavior that suggests a close psychological tie between earthly works and personal salvation. Perhaps the key difference between the two religions is the greater Amish emphasis on membership in Christian community, which they see as a prerequisite for redemption. Amish theology, far more than Calvinist, provides religious incentives for channeling industry into social solidarity.

Conversion is an uphill battle, however. One youth I know who was about 22 when he entered the local Amish church became dismayed at the amount of catching up he had to do, and not just in the area of know-how and skill. His Amish friends received help and money from their parents, enabling them to buy buggies and save up for marriage and a farm. Without similar support from his own parents, who lived several hundred miles away and had none of the same resources, he languished. He finally left—but not without heartache, for he had grown attached to the community and its ideals.

The Amish stress on religious dogma is another reason why converts may leave. The list of possible complaints, to be sure, can be daunting: that Amish religious indoctrination of the young generally leaves little room for critical or creative thought; that the traditionalism informing every sphere of Amish activity can inhibit even common sense; that the practice of shunning adult backsliders is unduly harsh; that baptism into the Amish faith, while ostensibly voluntary, is expected by parents when a child comes of age and virtually foreordained; and that, in short, the survival of Amish culture can be attributed less to anything positive than to inertia and the fear of change and reprisal. Amish expert John Hostetler, quoting a formerly Amish woman, notes that "in Amish society one is always conscious of keeping as many rules as one can, and each person tends to feel an obligation to see that other members of the family also keep these rules."

On the other hand, it is worth remembering that it was the Amish themselves who set up the rules, and that their efforts have borne fruit in a flourishing handicraft civilization and an oasis of religious sentiment. As Hostetler himself remarks, "conformity implies that persons are committed to the goals and means of their culture, voluntarily accept them, and live in a state of commitment to the appropriate rewards."

Yet while religion is inextricably interwoven with Amish practice for the Amish themselves, for the rest of us such issues can be kept more distinct. My wife and I are in the midst of an attempt to initiate an alternative to Amish community that is less enmeshed with dogma. Chary of religious codes of technological enforcement, we nonetheless appreciate the need such measures address: that of conscious boundary setting, backed by some kind of clout. Without some such boundaries, seekers of alternative lifestyles could easily drift and re-merge with the practices they have set out to avoid. Several other communities already in existence in this country employ one or another method of common discipline: land trust or coop or commune. For our part, we are trying to translate the Amish religious codes into deed restrictions, aiming for a kind of low-technology zoning. By legally limiting electric and telephone lines, as well as motorized farming and wood-cutting devices, we aim to help establish on a circumscribed space nonsectarian guideposts for the kind of social stability the Amish enjoy.

In other words, we hope for the best of both worlds: a reliable structure for neighborly solidarity without the pincers of theological sanction. This is not to say that we have disallowed religion. In fact, we are practicing Catholics ourselves. But we do not wish to make religious belief the basis of technological norms. We welcome anyone who shares a desire for technological simplicity.

Other ways to further the same ends might include a summer internship program in homesteading for college students, or a school for homesteading apprentices that would complement the conventional liberal arts curriculum. Plunging into the nitty-gritty versatility our ancestors once shared, students would pick up new skills, enhancing, shall we say, their career options. The Cabin Fever University at Celo, N.C., the Shelter Institute of Bath, Maine, and the Christian Homesteading Movement of Oxford, N.Y., are already pursuing this idea. I hope one day to inaugurate something similar here on our own homestead—if I ever know enough to do it.

But even in the absence of such opportunities, the mere example of Amish settlements is instructive. In a world whose rapid changes give credence to the notion of unstoppable "progress," the Amish remind us of what may have helped to propel much of that progress: Francis Bacon's dictum that knowledge is power. Amish successes are an index of what we can achieve by attempting to integrate practical knowledge and social intimacy. They teach us not that the modern world can or should somehow be ignored, but that within it—even as part of its own program—something vital from the past may persist. And perhaps even nose ahead of the present.

Seven Sustainable Wonders

Alan Thein Durning

Alan Thein Durning is a contemporary activist, urging us, in the spirit of Thoreau, to adopt a more sustainable lifestyle, one based more on quality of experience than on quantity of possessions. In the following short piece, he offers a list of useful items that can form the basis of such a sustainable way of life. Some of Durning's current interest has been in preserving the Pacific Northwest.

At the *Visions* Web Site Links to an interview with Durning, "On Creating a Sustainable Regional Future"; to groups advocating voluntary simplicity.

READING PROMPT

Write up a specific list of rules, serious or otherwise, that items must meet in order to qualify for Durning's list of "Sustainable Wonders."

I never have seen any of the Seven Wonders of the World, and to tell you the truth, I wouldn't really want to. If I ever got to the pyramids, I'm sure I'd only think about how many slaves must have died cubing boulders and shoving them about. I once came close to seeing the Taj Mahal, which, though it came too late to register as an original wonder, surely would qualify on a new ranking. Everyone we met in India said we had to go see it, but my wife and I balked when we figured out how early we would have to arise to do so. Who wants to get up before dawn for a hell ride through traffic (Hindus are especially bad drivers because they believe in reincarnation) to see a tomb built by a long-dead imperialist? We slept in, then spent the day riding bicycle rickshaws and smelling the jasmine in Delhi.

To me, the real wonders are all the little things that work, especially when they do it without hurting the Earth. Here's my list of simple things that, though they are taken for granted, are absolute wonders. These implements solve everyday problems so elegantly that everyone in the world today—and everyone who is likely to live in it in the next century—could have and use them without Mother Nature ever noticing.

The bicycle. The most thermodynamically efficient transportation device ever created and the most widely used private vehicle in the world, the bicycle merits first place on the list of the sustainable wonders of the world. Invented just a little over a century ago, bikes let individuals travel three times as far on a plateful of calories as a person could walking. Moreover, they are 53 times more energy efficient—comparing food calories with gasoline calories—than the typical car. They don't pollute the air, lead to oil spills (and oil wars), change the climate, send cities sprawling over the countryside, lock up half of urban space in roads and parking lots, or kill a quarter-million people in traffic accidents each year.

Bikes also are cheaper than any other vehicle, costing less than $100 new in most of the Third World. Mine, a zippier model, still ran less than $400. (Fluorescent spandex tights, of course, are extra.) Bicycles do take steel, aluminum, and rubber to manufacture, and making these has an environmental cost, as a glance at any iron mine or rubber factory will demonstrate. However, they are lightweight of necessity, so they require small amounts of materials compared to other vehicles. Their simplicity makes repair relatively easy. Where bicycles are prevalent, repair is a large, decentralized industry. Pedal a couple of blocks on any thoroughfare in India and you likely will pass at least one bike-fixer sitting on a mat, his tools spread around him.

The world doesn't have enough bikes yet for everybody to ride, but it is getting there quickly. Best estimates put the planet's expanding fleet of two-wheelers at 850,000,000—double its auto population. Americans have no excuses on this count: We have more bikes per person than China, where they are the principal vehicle. We just don't ride them much.

The ceiling fan. Appropriate technology's answer to air conditioning, ceiling fans cool tens of millions of people in Asia and Africa. A fan over the bed works like a charm in sweltering climes, as I've had plenty of time to reflect on during episodes of digestive turmoil in cheap hotels in the tropics.

With Americans moving to the Sunbelt and most of the Earth's people concentrated near the equator, air conditioning is a surging sector of the global economy. In 1960, 12% of US homes were air conditioned; now, two-thirds are. Air conditioning, though, consumes disproportionate amounts of energy and is the bane of the stratospheric ozone layer, because of its chlorofluorocarbon coolants. Ceiling fans are simple, durable, repairable, and take little energy to run.

The clothesline. A few years ago, I read about an engineering laboratory that claimed it had all but perfected a microwave clothes dryer. The machine, the story went, would get the moisture out of the wash with one-third the energy of a conventional unit and cause less wear and tear on the fabric.

I don't know if they ever got it on the market, but it struck me at the time that, if simple wonders had a public relations agent, there might have been a news story instead about the perfection of a solar clothes dryer.

It takes few materials to manufacture, requires absolutely no electricity or fuel, and even gets people outdoors where they can talk to their neighbors. "A team of government scientists," I imagined the article reporting, "have found such a technology in an all-but-abandoned device called the clothesline." The apartment building I lived in for four years had a ban on clotheslines, as do many apartment buildings. Apparently, laundry flapping in the breeze grates on people's sensibilities. "Looks like Appalachia," I've overheard folks say, and "Reminds me of a tenement." This aversion to frugality is strange. It's as if using the sun to dry clothes is a retrograde step, a betrayal of the American Dream.

Maybe clotheslines need Madison Avenue's help. Imagine grainy images of hard-body youths in tight close-ups. The copy might read: "Sun-dried clothes. They're hot like you."

The telephone. The greatest innovation in human communications since Johannes Gutenberg's printing press, telephone systems take small, and shrinking, quantities of resources to manufacture and operate. Hype of the information age notwithstanding, I'll wager that they never lose ground to other communications technologies. Unlike fax machines, PCs and computer networks, television sets, VCRs and camcorders, CD-ROM, and all the other flotsam and jetsam of the information age, telephones are a simple extension of humans' time-tested means of communication—speech.

The Earth can afford for everyone to have a phone. The only hitch is that the Earth can not afford for everyone to have a phone book with everyone else's number in it.

The public library. Public libraries are about the most democratic institutions around. Think of it! Equal access to information for anybody who comes in; a lifetime of learning, all free. Libraries foster community, too, by bringing people of different classes, races, and ages together in that endangered form of human habitat—noncommercial public space.

Although conceived without any ecological intention whatsoever, libraries are waste reduction at its best. Each saves a forestful of trees by making thousands of personal copies of books and periodicals unnecessary. All that paper savings means huge reductions in energy use, as well as water and air pollution. In principle, the library concept also can be applied elsewhere, further reducing the number of things society needs without reducing people's access to them; for example, cameras and camcorders, tapes and CDs, cleaning equipment, and extra dining chairs. The town of Takoma Park, Md., has a tool library, where people can check out a lawn mower, ratchet set, or sledgehammer.

The campus interdepartmental envelope. Those old-fashioned slotted manila envelopes bound with a string and covered with lines for routing papers to one person after another put modern recycling to shame. The other day, I took a parcel to Federal Express, carrying along one of those indestructible Federal Express envelopes I had received. In the bottom corner it said, "This PAK contains 25% recycled material. FEDEX uses only non-toxic inks, varnishes, and non-toxic adhesives." Great, I thought, they're on my side.

Yet, when I asked the clerk for some tape to seal my papers into the used "pak," he informed me that company policy says "no resealing." At the post office, it's the same story. The postal code says business reply envelopes cannot be used for other addresses, even if you put on your own stamp and completely cover all identifying information. (Anyone want to wager on whether envelope manufacturers had anything to do with getting that rule written?) It seems like a great target for grassroots-type action, perhaps a letter-writing campaign?

The condom. Among the oldest of contraceptives, the condom now is the preferred form available, protecting against both pregnancy and sexually transmitted disease. It is a remarkable device—highly effective, inexpensive, and portable. A few purist environmentalists might complain about its disposability and excess packaging, but these are trivial considering the work it has to do—battling the scourge of AIDS and stabilizing the human population at a level the Earth can support comfortably.

chapter 1 Explorations

1a Simple Things

Collaboration. In the spirit of Durning, make a list of five or more other simple or low-tech objects that play a helpful role in your life, then compare lists with your classmates.

Web work. For more on everyday tools, check out the Web sites of some of the better-known hardware chains, such as Ace Hardware, especially the Best Buys page; Lowe's Home Improvement Warehouse; and TrueValue. Yahoo! lists dozens of others under Business and Economy > Companies > Home and Garden > Hardware.

Stanley Work prides itself on its well-crafted hand tools, devoting a page on its site to its history.

One writing topic. Compose an essay based on one or more simple or low-tech objects that have played an important part in your life. Use one or more of the authors from this chapter to provide an effective outside (introduction and conclusion) to your essay.

1b Glamorous Things

Collaboration. We usually think of glamour in terms of people, not objects; yet one of the common features of the objects on Alan Thein Durning's list of seven sustainable wonders or of Thoreau's furniture is their lack of glamour. Can you make and share a comparable list of glamorous products? How would you define *glamour* in such a way that it applies to such objects?

Web work. The Sharper Image and Brookstone are two retailers that have made their national reputations locating and marketing tools with that something extra (glamour?). How do products presented at these Web sites differ from those at traditional hardware stores?

The Industrial Designers Society of America (IDSA) uses its Web site to promote what it sees as the best in product design, with an annual list of the year's best-designed products. In what ways are most of these winners glamorous as well as practical?

One writing topic. Compose an essay on dullness and glamour in objects. What makes an object suitable as a gift?

1c Packaging

Collaboration. The simple and straightforward objects that Durning likes might seem dull to many others. One way to add glamour to dull things is to put them into new packages. Make and share a list of fancy new ways of packaging or marketing any of Durning's seven objects or any of the simple objects you discussed in number 1a above and share these with your classmates. Try going to a grocery store and observing the different ways that essentially the same products are packaged. Bring some packages to class to share. Which packages do you find most effective, and why? What role does honesty or informativeness play versus some other appeal?

Web work. Check out the Web site for Packaging Online to see what's up in the container industry.

The liquor industry, its distilled spirits component especially, spends huge sums of money to attach a certain kind of prestige, akin to glamour, to its products, in no small part to justify the large differences in costs between high-priced and bargain brands. Perhaps no product in history has been as successful in using print ads and packaging to enhance its image as Absolut Vodka, whose Web presence is collected at Yahoo! under Business and Economy > Companies > Drinks > Alcoholic > Vodka > Absolut.

One writing topic. Write an essay on different ways the same or similar products are packaged; discuss how that packaging affects how you or others feel about the product.

1d Bikes

Collaboration. Make and share two lists: all the different ways you have used bicycles in your life before and after you were old enough to drive a car.

Web work. The Bike Site is just one of numerous bike sites on the Web, with lots of information and links on high-tech biking (where the money and hence the commercial interest in the biking industry seems to be). Mountain biking certainly represents one high-tech area, well represented on the Web by the British site ChainSmoke. The Web site of the Classic and Antique Bicycle Exchange Newsletter obviously takes a different slant on cycling, although there seems to be quite a lot of money involved here as well. The Bike-Friendly Berkeley Coalition is just one of a number of local advocacy groups trying to improve the lives of bicyclists in urban areas; at Yahoo!, look under Recreation > Sports > Cycling > Advocacy > Organizations for an updated list of the others. Meanwhile the American Discovery Trail (ADT) describes a multiuse transcontinental trail—for hiking, biking, and horses.

One writing topic. Do research on your own to see what is being done to make your campus and the surrounding community more bike-friendly. Write a report describing what you learn and offering suggestions for doing more; base your recommendations in part on what you discover from your Web work above.

1e Clutter

Collaboration. Thoreau and Carlin both discuss clutter: Make and share a list of different ways that clutter affects your own life. How do objects often get in our way? Describe the efforts you have made to reduce clutter. Were you successful? Why or why not?

Web work. How would Thoreau feel about the suggestions offered by the good people at the Container Store, a Dallas-based company? How would George Carlin? How do you feel?

There are many other similar businesses organized by Yahoo! under Business and Economy > Companies > Home and Garden > Storage.

One writing topic. Discuss clutter as both a personal and a universal phenomenon. One way to do this is to place either your personal problems or those of Carlin or Thoreau on the inside or the outside of the essay.

1f Organizers

Collaboration. How do you keep track of your daily schedule? What method do you now use to note assignments, appointments, concerts, trips, and so on? What methods have you tried in the past? Share your thoughts with your classmates.

Web work. The Dayrunner Corporation is in business to help simplify your life with paper products. Yahoo! organizes similar companies under Business and Economy > Companies > Office Supplies and Services > Calendars and Personal Organizers.

Pacific Business Systems has a Web site to help you select what may be the ultimate high-tech organizer, a "personal digital assistant" or PDA.

One writing topic. Discuss the widespread problem of organizing our lives. One way to do this is to place your personal problems on either the inside or the outside of the essay.

c h a p t e r **2**

Toys, Simple and Not

READING QUESTIONS

- Which of your childhood toys hold the greatest value for you today? Why?
- What games do you still play today?
- Were you a parent, what guidelines would you use in selecting toys for your kids?

QUOTATIONS

The Child's Toy & the Old Man's Reasons
Are the fruits of Two seasons.

— *William Blake, British visionary poet and artist, 1757–1827*

The one who dies with the most toys wins.

— *Wall Street T-shirt*

People dream they can fly
Birds dream they can skate.

— *Rollerblade slogan*

As men get older, the toys get more expensive.

— *Marvin Davis, on buying the Oakland A's baseball team in 1979*

Basically, *Doom* is a (violent) 3D arcade game where you run around in a maze and kill things with shotguns and chainsaws. . . . After you get tired of killing things, you can run it over a network and kill things together with your friends. After you get tired of that, you can kill your friends.

— *Frequently Asked Questions:* Doom

Life is too important to take so seriously.

— *Corky Siegel, blues musician*

Bedtime Barbie

Ann Hulbert

Ann Hulbert's short piece touches on any number of topics considered in this chapter: the history of Barbie; boy versus girl toys; novelty versus traditional toys; and, perhaps most important of all, a comparison of the mass-market retailer Toys "R" Us and the upscale newcomer Learningsmith.

Edvard Munch's 1893 painting *The Scream* (referred to in Hulbert's essay) has become one of the most recognizable images of the twentieth century: One expects to see it almost anywhere, although perhaps not in a toy store. For more on Barbie, see Topic 2f.

At the *Visions* Web Site Links to Toys "R" Us and Learningsmith; to Edvard Munch's 1893 painting, *The Scream*.

READING PROMPT

What makes a good toy in Ann Hulbert's view? Do you agree or disagree with her? How does her thinking influence her opinion about the Learningsmith stores?

Barbie is 35 this year [1994] and along with a special-edition vinyl reproduction of the original 1959 model, Mattel has launched a new Bedtime Barbie. Not a Barbie fan, I immediately imagined the worst (fuchsia teddy) and couldn't resist a classic parental shudder. Could there be a better symbolic indictment of juvenile consumer culture? Casting three decades of coyness aside, the fashion doll had, to judge by her name, finally exposed her disreputable origins. (Barbie was inspired by salacious Lilli, a German gag item for men.) The fact that "Bedtime Barbie" shared Christmas season pre-eminence with Mighty Morphin Power Rangers clinched the diagnosis of advanced decadence in the playroom.

But Bedtime Barbie, I discovered on my holiday shopping expedition to Toys "R" Us, is actually dressed like every little girl's grandmother. She wears a long fuzzy pink nightgown, with demure white lace at the throat and cuffs. The bigger surprise is the look of "the first soft body Barbie doll you can sleep with" underneath her nightwear. I peeked, eager to see (and feel) a cushiony version of the mighty bust. Bedtime Barbie turns out to be barely distinguishable from the refined-looking rag dolls every little girl's great-grandmother stitched delicate clothes for, except she has just a little extra stuffing above the waist.

The cuddly Barbie was a calming influence on me: suddenly the mid-'90s panic about dangerously mindless kid merchandise seemed like, well, panic. Even a couple of hours in a mass-market madhouse like Toys "R" Us didn't revive alarmism. Sure, there's too much of everything (15,000–20,000 kinds of toys on the shelves, out of a total market of 120,000). But just look at all the floor space devoted to wholesome sports equipment. Construction toys clutter the shelves. Kitchen play paraphernalia is made of the sturdiest plastic—not all of it pink. And the prices, as Toys "R" Us guarantees, can't be beat. As for the junk, I asked for Jibba Jabber, the most ludicrous sounding toy on the list of "hot" Christmas items from Toy Manufacturers of America Inc., the industry trade association. No salesperson had a clue of what I was talking about.

Above all, a cruise down the aisles undermined my support for the favorite reform currently urged on American toy manufacturers: liberate girls from their doll ghetto and give them the kind of dynamic toy opportunities boys enjoy. A disproportionate share of the 5,000–6,000 new playthings introduced into the toy market every year are aimed, unfairly it is claimed, at boys. It is unfair—for the boys. Roughly 80 percent of those novelties are flops, or quick fizzles, and a glance at the crummy combat- and sci-fi-oriented miscellany makes clear why. If diversifying for girls means forsaking loyalty to Barbie and sturdy Little Tyke kitchens in order to flirt with faddish T.V. tie-in action figures—or, say, Doctor Dreadful Food and Drink Labs (serving up Monster Warts and Putrid Potions)—that's hardly progress.

On my way out, I finally found Jibba Jabber ($12.99). Following the instructions, I grabbed the creature's long neck and shook it. Strangulated burbling ensued. The perfect release for a parent after a hard day in toyland, but happily I didn't need it.

Not then. It was the trip to the opposite pole of kid consumer culture, Learningsmith, that was too much. All blond wood, warm lighting and soft carpets, the three-year-old educational toy store (now in twenty-five locations) has taken media-saturated recreation a step higher: here, the T.V. tie-ins have a college education. Learningsmith pays royalties to use PBS affiliate stations' call letters, gets credited as a sponsor for a few shows, and features program-related products among its several thousand items.

"A General Store for the Curious Mind," Learningsmith is a shrine to upscale pushiness. The products tout their high I.Q.s: "It's O.K. To Be Smart!" "To Get a Smart Start," "Creativity for Kids," "Kindle Creativity." Instead of aisles there are "discovery centers" with names like "Tower of Babel" (for flashcards, etc.) and "Socrates Sandbox" (for the preschool set). Computers are available, to make shopping an "interactive experience."

In a place like this the curious mind quickly turns off. "The Backyard," a CD-ROM computer game for ages 3–6 that is on Learningsmith's shelf of recommended items, is promoted as "a fun place to explore nature, map reading and more." Hey, why get dirty? For older children, Broderbund's Aesop's Fable, "Tortoise and the Hare," gets the seal of approval. This "Living Book version includes dozens of hilarious new twists. Along the way, you'll meet a score of delightfully animated characters—including Simon, the story-telling bird." He's purple, with red sneakers, and evidently loquacious. So much for aphoristic Aesop.

Creativity turns out to mean mostly creative packaging on the part of the manufacturer, designed to spare consumers any act of ingenuity or, for that matter, common sense. Why buy a $5 hunk of modeling clay and spread out an old newspaper when for $29 you can buy Claymation Clay Fun Pack? Inside the large, attractive box you'll find a modest lump of clay, a smaller attractive box to keep it in, a plastic workmat and a modeling tool.

Leaving Learningsmith, I was accosted by a large inflatable toy called The Scream, for $27.95. Edvard Munch's figure, blown up, is fifty inches tall. "It's therapeutic," the label promises: "The Scream will be the one who understands you when no one else does." A swift punch didn't bring half the pleasure of throttling Jibba Jabber.

Playing with Dark-Skinned Dolls

bell hooks

bell hooks is the author of numerous books analyzing the interplay of race, gender, art; and politics. The following excerpt is from hooks's 1996 memoir, *Bone Black: Memories of Girlhood.*

At the *Visions* Web Site Links to online interviews with bell hooks, including an April 1996 interview from *The Washington Ripple,* an independent journal of politics produced at Washington University, an interview by Orlo in *Bare Essentials* (winter 1994), and a conversation between hooks and cyber-activist John Perry Barlow (Chapter 5), published by *The Shambhala Sun,* a Buddhist-inspired journal; to Barbie sites.

READING PROMPT

Describe hooks's complex attitude about Barbie dolls.

We learn early that it is important for a woman to marry. We are always marrying our dolls to someone. He of course is always invisible, that is until they made the Ken doll to go with Barbie. One of us has been given a Barbie doll for Christmas. Her skin is not white—white but almost brown from the tan they have painted on her. We know she is White because of her blond hair. The newest Barbie is bald, with many wigs of all different colors. We spend hours dressing and undressing her, pretending she is going somewhere important. We want to make new clothes for her. We want to buy the outfits made just for her that we see in the store, but they are too expensive. Some of them cost as much as real clothes for real people.

Barbie is anything but real, which is why we like her. She never does housework, washes dishes or has children to care for. She is free to spend all day dreaming about the Kens of the world. Mama laughs when we tell her there should be more than one Ken for Barbie: There should be a Joe, Sam, Charlie, men in all shapes and sizes. We do not think that Barbie should have a girlfriend. We know that Barbie was born to be alone—that the fantasy woman, the soap-opera girl, the girl of True Confessions, the Miss America girl, was born to be alone. We know that she is not us.

My favorite doll is brown, brown like light milk chocolate. She is a baby doll, and I give her a baby doll's name, Baby. She is almost the same size as a real baby. She comes with no clothes, only a pink diaper fastened with tiny gold pins and a plastic bottle. She has a red mouth the color of lipstick, which is slightly open so that we can stick the bottle in it. We fill the bottle with water and wait for it to come through the tiny hole in Baby's bottom. We make her many new diapers, but we are soon bored with changing them. We lose the bottle, and Baby can no longer drink. But we still love her. She is the only doll we will not destroy.

We have lost Barbie. We have broken the leg of another doll. We have cracked open the head of an antique doll to see what makes the crying sound. The little thing inside is not interesting. We are sorry, but nothing can be done—not even Mama can put the pieces together again. She tells us that if this is the way we intend to treat our babies she hopes we do not have any. She laughs at our careless parenting. Sometimes she takes a minute to show us the right thing to do. She, too, is terribly fond of Baby. She says that she looks so much like a real newborn. Once Mama came upstairs, saw Baby under the covers, and wanted to know who had brought the real baby from downstairs.

She loves to tell the story of how Baby was born. She tells us that I, her problem child, decided out of nowhere that I did not want a White doll to play with. I demanded a brown doll, one that would look like me. Only grown-ups think that the things children say come out of nowhere. We know they come from the deepest part of ourselves.

Deep within myself I had begun to worry that all this loving care we gave to the pink-and-white flesh-colored dolls meant that somewhere left high on the shelves were boxes of unwanted, unloved brown dolls, covered with dust. I thought that they would remain there forever, orphaned and alone, unless someone began to want them, to want to give them love and care, to want them more than anything. At first my parents ignored my wanting. They complained. They pointed out that white dolls were easier to find, cheaper. They never said where they found Baby, but I know. She was always there high on the shelf, covered with dust—waiting.

Toys That Bind *Douglas Coupland*

Canadian author Douglas Coupland may be best known for coining the phrase *Generation X,* the title of his 1992 novel. Coupland's work often deals with people caught up in new technology, as is the case with his 1996 satire *Microserfs,* about the desperate lives of programmers slaving away for a giant corporate software company.

At the *Visions* Web Site Links to Lego sites; to Coupland's Web site (www.coupland.com).

READING PROMPT

Explain how both Coupland's essay and his feeling about Legos are built on alternating concerns with construction and destruction.

Legoland is a real place—the Orlando of Europe—located in the center of Denmark's Jutland peninsula in a region that was once an atopian potato and dairy farm nowhere but is now a powerful utopian tourist somewhere. Lego, the modular plastic toy that has become a playtime cultural staple in the West, was first manufactured in 1947 in the nearby town of Billund—hence Legoland's somewhat random-seeming location. When families in Antwerp, Nancy, Linz or Bern discuss holiday locales, Legoland is the focus of untold Eurobrat tantrums.

Legoland, which opened in 1968, is a big attraction. Last year it took in 1,284,831 visitors (touché EuroDisney). Legoland consists of 42 million bricks configured into thousands of waist-high exhibits covering 120,000 square meters. There is a Lego Neuschwanstein (236,000 bricks) and a Lego Bergen, Norway, with a teeny funicular railway. There is a Lego Amsterdam (sans Lego squatters) and there is a "Legoredo" American Wild West theme park, which has little if anything to do with Euroculture, but seems wildly popular nevertheless (See Mount Rushmore!, 1,500,000 bricks). Other attractions include: a Lego Taj Mahal, a Lego Statue of Liberty and even a Lego US Capitol Building. A new Legoland is currently planned for Carlsbad, California, just north of San Diego.

I came to Legoland from Silicon Valley, where I had been spending time with scores of highly gifted, successful computer hardware and software engineers. In meeting these people, it quickly became apparent that every one of them had spent his or her youth heavily steeped in Lego and its highly focused, solitude-promoting culture. While a few made youthful excursions into alternate construction systems such as Meccano, Lego was their common denominator toy. As computer engineering culture is dense with any number of complex assembly languages, some sort of linguistic connection sprang to mind.

Now, I think it is safe to say that Lego is a potent three-dimensional modeling tool and a language in itself. And prolonged exposure to any language, either visual or verbal, doubtless alters the way a child perceives his or her universe. So let us examine the toy briefly.

First, Lego is ontologically not unlike computers. This is to say that a computer by itself is, well, nothing. Computers only become something when given a specific application. Ditto Lego. To use a Lotus 1–2–3 spreadsheet or to build a racing car—this is why we have computers and Lego. A PC or a Lego brick by itself is inert and pointless: a doorstop; litter. Made of acrylonitrile butadiene styrene, Lego's discrete modular bricks are indestructible and fully intended to be nothing except themselves.

Second, Lego is "binary"—a yes/no structure; that is to say, the little nubblies atop any given Lego block are either connected to another unit of Lego, or they are not. Analog relationships do not exist.

Third, Lego anticipates a future of pixilated ideas. The forms it builds are digital. The charm and fun of Legoland stems from seeing what was once organic reduced to the modular: a life-size zebra seemingly built of little cubes; cathedrals as seen through the *Hard Copy* TV lens that converts the victim's face into small squares of color; a statue of Hans Christian Andersen as constructed by Chuck Close.

The Lego-built universe thwarts entropy, and that can be both frightening and cute. This is unexpected. For North Americans, Legoland seems cloyingly European and tame. At times it makes Disneyland seem almost noir. A North American is left pining for a burning Lego Los Angeles; a Lego DC-10 crashed in an Iowa cornfield; a Lego earthquake-collapsed 880 Nimitz freeway—modernism facing and dealing with crisis rather than mummifying in eternal stasis. In the same manner that one searches for irregularities among Andy Warhol's multiples, one also searches for Lego bricks that are bleached and cracked by exposure to Scandinavian winters—errors in the system; decay; disorder. Legoland highlights this century's continual battle between linearity and chaos: witness Sarajevo's bombed-out Philip Johnson–like office towers, which still seem strangely intact.

Also, amid Legoland's crenellated minibuildings and railways are thousands of hand-high, Lego-built citizens, most of whom emulate shoppers, pedestrians, restaurant-goers and other Eurocitizens indulging in harmless enough middle-class pursuits. Yet there are a few small figures who, either through strong wind or simple misplacement, seem to have had their narratives stripped from their lives or had darker narratives imposed on them. For example, there are six gray bodies crawling across a lawn outside the Amsterdam-plex, past a field of digital-looking black and white cows, looking, for all the world, as though they had survived the fires of Waco. In Bergen, a character holding a bike outside the window of a house looks like a Peeping Tom. The brain willfully rejects the niceties presented by Lego's scamless psychic ecology.

Before any idea occurs, there must first be a dream. The Apollo rocket designers and the NASA engineers of Houston and Sunnyvale grew up in the 1930s and 1940s dreaming of Buck Rogers. And so when this aerospace generation grew old enough, they chose to make those dreams real. The children of Lego grew up dreaming of another world—a seamless world of HAL 9000, of *I Robot,* modularity, indestructibility, sound bites, acrylonitrile butadiene styrene. And now this generation, working in high-tech "campuses" that resemble suburban rumpus rooms, is silently choosing to build CD-ROM drives and code technologies that will help bring the world closer to a vision of Legoland. The darker narratives will always take care of themselves.

Girl Games | *G. Beato*

G. Beato is a San Francisco–based freelance writer. The following article appeared originally in the April 1997 issue of *Wired* magazine. Beato's article features extensive reference to video-game designer Brenda Laurel, a person with a wide Web presence of her own.

At the *Visions* Web Site Links to girl-oriented electronic games; to information by and about Brenda Laurel.

READING PROMPT

According to reports cited by Beato, what do experts see as the differences between the computer-mediated play of boys and that of girls?

The toughest computer game ever? It has to be *Doom,* right, with its endless toxic corridors and fidelity to the aesthetics of terminal carnage? Or maybe *Duke Nukem* and its Mark Fuhrman-on-acid Pig Cops? Or how about the gentler trials of *Myst,* all those cryptic machines and contraptions, and not a user manual in sight?

Those games are tough, sure, but there's one that's even tougher. It has mazes riddled with conundrums, inscrutable adversaries whose unshakable indifference to your presence leaves you wondering if it's even worth the effort to attack. And there are no cheat codes to bail you out when nothing's going right. Your goal? To reach the testosterone-spattered war rooms of the interactive entertainment industry and persuade the pasty knuckle-draggers who reside there to conceive, develop, and deliver games for girls.

Call it *Woom.*

For the last several years, men and women throughout the software industry have been playing this real-life game. And usually not winning.

But now, after years of disregard and sporadic, sometimes ludicrous attempts to serve the female market, the industry's game boys are experiencing a change of heart. Companies like Mattel Media, Hasbro Interactive, Sega, DreamWorks Interactive, Starwave, R/GA Interactive, Broderbund, and Philips Media have all introduced or are developing products targeted to girls. Start-ups such as Her Interactive, Girl Games Inc., Cybergrrl Inc., Girl Tech, and Interval Research's Purple Moon are doing the same. Some expect 200 new girl games to reach store shelves or go online this year—a tenfold increase from 1996.

Why the sudden interest?

Blame it on Barbie. Last November, the CD-ROM industry received a wake-up call when the runaway best-seller turned out to be *Barbie Fashion Designer.* In its first two months of sales, Mattel's digital incarnation of the oft-denigrated but remarkably enduring role model sold more than 500,000 copies, outstripping even popular titles such as *Quake*—and leaving the rest of the industry wondering how to cash in on this newfound wellspring.

The surge of girl game activity also reflects demographics—there are simply more women developing games and using computers today than five, ten years ago. And these women are bringing their perspectives to the development process. As girl game developer Brenda Laurel points out, "The game business arose from computer programs that were written by and for young men in the late 1960s and early 1970s. They worked so well that they formed a very lucrative industry fairly quickly. But what worked for that demographic absolutely did not work for most girls and women." While initial attempts at creating girl games amounted to little more than painting traditional titles pink, the current crop of developers understand that it's a far more complicated business than that.

THE LAND OF SWEEPING GENERALIZATIONS

That there's consumer demand for girl games comes as no surprise to Brenda Laurel. A pioneer in developing virtual reality, Laurel first zeroed in on the market at a conference in 1992, where she met David Liddle, who was co-founding Palo Alto, California-based Interval Research with Paul Allen. "We started talking about how the industry had consistently missed opportunities to get girls involved in technology," she recalls. "We asked, What would it really take to get a large number of girls using computers so often that the technology became transparent to them? In the end, we both had to admit that neither one of us really knew the answer."

Inside Interval's standard-issue gray-and-white Silicon Valley offices, Laurel leans back in her chair and smiles. In contrast to the high-collared, Vulcan-diva persona she assumed in the publicity photos for her book *Computers as Theatre,* in person she is warm and engaging, her curly auburn hair falling to her shoulders, her everyday earthwoman's garb giving her a decidedly human appearance. Compared with her forays into virtual reality, developing engaging content for girls had a relatively low fetish-factor, but Laurel eagerly accepted Liddle's offer to pursue the project at Interval—even though it came with a string attached. "I agreed that whatever solution the research suggested, I'd go along with," laughs Laurel. "Even if it meant shipping products in pink boxes."

To figure out the kinds of interactive entertainment girls would really find compelling, Laurel launched a major research campaign. "We took a three-pronged approach," she explains. "We did hundreds—maybe thousands—of interviews with 7- to 12-year-olds, the group we wanted to target with our products. We watched play differences between boys and girls. We asked kids how they liked to play; we gave them props and mocked-up products to fool around with." Laurel and company consulted experts in the field of children's play: toy store owners, teachers, scout leaders, coaches. Finally, they looked at all the research literature they could get their hands on, including material on play theory, brain-based sex differences, even primate social behavior—all with the goal of seeing how it might carry over into the realm of interactive entertainment.

Now, before we move into the Land of Sweeping Generalizations, a disclaimer: there are girls and women who like to slaughter mutant humanoids as much as any man does, and whose only discontent with *Duke Nukem* is that the bloodbaths it facilitates are simply too tepid; on the other hand, there are boys and men who don't immediately turn into glassy-eyed alien snuff zombies when presented with the latest *Doom* level. That said, Laurel's research did reveal certain patterns and tendencies.

"Girls enjoy complex social interaction," Laurel says. "Their verbal skills—and their delight in using them—develop earlier than boys'." Laurel further found that while girls often feel their own lives are boring, and thus have an interest in acting out other lives, they like to do so in familiar settings with characters who behave like people they actually know.

"We also learned that girls are extremely fond of transmedia," Laurel continues. "Things that make a magical migration from one media to the next. Or things that can appear in more than one form, like those Transformer toys." As it turns out, Transformers—the plastic contraptions that lead dual lives as robots and heavy artillery—offer a vivid example of how girls and boys tend to approach toys differently: whereas boys are apt to use them as a means of demonstrating mastery, concentrating on the ability to transform them as quickly as possible, girls focus on their magical quality, taking delight in the fact that the toy has a secret.

Laurel may have been one of the first to try to crack the elusive girl's market, but she wasn't alone. Heidi Dangelmaier, a former doctoral candidate from Princeton's computer science program, left the school in 1992 to wage an outspoken campaign to get traditional developers to make titles for girls. Patricia Flanigan, an entrepreneur who'd previously specialized in children's furniture, started Her Interactive, the first company devoted exclusively to developing interactive entertainment for girls. Laura Groppe, a former movie and music video producer, started Girl Games Inc. Doug Glen at Mattel Media launched a multimillion-dollar effort to turn the company's successful brands into digital designs (see "Gender Blender," *Wired* 4.11, page 190).

These innovators were doing research of their own, and reaching conclusions that echoed Laurel's. "It all comes down to the nature of value," says Dangelmaier, who after brief bouts of corporate kick-boxing with Sega and other traditional developers ended up cofounding a Web development company called Hi-D. "What's worth spending time on? What's a waste of time? Females want experiences where they can make emotional and social discoveries they can apply to their own lives."

Sheri Graner Ray, a producer who left her job at Origin Systems when she grew frustrated with her colleagues' lack of interest in female players, agrees. What girls and women want, says Graner Ray, now director of product development at Her Interactive, is a game that allows them to create "mutually beneficial solutions to socially significant problems." By socially significant problems she means conflicts that happen in a social realm, that involve a group of people rather than a lone space commando going up against a ceaseless supply of enemies. In such a context, girls can use skills they tend to find more compelling than trigger-finger aggression—diplomacy, negotiation, compromise, and manipulation. "This doesn't mean there can't be fighting or combat in a game for females," adds Graner Ray. "There just has to be something beyond confrontation as the reward."

BUT ENOUGH WITH THE THEORIES . . . WHAT DO THE NEW TITLES FOR GIRLS LOOK LIKE?

In the case of Interval Research, we don't really know yet. While the firm recently announced the formation of Purple Moon Inc., a spin-off that will publish products informed by Brenda Laurel's research, none will hit the shelves until late 1997. And until they do, the company's keeping them under wraps. Still, Laurel, who will serve as Purple Moon's vice president of design, drops enough details to suggest that these products will differ vastly from the single-minded mayhem of the typical shoot-'em-up. Indeed, inside the company, they're referring to the titles as a whole new genre: "friendship adventures for girls."

Laurel says Purple Moon will launch with two multi-title product lines, which will focus on making friends and shared experiences. The lines will take place in different environments: one in a more social world, with settings like school and the principal's office and a focus on day-to-day issues; the other in a dreamier, neoromantic world of secret gardens and moonlit trails overlooking the ocean, where nature and reflection are emphasized. Both series will include a strong storytelling and narrative element and many of the same characters, but no clocks and no scores. "We want to let girls play in an exploratory, open-ended fashion, to let them have control over their environment," says Laurel. To extend these environments (and profit margins, no doubt) beyond the realm of the computer, a battery of offshoot merchandise is in the works.

Interactive stories—like Her Interactive's *McKenzie & Co.* and *Vampire Diaries,* or DreamWorks Interactive's *Goosebumps*—follow more traditional game models but include elements rare in the interactive entertainment world: teenage girl protagonists and plots that aren't based on killing someone, finding out why someone was killed, or taking over the world. Story lines focus on problem-solving, investigation, and communication with onscreen characters as a key to progressing through the drama.

And then there are titles like *Chop Suey* and *Mimi Smartypants,* the work of writer–producer Theresa Duncan, featuring nonlinear, fictional worlds to explore. With its sly whimsy and tactile, folk-art imagery, *Chop Suey* brings a whole new sensibility—quirky, poetic, almost bittersweet—to a medium that's often lacking in such nuance.

Finally, there's that feminist-nightmare blockbuster, *Barbie Fashion Designer.* Unlike almost every other interactive entertainment title, *Barbie* exists as a mere part of an overall play experience. "Instead of looking at the computer as a game machine, we looked at it as a power tool that makes things," explains Doug Glen, president of Mattel Media. Given that that's exactly how the computer is seen in many other application categories, this is hardly an earthshattering observation. And yet very few interactive entertainment titles employ this metaphor. In the case of *Barbie Fashion Designer,* girls can make clothes for their dolls by choosing styles, patterns, and colors onscreen, then printing the resulting outfits on special paper-backed fabric that can be run through an inkjet or laser printer. At that point, they can use color markers, fabric paint, and other materials that come with the package to further enhance their designs. Like so many of the toy industry's most successful "interactive entertainment" products—think Lego, Lincoln Logs, even Barbie herself—*Barbie Fashion Designer* is designed to let the user's imagination become the most important part of the play experience. In so many children's titles—and to a lesser extent in CD-ROM games aimed at older audiences—this simply doesn't happen. "It's CAD software for kids," says Ann Stephens, president of the high-tech analysis firm PC Data. "Barbie did so well because it's a very good product that incorporates girl play models and a strong franchise."

Now, titles that emphasize fashion and makeup might sound like a conspiracy hatched by Rush Limbaugh to turn prospective riot grrrls into complacent, pretty little consumers. But if the product's intended audience likes it, and if it introduces them to the world of technology, then why complain? This, at least, is how Her Interactive's Patricia Flanigan responds to critics. Besides, she points out, her company surveyed 2,000 girls before embarking on development of *McKenzie & Co.* and found that makeup, fashion, shopping, and boys were subjects girls wanted to see.

In addition, *McKenzie* and similar titles have real utility; they let girls experiment—in a comfortable way—with identity, appearance, and communication at an age when these things are extremely important to them. They also familiarize girls with interface and interactive media conventions. Indeed, diary-style titles like Girl Games's *Let's Talk About Me* and Philips Media's *The Baby-Sitter's Club Friendship Kit* are practically full-blown personal information managers, with address books, calendars, daily planners, diaries, and other pre-Office features built into them.

But however individual developers feel about selling stereotyped girl themes, most in the interactive entertainment industry are overjoyed by *Barbie Fashion Designer*'s success. In one fell swoop, *Barbie* cracked open the market for girl games. Purple Moon vice president Nancy Deyo has nothing but praise for the title: "We're thrilled to see Barbie do so well," she says. "We're going to enter a retail market that simply didn't exist six months ago."

THE FUTURE OF GIRL GAMES

The greatest potential for girl games still lies largely untapped. At a time when many companies view the Web as an all-purpose revenue enhancer, expected to add mouthwatering zest to even the blandest business plan, the firms focusing on interactive entertainment for girls seem to have reason to be licking their chops. As Brenda Laurel says, "The Web has an innate sociability—so there's loads of potential for activities that appeal to girls' social intelligence, their penchant for narrative play."

For many girls, the online world has already begun to supersede that sacred tool of female adolescence, the telephone. According to Aliza Sherman, creator of the popular Cybergrrl Web site, "Girls want to meet other girls their age and they really want to chat. When we held a Team Webgrrls event to teach 25 girls age 5 to 15 to learn to surf the Web, they got the biggest charge out of the CU-SeeMe and IRC instead of the Web sites themselves. They wanted to make contact and interact."

Ellen Steuer, a 20-year-old sophomore at Mills College in Oakland, California, first started going online when she was in high school. "My brothers wanted me to get a computer because they said I'd need one in college," explains Steuer. "I didn't really have any interest in computers until I discovered chat. For me, it's all about people—I've become friends with so many people I never would have met except online. And along the way, I really learned some interesting stuff."

So much stuff, in fact, that soon she was switching from her initial AOL account to an ISP and creating her own Web site. Today she has a job as a technical assistant in her college's information technology department and is planning to pursue a career that involves the Internet. In short, she's a perfect example of a girl whose introduction to technology has had a major impact on her life. By the time she finishes school, she'll have more than six years of experience using interactive technologies.

Thousands of other girls are creating their own Web pages and chat rooms, forming alliances to promote each other's pages, and sometimes even starting secret clubs that require a password to view other members' sites. For developers, then, the question is this: How can we create products and services that can add to what girls are already doing themselves online? In their efforts to answer this question, Her Interactive and Girl Games Inc. have created community-oriented Web sites, with bulletin boards, advice columns, contests, pen pals, interviews with mentors, and online games. None of these sites is exactly cutting edge, but compared with the brochure-style sites that many traditional game developers have put up, their grasp of basic cyberspace principles is quite apparent.

In addition to creating its own site, Girl Tech, a start-up targeting 6- to 14-year-old girls, has several other Web projects in development. The company's trying to negotiate a deal with a major search directory for use of its "girl-friendly" rating system, and it's also created a book called *Tech Girl's Internet Adventures*. Along with site reviews and basic how-to information about the Web, the book includes a CD-ROM with software and clip art that girls can use to develop their own Web pages.

But how much interest do 6-year-old girls have in the Web? Girl Tech's founder and CEO Janese Swanson tells the story of how she helped her daughter have an online conversation with another girl on the other side of the country—who, in turn, was being helped by her dad. "My daughter loved learning about this little girl from a different part of the country," says Swanson, who as a product manager at Broderbund helped create *Where in the World Is Carmen Sandiego?* "We took out maps and asked questions about what it was like there."

As the Net continues to develop as a platform for interactive entertainment, look for girl game developers to be at the forefront. "So much of the Web's power comes through orchestrating human interactions," exclaims Heidi Dangelmaier. "The key to success in this medium lies in communication, human interaction, participation, and emotional impact."

In other words, all the things girl game evangelists have been thinking about for years.

Dip into the Future, Far as Cyborg Eye Can See—and Wince

The following piece from the British weekly *The Economist* delves into the violent and, to nonplayers, probably strange world of video games. For more on video games, see Explorations topic 2c. U.S. Kids TV is a site and organization dedicated to stemming the amount of violent entertainment directed toward children.

At the *Visions* Web Site Links to manufacturers of electronic games; to manufacturers of high-tech weapons.

READING PROMPT

What connections does this essay present between video games and real life—and what connections do you see? Compare the view of science fiction video games offered here with that of science fiction and feminism offered by Kathe Davis ("What about Us Grils?") in Chapter 15.

Ever wonder why your kids have so little hope for the future? Look at the games they play. Nowhere is tomorrow more vividly alive than in video games. These things are nearing a film-like realism in which robots, aliens and humans battle in lovingly detailed three-dimensional settings. The world's homes contain more than 150 million video game machines and multimedia computers. A big chunk of them, often for several hours a day, are time-warping those who play games on them several centuries forward, establishing a lore of the future far more ingrained than any science-fiction literature. The overwhelming theme: one way or another, the world is toast.

Video games and the future have been locked together from the beginning. The father of them all was *Space War,* which was written in 1961 on one of the first interactive computers, the Digital PDP-1, as an exercise in finding something interesting for a computer to do. The answer: move two little outline rocketships against a background of stars. It seemed only natural to allow them to shoot torpedoes at each other. Thus was born the video game. Why were the spaceships shooting at each other? Because that is what future spaceships do. But *Space War,* in a sense the first video game, was also one of the last to arrive without at least a bit of hokey sci-fi plot explaining why, exactly, you need to blow away everything that moves on the screen.

IF YOU DIDN'T BELIEVE IN ORIGINAL SIN

All novels are said to be essentially a re-telling of one of seven possible stories. Futuristic video games have about four. Man v Merciless Aliens. Man v Merciless Machines. Good Man v Merciless Legions Of Bad Men. And the catch-all category: ordinary sports or racing games or puzzles placed, on thin pretext, in the future because it looks cooler that way. In each case it all ends in tears. Today most of us may coexist relatively peacefully, but within a decade the world (what's left of it) will be at war. And the bad guys are going to win, unless someone presses just the right buttons.

To be sure, science-fiction books are no stranger to apocalyptic cliché either. Drama requires conflict of some sort, and the notion that things might actually be rather nice in a hundred years or so is no way to sell books. But written sci-fi is at least able to flesh out a vision of a strife-filled future with some chat, even a romance or two, between the laser blasts. Video games have no time for this soft side of humanity: the next wave of aliens is already on its way.

Why such poverty of plot? Unlike ordinary, old-fashioned fiction, video game plots are constrained by some tough rules. There is usually just one player, and all action must happen around him (women play plenty of video games, too, but for some reason those set in the future are aimed almost entirely at boys and men). To pass more than a minute or two without killing something or otherwise scoring is boring. Chaos must reign, or else such rampant bloodshed would seem unjustified. But parents get upset if games kill too many humans without good reason, so the characters at the receiving end of the mayhem must be made aliens or robots or savage animals, or at least something demonstrably evil on the fringe of humanity (terrorists, invading hordes, prison guards, or at least members of an opposing tribe).

For all the vast noise emerging from game machines, their characters might as well be mute. Apart from grunting, most video game personalities have no voice but the chatter of their machinegun. With nothing but a joystick to control them, figures can move speechlessly, or utter a few pre-programmed remarks. As a result, they tend to shoot first and ask questions never, which suggests a future where summary justice is the only kind.

The explanations for all this mayhem are generally cobbled together as an afterthought. Why, in the game *Pod*, are cars racing around an empty city of the future and shooting at each other? Because, you see, a virus has taken over the planet, the last shuttle is about to leave, and there is just one seat left. Why in *Darklight Conflict* are the pilots of battling alien ships speaking English? Because they are really terrestrial fighter aces kidnapped by aliens to fight their wars. It is up to the poor saps in the rendering rooms to come up with a few minutes of computer video footage that establish these story foundations.

LOVE YOUR ENEMY

Look at *Doom,* a classic. A lone marine gets a message from a space outpost that the energy stream it has been tapping turns out to be a highway to hell, and a rush-hour of demons is flooding into the ship. By the time the marine gets there, everyone is dead or zombified. So, when the marine blows away other marines, that is okay, because they are really zombies and he is saving them from a fate worse than death. Blowing away overt aliens hardly needs explaining (aliens are, *ipso facto,* Evil).

Although the plot of *Doom* is tailored to game play, it is dictated even more by computer technology. The advantage of fighting in a space outpost is that it all happens inside something, usually dark corridors. This is easy for the computer to draw in three dimensions, being largely an elementary-school exercise in vanishing points and perspective, with the additional advantage of a very limited field of view. No need to draw the wide-open world or even the next corridor: one set of walls with a few internal features is fine, and imparts a great closed-in terror to the game, too. Because it is a futuristic scene, the walls can be largely undecorated (the future, we assume, will be minimalistic), which saves time for the "paint monkeys" in software companies who must put flesh on the game sets.

After the success of *Doom* came a wave of *Doom-clones,* creating an industry-standard claustrophobic vision of the future. *Duke Nukem 3-D* took the model to a post-apocalyptic Los Angeles (blighted concrete jungles have the same game-design advantages as space outposts: lots of featureless walls, maze-like paths, few hard-to-draw horizons, and no guilt about blowing them all to smithereens), as well as to a space station and the Grand Canyon (same deal, although the walls are a bit farther away). And plenty of other variations emerged—*Descent* added a vertical dimension, and *Tunnel B1* turned the marine into a car—but the message stayed the same: in the future, battles will be fought indoors, and a notably gloomy indoors.

Computer technology now provides enough graphics horsepower to let games move outdoors, but the future has not become much brighter. Games such as *MechWarrior 2* (guide your giant robots into battle with those from the opposing clan) are set mostly on barren planets or deserts, or in cities that might have been made of building blocks—all to cut down on the number of three-dimensional objects the processor is required to draw.

Even those games, such as *Turok: Dinosaur Hunter,* which attempt to add some grass, trees and other real-world detail to their landscapes often have to use other tricks to make up for the difficulty of doing anything of the sort. "Fogging"—the notion that the farther away something is from the viewer the more it should take on the horizon's gray color—not only ensures that the number of objects to be drawn is kept to a manageable minimum but also paints its world as a misty, smoky place. It is much like early sci-fi films, when a liberal use of dry ice concealed the smallness of a Hollywood set.

Now, just as effectively, Nintendo's software tricks hide the limitations of its processor.

Populating these barren futurescapes are legions of creatures that paint a pathological portrait of life elsewhere in the universe. Those that have not risen straight from the bowels of hell have arrived, guns blazing, from a planet where evolution has taken a distinctly nasty turn. Apart from tentacled and fanged things of every description, there are flesh-throwing zombies and chainsaw-wielding fiends (*Quake*), vicious police pigs and rocket-launching mounds of blubber (*Duke Nukem*), huge mutated insects, evil spiders, cyberdemons, imps, flying flaming skulls, towering Barons of Hell, even a revenant skeleton.

J. C. Hertz, author of *Joystick Nation* (Little, Brown, 1997), by far the best analysis of the video game culture, explains the appeal of such foes:

> The *Doom* universe gives you fire and brimstone by way of cyberpunk—everything that flies at you seems to combine medieval demonology with advanced robotics. Religion doesn't provide depictions of evil this vivid any more. We miss those honest-to-God, pitchfork-carrying, cloven-hoofed, Lake-of-Fire, shit-kicking devils. *Doom* fills that niche.

There are no behavioral ambiguities or moral dilemmas in this sort of video game, Ms. Hertz observes. If it moves, shoot it.

You get to visit a place where there is no way to humanize the enemy because the enemy is, by definition, Evil. Not just bad. Not misunderstood. Not the victim of childhood abuse, ethnic discrimination, faulty anti-depressants or low self-esteem. Not a belligerent race of aliens on *Star Trek* with whom you have some responsibility to negotiate and understand. It's the devil, OK? It's printed right there in the instruction booklet: "They have no pity, no mercy, take no quarter and crave none. They're the perfect enemy." Yum.

Perfect enemies are a scarce commodity in today's world. Hitler was that (and the predecessor of *Doom, Wolfenstein 3-D,* pitted you against him and the SS), but they have paled recently. Even Saddam Hussein cannot quite touch the aliens of *Fade to Black,* with their crab legs for lips.

BLAME CAPITALISM, THE MILLENNIUM

Video games' nightmarish view of the future often reflects the paranoia of their day. *Missile Command,* a 1980 classic in which a player aimed anti-missile missiles to stop an incoming shower of nukes, was a product of the cold war and the yearning for a recovered security that led to the Strategic Defense Initiative. *Robotron* and *Tron,* both 1982 games (*Tron* based on a film of the same name), were built around the notion of computers or robots run amok, tapping a fear of technology that was arising as computers entered daily life. *Balance of Power,* in the mid-1980s, let a player be the leader of the free world, trying to avert nuclear war between the superpowers. In the late 1980s and early 1990s fighting games, such as *Mortal Kombat,* brought graphic bloodshed to the video screen just as a jaded generation was seeking new thrills.

Today's games, deprived of superpowers, the space race, cold wars and (for the moment, anyway) an imminent nuclear or bacteriological missile threat, have had to look elsewhere for their menace. Rather than a fight between warring states, many today imagine a world dominated by giant companies or a cabalistic mega-government. Religion battles against capitalism, or science. Computers and machines turn on their masters. Sometimes mankind itself has become malignant, and the peaceful aliens must defend themselves from it. The economy is in ruins, pollution has devastated the environment, the earth's resources have run dry, disease is spreading, and the population is either bursting at the seams or down to a lone, huddled band.

Take *G-Police,* a great game but a dire glimpse of tomorrow. The year is 2097. The earth's mineral resources have been depleted, governments and society are in tatters, and powerful multinationals are racing to plunder the solar system for ore. Fuelled by massive profits, they seize control, allowing token governments to set up a police force to keep order in *Blade Runner*–like domed cities, where smokestacks belch flame and it is always night. But the corporation still really runs the show, and it's crooked, naturally enough. A few brave cops in cool helicopter gunships have to untangle the mess. Huge explosions ensue.

In *Syndicate Wars,* besuited businessmen from the Syndicate run what amounts to a dictatorship, blinding the masses to the squalor of their cities by putting a Utopia chip in their brains. But the rogue Church of the New Epoch inserts a virus into the Syndicate's computers and law and order begin to break down. Unguided citizens riot and loot, and zealot-filled churches sprout. You can play the Syndicate side and restore order or play Church and try to bring on Armageddon. Either way, it's shoot to kill. In *MDK* (Murder Death Kill) huge alien ore-eating robots are digesting the world's cities. It is up to one man with a clever gun to stop them; every minute he dallies is a few million lives lost.

In *Epidemic,* the human race has moved underground to escape a virus. The Biflos Corporation runs the subterranean Neurol City, oppressing its people with the power of its central computer as it searches for the perfect human DNA that will ensure the survival of the race, albeit without the nobler traits of humanity. You are part of a rebel team that seeks to undermine the corporation. Man your Mech and fire away.

Blame millenarianism, which has given software developers license to repackage any game concept as a post-apocalyptic duel to the death. Or blame a paucity of imagination: thanks to *Blade Runner,* a bit of blighted-future hand-waving conjures up a vivid enough image to justify the absence of any explanation of why, for example, you need to kill replicants. Or perhaps just blame a demand for 3-D games that has exceeded the ability of the hardware to deliver anything like the rich textures of reality, thus limiting most worlds to easy-to-render blasted moonscapes or the usual corridors of death.

A HOPE TO CLING ONTO

Whatever the reason, today's game-players are getting a pretty stiff dose of gloom. Is this sort of entertainment going to produce a generation of little nihilists, as ethically vacuous as they are aimless? As the first wave of arcade video games mesmerized teenagers in the early 1980s, parents and politicians lamented that the impressionable button-punchers would emerge from their video trance as moral zombies, who could barely distinguish between zapping aliens and other kids, or care much.

All this generated a 1991 book, *Video Kids,* in which Eugene Provenzo, a professor of education, said that video games were corrupting the youth of the world by failing to foster a sense of community: "Each person is out for himself. One must shoot or be shot, consume or be consumed, fight or lose." By 1993 America's Congress, horrified by the violence of *Mortal Kombat* and its kin, was debating whether it ought to restrict such games. Some American towns banned video games. In the end, the video game industry dodged a bullet by agreeing to a voluntary ratings system. But in the mind of every parent watching Junior gun down another wave of photorealistic humanoids in a radioactive wasteland, the worry undoubtedly remains.

It is right to worry about the violence; but the wider and vaguer bleakness seen in most of these games is so silly that even children are probably able to smile at it. On screen it is always twilight. What plot there is gets forgotten within seconds of the first encounter with something fanged. Who cares why the Krakov Corporation is plundering the earth's resources? Line them up in the sights and pull the trigger.

Cheerier parents, looking past the grim storylines, may see video games as a classroom for 21st-century skills. As everything from share-trading to manufacturing becomes increasingly computerized, it gets very hard to tell the difference between a game and many jobs (except that the games are usually more fun). Modern armies use video game–like tank simulators to train tank crews, aircraft simulators to train pilots, and war games to train generals.

Ronald Reagan presciently said that today's joystick jockeys would be tomorrow's soldiers. The Gulf War, with cruise missiles guided like the rocket-propelled grenades in *MDK,* made his point. "Concerned mothers can now rest assured that their children have a mandate, if not a moral obligation, to play as much *Virtua Fighter* as possible," writes Ms. Hertz firmly. "It's in the interest of national security."

chapter 2 Explorations

2a Toys, Mobile and Stationary

Collaboration. Share with classmates two lists of toys you have had with brief descriptions: one list of toys, such as a yo-yo, that are designed to move through space (either on their own or propelled by you) and a second list of earthbound toys. Discuss the particular virtues of each kind of toy, mobile and stationary.

Web work. For more than you ever wanted to know about yo-yos, try Dave's Wonderful World of Yo-yos or the Yo-yo Guy, where you can participate in a Web discussion area.

There is much less on the Web about tops or marbles, or about that toy sensation of the 1950s that moved only with the greatest of effort: the hula hoop.

One writing topic. Write an essay on the role motion (or the lack of motion) plays in our childhood experience with toys or with objects we make into toys. How may toys like yo-yos and hula hoops fare in the twenty-first century?

2b Traditional Toys

Collaboration. Share with classmates a list, with brief descriptions, of toys you have encountered that were in existence when your parents were children.

Web work. To learn more about traditional toys, visit the London Toy Museum. Not all toys your parents or even your grandparents played with are quite traditional enough to earn a place in the London Toy Museum. Two other sites to visit for classic toys are the Ohio Art Company, home of the Etch-a-Sketch, and the Official Lego Site.

One writing topic. Write an essay on the generational value of toys: What toys from an older generation did you enjoy when you were young? Why? What toys from your generation would you think your children or grandchildren might enjoy? What impact do you imagine the toy business has on the survival of traditional toys?

2c Video Games

Collaboration. Share with classmates a brief account of the first video game you remember encountering and of your favorite. Discuss what aspects of these games make them successful and addictive.

Web work. As might be expected, there is a huge amount of Web material on video games. Two places to begin: The History of Home Video Games and GamePen.

One writing topic. Write an essay on the attraction that video games had or still have for you. What, if anything, gives video games an addictive power?

2d Board Games

Collaboration. Share with classmates descriptions of a handful of favorite board games, including puzzles and card games. Discuss what made these games so enjoyable. Which of these games can be played alone?

Web work. Hasbro has sites dedicated to three of the most popular board games of all times: Monopoly, Scrabble, and Trivial Pursuit.

One writing topic. Write an essay on the difference between games played with others and games played alone.

2e Icing, Sweets, and Other Unnecessary Trifles

Collaboration. Most people associate candies, desserts, and sweets with strong feelings—usually with rewards and comfort, although sometimes with guilt. Make and share a list of memorable experiences associated with sweets.

Web work. The Web is a haven for anyone with a sweet tooth. For candy, check out Hershey's site, the exclusive M&Ms site, or England's Cadbury site. To combine an interest in candy with the earlier topic of glamour, have a look at such upscale chocolate sites as Godiva's. Cakes of Elegance in Dallas, Texas, allows you to view their wedding cakes online.

One writing topic. Write an essay classifying the major sweets in your life, either by the types of sweets you most enjoy or by the different times and ways and reasons you enjoy them. What would your life be without sweets?

2f Toys and Gender

Collaboration. Share with classmates experiences with toys that illustrate toys' connection with gender. What toys did you have that most strongly indicated gender, whether your gender or not? Which toys did you have that did not indicate gender at all?

Web work. Mattel hosts the official Barbie site; Barbienet is a good unofficial site. Hasbro hosts the parallel GI Joe site.

One writing topic. Write an essay on your sense of how toys helped shape your current attitude about gender, and reflect on what you would do differently or the same with children born today.

chapter 3

Gadgets and the Flight from Simplicity

READING QUESTIONS

- What are some gadgets you have used in the past, still use today, or are eager to start using? What are some you have collected but never used?
- What are the joys of things that don't work well or efficiently, of useless things, or of things totally out of order?
- How do you feel about new technological solutions to small, everyday tasks—the more the better? Or the less the better?

QUOTATIONS

Americans . . . attach such a fantastic importance to their baths and plumbing and gadgets of all sorts. They talk as if people could hardly be human beings without all that; we in Europe are beginning to wonder if people can be human beings *with* it.

— *Ann Bridge (1946), British author, 1891–1974*

It's kind of fun to do the impossible.

— *Walt Disney, American filmmaker, 1901–1966*

Without order nothing can exist—without chaos nothing can evolve.

— *Anonymous*

A computer lets you make more mistakes faster than any invention in human history—with the possible exceptions of handguns and tequila.

— *Mitch Ratcliffe (1992), media commentator*

Creativity is the sudden cessation of stupidity.

— *Attributed to Edwin Land, inventor of Polaroid photography*

Keeping Screen Doors Closed

Rube Goldberg

Rube Goldberg (1883–1970) was one of America's most celebrated cartoonists, hence a major satirist as well. Although most famous for cartoons depicting outlandishly complicated machines that perform simple tasks (now appropriately called "Rube Goldberg" devices), he worked in many styles and areas, winning the Pulitzer prize for political cartoons in 1948.

At the *Visions* Web Site Links to more Rube Goldberg cartoons.

READING PROMPT

Exactly what is Goldberg satirizing in his cartoon?

Professor Butts makes a parachute jump, forgets to pull the string and wakes up three weeks later with an automatic device for keeping screen doors closed. Houseflies (A), seeing open door, fly on porch. Spider (B) descends to catch them and frightens potato-bug (C), which jumps from hammer (D), allowing it to drop on pancake turner (E) which tosses pancake into pan (F). Weight of pancake causes pan to tilt and pull cord (G) which starts mechanical soldier (H) walking. Soldier walks to edge of table and catches his head in noose (I), thereby hanging himself. Weight in noose causes string to pull lever

and push shoe against bowling ball (J), throwing it into hands of circus monkey (K) who is expert bowler. Monkey throws ball at bowling pins painted on screen door thereby closing it with a bang.

The monkey is liable to get sore when he discovers that the bowling pins are phony, so it is a good idea to take him to a real bowling alley once in a while just to keep his good will.

Rubes in Training *Jeffrey Kluger*

Jeffrey Kluger's essay depicts an actual annual event, The Rube Goldberg Machine Contest, sponsored by the Purdue University chapter of a national engineering honor society. In Kluger's account of one year's contest, we can see a little of how play and invention can overlap. While the contest seems to take the form of a joke, there is finally something serious about the skill and ingenuity a student needs to cart home the trophy.

The contest described below is based on the world's longest-running gadget joke: how to screw in a lightbulb.

At the *Visions* Web Site Links to the official contest rules; to lightbulb jokes; to the College of Engineering at Purdue University.

READING PROMPT

How does the notion of uselessness figure in Kluger's depiction of collegiate humor? How do you explain the fascination with uselessness among students in active training as engineers, who are preparing to devote their lives to building useful things? Those interested in the relationship between useful and useless might want to look at Explorations topic 3c, Trivial Pursuits.

If you've gone shopping any time in the last decade or two, you know that lately Yankee ingenuity has been getting a little less ingenious. In a global economy becoming ever more competitive, American products have fallen further and further behind the merchandise of most high-tech, cutting-edge countries—like Japan, Germany, Taiwan, Togo, Uzbekistan, and Western Samoa.

Most economists agree that the first sign of Yankee decline came more than 20 years ago, with the introduction of the Ford Pinto—a car whose imaginative Blows-Up-When-Hit-From-Behind option proved surprisingly unpopular with finicky consumers—and Chevrolet's Chevette, a teensy-weensy economy car that doubled as an attractive tie tack. In subsequent years more and more American industries fell by the wayside, until, by the 1990s, the only piece of Yankee engineering to lead the world in innovation is the wildly successful SaladShooter, a product that has established US dominance in the exciting field of ballistic vegetables.

All is not lost, however. If the United States can no longer build marketable machines, we can still, indisputably, build funny machines. Having long ago set the global standard for joy buzzer and light-up necktie technology, America has maintained its dominance in the merriment market with such innovations as IBM's rib-ticklingly disastrous PC Junior and NASA's sidesplitting blueprints for the space station Freedom. Recently the country took one more step toward ensuring its technologically frolicsome future, with this year's edition of the National Rube Goldberg Machine Contest at Purdue University in West Lafayette, Indiana.

Rube Goldberg, as you know if you ever opened a Sunday comics section during the first two-thirds of this century, was a cartoonist who became famous for drawing fantastically complicated machines that performed fantastically simple tasks. A typical Rube Goldberg device could not perform a job as straightforward as, say, turning on a faucet without the assistance of pulleys, relays, switches, cables, fulcrums, mousetraps, and, when necessary, actual mice. In an era like ours, in which just this kind of technology must be used to install and program a VCR, such cartoons have lost some of their appeal. In Goldberg's era, however, things were simpler, and by the time the cartoonist retired, the term Rube Goldbergian had been enshrined in the language to describe anything characterized by excess complexity. (Such linguistic homage is a rare honor indeed, granted only by such terms as Stalinist, Orwellian, and the newly coined Clintonesque, which describes the act of running a large country while consuming your own body weight in Pop-Tarts.)

Since 1983 Purdue University and the Theta Tau engineering fraternity have held an annual contest in which engineering students from around the country invent Rube Goldberg–style machines to perform such tasks as unlocking a lock, sticking a stamp on a letter, and toasting a piece of bread. This year's competition called for a gadget that could screw a light bulb into a working socket. Tired of seeing America take it on the chin from countries that wouldn't know how to build a good whoopee cushion if you spotted them the whoop, I decided to visit Purdue and see for myself that the USA could still build the best darn hardy-hardware anywhere in the world.

The Rube Goldberg contest was held in the Elliott Hall of Music on the Purdue campus. The six machines that had qualified for the finals were displayed along the edge of the stage and, to the lay eye, were utterly impossible to figure out. According to the contest rules, the designers could build their gizmos out of virtually anything, but the machines had to meet some specific requirements: each had to measure five feet by five feet by six feet; each had to include at least 20 steps before the light bulb was screwed in; each machine was permitted to include flying objects and other projectiles as long as they remained within the five-by-five-by-six boundaries; each had to have some theme or story tying the components together; and no machine was allowed to use combustible fluids or explosives. Near as I could tell, all the contraptions met these criteria, especially, I noted with some relief, the part about the combustibles.

More impressive to me than the look of the machines was the look of the students who had built them. When I was in college, during the mid-1970s, few people I knew had the focus, discipline, or remaining EEG tracings to even consider becoming engineers. Most of our time was spent in darkened dorm rooms, listening to The Who at volumes that could cause tectonic plates to shift, while debating such topics as whether it was possible that Ringo was actually the walrus. By the time we graduated, we had done away with nearly all of the pesky brain tissue we had come to school with in the first place, limiting our "engineering" skills to such relatively low-tech tasks as driving to 7-Eleven to pick up more Ring-Dings.

The students here today, however, were a whole different breed. Hovering lovingly about their contraptions, they had the focused expressions of NASA gantry technicians moments before a launch. Of the six machines, the first to catch my eye was the one built by the University of Wisconsin at Milwaukee, the defending champion from last year. At a glance, Milwaukee's machine was an absolutely incomprehensible array of pulleys, rails, cables, and tracks, as well as what appeared to be a collection of every stuffed animal or doll manufactured in the United States in the last 20 years, from Buzzy the Bee to Oscar the Grouch to Oscar the Clinically Depressed. The name of the machine, Rube's Toyland, seemed more than apt.

"Our machine is sort of my prodigy," said Spencer Koenig, the captain of the team. "I came up with the toy theme, but all seven members of the team contributed. James was our relay master; Rick handled power; Matt grunted out the hardware."

I asked Spencer how the machine worked and, as I feared, he began by assuring me that it was actually pretty simple—a promise engineers always make just before they start speaking in tongues. Near as I could understand from his explanation, the contraption gets started with a ball bearing rolling down a chute, which sets into motion a chain of events in which one stuffed animal bumps into another stuffed animal, which swings to the other side of the machine on a trapeze, throwing a switch on a third, fourth, and fifth stuffed animal, which eventually phone out for a stuffed electrician to drop by and screw the light bulb in for them.

No fool, I could tell by the look on Spencer's face that I didn't quite get it yet, and that, with dozens of people clamoring to see his machine, he probably wished I'd get something else—like lost. Adjacent to Rube's Toyland, I noticed the entry from Hofstra University in Hempstead, New York, and immediately felt I'd have better luck with this one. Dubbed Fester's Finaglers, the machine was designed around an Addams family theme, with accouterments like guillotines, witch's castles, beakers of bogus blood, and the occasional flapping bat. The Addams family is enormously popular with college students in the 1990s (a huge improvement over the 1970s, when it was the vastly more ghoulish Loud family), and a crowd formed around the Hofstra machine that would make Herman Munster green—or greener.

"This is our first Rube Goldberg competition," said Hofstra junior Nick Croce, "and we wanted to do something memorable. We've been at work on this for months, and we put it together with parts from practically everywhere. One of the motors was taken from an oscillograph built in the 1950s. One of the plastic animals was actually a toy one team member had in 1976, when he was five."

Nick did not attempt to explain to me how the Hofstra machine worked but instead handed me a sheet of paper with all 27 steps clearly spelled out. The list took some puzzling out, but as close as I could figure it, the contraption works this way:

The action begins when "Gomez Addams," in the form of Nick or one of his Hofstra cohorts, pulls a tiny noose in one corner of the machine. The noose raises a barricade that sends a car speeding down a track. The car, equipped with a prong in front, pops a balloon which, when burst, slips through a ring that releases a toy guillotine. The guillotine blade falls onto a prone Pugsley doll, chopping off his head. The head rolls down a chute, landing on a switch that activates an elevator. The elevator rises and tips forward, releasing a ball that rolls down a track and lands between two relays, completing a circuit. The circuit triggers Spot the dinosaur (borrowed from the Munsters since "the Addamses didn't have a good pet," according to Nick). Spot walks into a toggle switch, turning on a pair of toy trains. The trains collide, causing one to roll off the track, fall into a basket, and depress a switch. The switch turns on a fan, which blows a magnetically levitated sail-car across a bridge. The car strikes a switch, releasing a plastic spider on a string. The spider falls, pulling a switch that releases marbles. The marbles trigger another fan, which rotates an Archimedes' screw, which in turn pulls a string on a mousetrap. The snapping trap knocks a support bar out from underneath a can of fake blood, which spills into a funnel. The blood runs through a length of plastic tubing that spells out Hofstra, eventually draining into a jar filled with plastic eyeballs. The eyeballs rise, striking a switch, which activates a toy witch. The witch pivots slightly, pulling a cord on a rattrap. The trap causes a gun to fire a rubber dart at a series of tombstones, which topple domino-style, striking yet another mousetrap. The mousetrap snaps, pulling a cable that releases a dead bolt from a spring-loaded box. The lid of the box opens, and an arm and hand—the Addams family's Thing—rises from it. Holding a light bulb, Thing begins rotating clockwise, rising toward a papier-mâché Uncle Fester head. When the bulb reaches Uncle Fester's mouth, Thing screws it in and the light goes on.

Fester's Finaglers had the look of a winner, and Nick was understandably pleased with it. According to rumors swirling around the contest, Pugsley-head technology had already been pirated by the Japanese and was even now being sold to countries throughout the Pacific Rim. For the Hofstra engineers, however, it was the eyeball-marinating jar that inspired the most pride. "It took us a while to get just the right number of eyes," Nick said confidentially.

While the entries from the other four schools—Oakland University, Lawrence Technological Institute, Western Michigan, and sentimental-favorite Purdue—didn't have the dazzle of Hofstra's, all were impressive. The Lawrence students chose an Operation Desert Storm theme for their machine, and their contraption bristled with plastic missiles and other memorabilia from the Gulf War. For the most part, however, the crowd seemed to be giving the machine a pass, possibly because Operation Desert Storm has long been yesterday's news (though a more topical reference, like Operation Comprehensive Health Care and Deficit Reduction, would probably have been less dramatic). Purdue selected a spy theme, in which the mission of the secret agents was to destroy missiles aimed at Purdue by the sinister organization Big Red, otherwise known as Purdue's arch-rival, Indiana University. The Purdue students managed to incorporate a decapitated figurine of Bobby Knight, IU's basketball coach, into the contraption. "It's not absolutely necessary for contestants to decapitate Bobby Knight in their machine," the master of ceremonies informed the crowd, "but it helps.") To its credit, Purdue was also the one school that managed to include female students on its team. The over-concentration of Y-chromosome contestants was a bit puzzling, since while the contest was sponsored by a fraternity, there certainly wasn't any No Gurlz Allowed policy, and as far as I know, none of the female students from any of the schools had hired nannies and neglected to pay their social security.

When the crowds had a look at all the entries, the master of ceremonies called the event to order. There would be two runnings of the machines, but it would be the first one that would make the biggest impression on the judges. As it turned out, for most of the teams this was not good news. The Oakland, Western Michigan, and Purdue machines all got their bulbs successfully screwed in, but all took less than 15 seconds to complete the job. This was a tribute, of course, to how efficiently the machines were designed—something that earns you no points at all in a Rube Goldberg contest.

The real excitement began when Hofstra's and Milwaukee's entries started to roll. The Toyland machine was a dramatic, choreographed swirl of dolls and stuffed animals resembling less a Rube Goldberg machine than *A Chorus Line* with floppy ears and paw pads. The Addams family machine was even more impressive. As cinematic giants like Sam Peckinpah learned a generation ago, you can never go wrong with fake blood, and when the red food coloring began to run through the Hofstra tubing and trickle into the eyeball jar, the audience was on the edge of its seats. When Thing finally screwed the bulb into Uncle Fester's mouth, the ovation could be heard in Indianapolis.

"They really have us on the theme," Spencer said disconsolately as he reset his Toyland components between rounds. "It's tough to compete with something as popular as the Addams family."

"It's that fake blood," grumbled junior John Cerone, across the stage with the Western Michigan team. "It takes forever to flow through that tubing. Those guys really build the suspense."

In the second round the trend only continued, and the judges retired backstage for just a minute before returning with their decision: third place and $250 to Purdue; second place and $300 to Rube's Toyland; and first place and $500 to Fester's Finaglers of Hofstra.

No sooner were the kudos offered and the trophies distributed than the students in the 1993 Goldberg contest were talking about the 1994 edition. The next competition will be held at Purdue in early spring, and nearly all the contestants here vowed to be back—most of them promising to take a hint from Hofstra and model their inventions after one of this year's popular movies (though, if Hofstra itself chooses *The Crying Game,* I for one don't want to see what part of the human anatomy emerges from its spring-loaded box).

The task next year is to invent a machine that can make a cup of coffee, but before the sponsors of the Goldberg contest take on any new challenges, perhaps they should try to perfect some of their stamp-licking or bread-toasting machines of past tournaments. What about a machine that can stick a postage stamp on an envelope and then see that the letter gets delivered within the half-life of a piece of uranium? What about a machine that can toast a slice of white bread and then keep it intact when you try to butter it with a pat of Land O' Lakes that's been refrigerated to the consistency of a Scrabble tile? A Goldberg gadget like these could be a gold mine. After all, there's a rube born every minute.

Not Available in Stores

Mark Kingwell

Mark Kingwell here writes about "infomercials," that new breed of television show that mixes advertisement and low-level audience-participation entertainment. The single person most responsible for creating this new genre is longtime television pitch person Ron Popeil, now with his own Web presence.

At the *Visions* Web Site Links to the official Ronco Web site; to more on infomercials.

READING PROMPT

In the 1960s the term camp *came into use in reference to the pleasure so many people seem to get from truly tasteless aspects of popular culture. To what extent is this term helpful in describing Kingwell's or your own response to infomercials?*

It begins like one of those cozy Women's Television Network chat shows, complete with bad lighting, fuzzy lenses, and warm looks. The host is an attractive, soft-spoken woman of a certain age. She purrs at the camera. She and her guests are here to tell you about what she chucklingly calls "Hollywood's breast-kept secret." Yes, it's true: Accents, the Plasticine bust enhancers favored by movie stars and models alike, are now available to you, the lowly viewer. No surgery. No hideous contraptions. You don't even have to leave home to get them.

And what a difference they make! Soon a line-up of gorgeous but slightly flat-chested women are being transformed before your eyes into jiggly supermodels or *Baywatch* lifeguards. These flesh-colored slabs of silicone gel that "fit into any underwire bra" and "within minutes warm to your natural body temperature" can actually be used in the swimming pool! At the end of the half-hour, the ever-smiling host and her guests admit that they are all wearing Accents themselves! Well, shut my mouth.

"Accents" is only the most outrageous of the current crop of television infomercials: those over-the-top attempts to hawk make-up, cleaning products, and ab-flexers under the guise of a genial talk show *(Kathie Lee Talks)* or breathless science program *(Amazing Discoveries!)*. Turn on your television late at night or on a weekend afternoon—even, these days, at midmorning—and the good-natured hosts, a has-been actress (Ali McGraw) or never-was celeb (Ed McMahon), are touting cosmetics or miracle car wax as if they are doing us a public service. Information + commercial = infomercial. Line up the word, and the phenomenon, next to those long advertising features in newspapers and magazines, often slyly imitating the publications' actual typeface and design, known as "advertorials."

Patently absurd, maybe, but if emerging trends continue, infomercials will not remain what they have been so far—a marginal and benign, if irritating, television presence. With the loosening of CRTC [Canadian Radio–Television and Telecommunications Commission] regulations, the explosion of cable channels, and the crude economics that can make them more lucrative than regular programming for network affiliates, infomercials are showing up in more and more places on the TV schedule, elbowing aside such popular quality fare as Sunday-afternoon sports, syndicated comedies, and old movies. They are also getting more and more sophisticated, as big-name companies with mainstream products—Ford Motor Co., Procter & Gamble, Apple Canada—enter the infomercial market.

And if, as enthusiasts in the business press insist, this is the future of TV advertising, then that is very bad news indeed for television and its viewers. But not because there is anything inherently wrong with infomercials, at least not as they have existed until now. The delicate pact between ads and shows that makes television possible has always been able to withstand the amateurish, ad-becomes-show genre they represent. But when infomercials are everywhere, and especially when they go high market, that pact is in danger of being overturned, and the thin line between entertainment and pitch may be erased for good.

Blame Ron Popeil. Blame him a lot, and at length. Blame him until his smiling, trout-like face is imprinted on your mind as the fount of all evil. Because Popeil is the one who started the sort of television hard sell that reaches its tacky terminus in today's infomercials. Founder of Ronco, restless inventor of the Popeil Pocket Fisherman, the Patti-Stacker, and other cheesy "labor-saving" devices too numerous to mention, Popeil is the guy who all but invented television shopping. In the 1970s he discovered that people got very excited, and very willing to spend, at the thought that you need never leave your couch to have the entire Ronco or K-Tel product line delivered to your home. His favorite author was the guy who came up with "Call this toll-free number now."

Popeil has recently come out from behind the camera to appear in his own convection-oven and pasta-machine infomercials. Looking like an also-ran from a professional tanning competition, he slops flour and water into slowly spinning machines that disgorge brightly colored goo for thirty minutes. Your own fresh pasta every night! Operators are standing by!

It isn't hard to decipher what makes these and other low-end infomercials so successful. Potential buyers are never made to feel bad, even as their baser desires are being pandered to. For example, we are told at least four times that Accents "are shipped confidentially" and arrive at your door in (get this) "a beautiful designer chest that will look great on your vanity." The Accents people even muster expert opinion, the sine qua non of the TV hard sell. In this case, it's a panel of Hollywood make-up artists and photographers. "I tried everything," says one. "Foam pads, wires, push-up bras, duct tape. Nothing works like Accents." (Duct tape?)

The same forms of reassurance are visible on all the successful infomercials now airing, from The Stimulator to the Ab-Roller Plus. The Stimulator—a small syringe-like device that is supposed to kill pain by means of mild electric shock, a sort of mini stun gun—also produces what has to be the funniest infomercial moment of all time. Evel Knievel, the all-but-forgotten daredevil of the 1970s, shares, over footage of his famous Caesars Palace motorcycle crash, his belief in the pain-relieving properties of The Stimulator. "If it hepped me," Knievel twangs, "it can hep you." Now that's expert opinion.

This is so silly that it is easy to imagine a kind of self-parody operating, of the sort in the hilarious *Money Show* spots on CBC's "This Hour Has 22 Minutes": "Gus, I want to pay less in taxes, but I'm not sure how." "Marsha, it couldn't be easier; stop filing your returns!" But that would misread the intentions of the makers—and the attitudes of the audience, whose response to infomercials has been wholehearted. Canadians spent $100 million on infomercial products in 1995, up thirty-four per cent from 1994. One Ontario company, Iona Appliances Inc., quadrupled annual sales of its "dual-cyclonic" vacuum cleaner when it started marketing via infomercial.

In fact, the point of infomercials has so far been their lack of sophistication. The niche is still dominated by the charmingly inept likes of Quality Special Products, the Canadian company responsible for such thoroughly trailer-park items as the Sweepa ("The last broom you'll ever have to buy!") and the Sophist-O-Twist hair accessory ("French braids made easy!").

Most current efforts eschew the cleverness and quality visible on more traditional commercial spots in favor of the lowball aesthetic of public-access cable. Instead of competing with shows for our attention—and therefore being pushed to find better writing, multimillion-dollar budgets, and gilt-edged directorial talent—infomercials become the shows. Yet they do so in ways so obviously half-hearted that nobody, not even the quintessential couch-potato viewer, could actually be fooled. The talk-show cover story is really nothing more than a tacit agreement between marketer and viewer that they're going to spend half an hour in each other's company, working over a deal.

And this is what many critics miss: most infomercials, as they now appear, aren't really trying to dupe the viewer. They are instead the bottom-feeding equivalent of the irony observable in many regular commercials. Bargain-basement infomercials offer a simpler form of customer complicity than the crafty self-mockery and self-reference that appeals to young, kitsch-hungry viewers. Infomercials are a pure game of "let's pretend," taken straight from the carnival midway.

That's why the entry of high-end marketers into the field is so alarming. Big-money companies are not content to maintain the artless facade that now surrounds infomercials. They break the carny-style spell of cheap infomercials, where we know what we see is fake, but we go along anyway, and offer instead the high production quality, narrative structure, and decent acting of actual shows.

A recent Apple Canada effort, for example, which aired last year in Toronto, Calgary, and Vancouver, is set up as a saccharine half-hour sitcom about a whitebread family deciding to buy a home computer ("The Marinettis Bring Home a Computer"). It is reminiscent of *Leave It to Beaver* or *The Wonder Years*, complete with Mom, Pop, Gramps, the family dog, and an annoying pre-teen narrator named TJ. Gramps buys the computer, then bets grumpy Pop that the family will use it enough to justify the expense. Soon TJ is bringing up his slumping math grades, Mom is designing greeting cards for profit, and Gramps is e-mailing fellow opera buffs. It's nauseating, but effective. Heather Hutchison, marketing communications manager for Apple Canada, explains the company's decision to enter the infomercial universe this way: "Having produced something of higher quality," she says, "there's a recognition at—I hesitate to use the word 'subconscious,' but at a lower level—that it says something about the quality of the product. The Canadian market responds well to this kind of softer sell."

We all know that television, as it now operates, is primarily a vehicle for the delivery of advertising. That is, we know that if it weren't for ads, nobody would get to spend a million dollars on a single episode of an hour-long drama or employ some of the best dramatic writers and directors now working. True, this symbiosis is uneasy at best, with good shows all but free-riding on the masses of dreck that keep the advertisers happily reaching their targets. That's fine or at least not apocalyptic. We can accept that advertising is the price we have to pay (every seven minutes) for good television.

But slick infomercials, unlike their cheapo forebears, threaten to destroy this shaky covenant. Only a moron could mistake a low-end infomercial for a real show. (And only a condescending jerk could think that all people who buy Sweepas and Abdomenizers are, in fact, morons.) Up-market infomercials have a much greater potential to muddy the waters between advertising and programming. It may be that, without the cheesy aesthetics and side-show barker style, these new infomercials won't find an audience. But it's more likely that big companies with big budgets and top advertising talent will be able to suck even non-morons into these narrative ads that masquerade as entertainment. The new corporate offerings, in other words, may actually do what Ron Popeil couldn't: strip TV of extraneous effects like quality programming so that it finally reveals its essential nature—selling things, selling things, and selling things.

When that's true, maybe it's time to turn the damn thing off for good.

Technophiles and Technophobes

Susan Mitchell

Susan Mitchell attempts to provide a real service by classifying each of us into one of two large groups—those who are especially attracted to technology (technophiles) and those who fear it (technophobes)—even though many of us may find ourselves caught between the two groups. Not surprisingly, there are many more Web sites devoted to technophiles than to technophobes.

At the *Visions* Web Site Link to *Popular Mechanics* magazine, for years a print-based rallying point for techies.

READING PROMPT

What are the major characteristics of technophiles? Of technophobes?

Your personal communicator beeps to alert you to an incoming message. Its palm-sized screen tells you that your "smart home" needs instructions. You tell your portable computer to dial your home-control center, which relays an image of the problem. You left the rice cooking. Inserting your hand into a virtual-reality glove, you control the movements of a home robot to turn off the stovetop.

When you read the above, do you: (a) abstractly ponder what it would be like to live in such a world; (b) jump up from your chair and begin making animated hand gestures while screaming, "Yeah, yeah! That's great!"; or (c) feel the need to lie down? If you answered (a), you're a regular Jane or Joe. If you answered (b) or (c), you are probably a technophile or technophobe.

Technophiles are people who can program their VCRs. But more than that, they are "early adopters," people whose interest extends beyond the practical use of technology. They get excited about technology itself, whatever its purpose. "These are the customers who bought the original PCs in the 1970s," says Charles Humphrey, group vice president of Computer Intelligence InfoCorp of Santa Clara, California. "They are technically curious people, whether they are dealing with microwave ovens, entertainment systems, or information technologies."

But the original personal computer is like an abacus when compared with new technologies on the horizon. Americans will soon have the chance to buy interactive television, virtual-reality toys, and wireless personal communications. The world is divided into those who will eagerly embrace these new products, those who will shrug at them, and a large group who actively fears them. Understanding what makes each group tick is vital to selling any new technology and the services that go with it.

TECHTHUSIASTS

One in five American adults is a "techthusiast" who is most likely to purchase or subscribe to emerging technological products and services, according to a study by Backer Spielvogel Bates (BSB) of New York City. These 37 million adults are younger, more affluent, and better-educated than the average American. Despite their technical bent, they watch less television than average and read more magazines and newspapers.

With a median household income of $56,500, techthusiasts can afford to take a chance on new gadgets. They can also make sense of new technologies because they have completed a median of slightly more than 14 years of school, compared with less than 13 for the general population. Techthusiasts are dispersed fairly evenly throughout the U.S., but they tend to concentrate in cities that house major universities and technology-related companies, such as Boston, Houston, San Diego, and Seattle.

Techthusiasts are younger than the average American adult, with a median age of 38. "Younger people are a bit more receptive to the changes these technologies may bring," says Craig Gugel, senior vice president and executive director of media research and technology for BSB. "If you look at the learning curve and the fact that many older adults were not brought up with this technology, you find that older adults tend to be technology-averse."

For evidence of the generational technology gap, look no further than the nation's capital. When boomer president Bill Clinton moved in, he found a White House frozen in time. Telephone operators had to plug cords into the switchboard to connect callers. There was one lonely fax machine in the communications department. Clinton could barely grasp how anyone could work in an environment so low-tech that it didn't have e-mail. Clinton's young staffers quickly brought in new technologies. Now, if you want to reach the president, you can send an e-mail message to president@whitehouse.gov.

Sizable shares of adults of all ages are interested in various new technologies, according to the BSB study. But techthusiasts, by definition, find new gadgets even more appealing. For example, one-third of all adults are interested in owning a CD-ROM recorder/player, but well over one-half of techthusiasts would buy one.

Another place to look for technology lovers is among cable-television subscribers. Almost 14 percent of all adults fall into a category called "Electronic Consumer Innovators" by Mediamark Research Inc. of New York City. Yet nearly one-fourth of people living in homes with cable TV qualify for the label. Electronic Consumer Innovators are good prospects for "movies on demand," because they are more likely than other cable viewers to order pay-per-view and subscribe to premium channels, according to Mediamark. The catch is that, like the Techthusiasts identified by BSB, Innovators are not heavy cable-television viewers.

Companies that make high-tech consumer products often expect technophiles to buy whatever they produce, which may explain why few market directly to this group. But the importance of technophiles goes beyond their own purchasing power. "Computer-comfortable consumers are making buying decisions for their work groups, friends, and families," says Computer Intelligence InfoCorp's Humphrey. People who buy personal computers usually do a lot of research first, according to a survey by Computer Intelligence InfoCorp. Publications are the number-one source of information, followed by knowledgeable friends and relatives.

THE TECHNOPHOBIC CHALLENGE

Some people wouldn't touch a computer if you paid them. Technology is to them as snakes, spiders, and heights are to other fear-filled people. Fear of technology may be the phobia of the 1990s, according to a survey by Dell

Computer Corporation of Austin, Texas. As technophiles run gleefully into the 21st century, many Americans run away.

One-fourth of American adults have never used a computer or programmed a VCR. A similar share are not comfortable using a computer on their own. More striking is the nearly one-third of adults who are so intimidated by computers that they are afraid they will break them. In this group are found the most serious technophobes: nearly one-fourth of the phobes, or about 8 percent of all adults, are uncomfortable setting a digital alarm clock.

The age gap shows up here. Nine in ten teenagers are comfortable using a range of technical devices, from answering machines and VCRs to compact-disc players and computers. Only 74 percent of older adults can claim this level of ease, according to the Dell survey.

Technophobes are not necessarily opposed to the onward march of technology. But they are seriously uncomfortable using new technologies. Overcoming their fears is the key to growth for the cottage industry of high-tech education. A few years back, Dan Gookin wrote a self-help book for people who get hives looking at a "c prompt." *DOS for Dummies* has sold more than 1 million copies.

Auto manufacturers would never try to sell a car that requires drivers to understand the workings of an internal combustion engine. Yet many businesses that sell new technologies still do not realize that many consumers just want to turn it on and go.

Apple Computer is the leading exception. For years, it has touted ease of use as a major selling point. Print ads for its Performa, "The Family Macintosh," deliberately invoke techno-anxiety: "All it takes is a path command error or an AUTOEXEC.BAT file to send you or your kids back to reruns on TV." Performa has "no codes to confuse you." Apple's Newton can be carried in a pocket and runs on handwritten commands scribbled on a small screen. Hey, you don't even have to know how to type!

TECH FOR THE MASSES

Some products may succeed by selling only to technophiles. Others benefit from overcoming technophobe fears. Yet many new technologies need mass appeal to survive.

Some high-tech products have a serious image problem. In fact, they have no image at all. Roper Starch Worldwide of New York City asked people if they had heard of 15 new or emerging home-electronics products. Only four products were familiar to at least half of the general public—car phones, fax machines, laptop computers, and laser discs. It's hard to build consumer demand for products the public doesn't even know exist.

Awareness is just the first step, because people do not always stampede to buy the things they do know about. Only 10 percent of Americans who are familiar with car phones want to buy one in the next year or two. Just 7 percent want a laptop computer, 5 percent want a fax machine, and 3 percent want a laser disc. It's not clear whether fear or sheer confusion keeps people from buying certain high-tech products. But high-tech success stories can provide some clues about who buys high-tech and why they buy.

One way to measure a potential market is to gauge interest. People aged 18 to 44 are most likely to say they would be interested in purchasing one or more of the new high-tech products, according to Roper. On the other hand, true to their techie image, people who live in the West are significantly more likely to be interested in buying the latest high-tech gadgets.

Even if people want a product, they can't buy it unless they have the money. Young adults, for example, have never known a world without computers, but Roper finds that only 20 percent of households headed by people aged 18 to 29 own a home computer, compared with 28 percent of those headed by 30-to-44-year-olds. Likewise, younger-adult households express the strongest interest in owning a CD player that records, but they are no more likely than better-salaried middle-aged households to actually own one.

THE FUN FACTOR

Eighty-one percent of American households now have microwave ovens, according to Roper. Sixty-eight percent own VCRs, 32 percent have Nintendo or other electronic TV games, 21 percent own compact disc players, 20 percent own home computers, 15 percent have video cameras, and 11 percent have video-disc players. You are most likely to find these techno-toys and products in baby-boomer households headed by people aged 30 to 44. Households headed by people aged 60 and older are least likely to contain electronic goodies.

It's no surprise that income is a major factor in ownership of electronic items. Only 10 percent of households with incomes under $15,000 own a home computer, compared with 41 percent of households with incomes of $50,000 or more. Married-couple households with both spouses working are also more likely to own a variety of electronic items, including home computers, than are those with one earner.

"For a long time, the computer industry tried to sell PCs to the average consumer based on marketing messages like 'buy one of these things so your kids don't become a burden on society'," says Humphrey. "But the average consumer looked at the price and the message, and there was a disconnect." Now, prices for home computers have plummeted and pent-up demand is driving sales. Sales of home computers are expected to have risen to 5 million in 1993, an increase of 11 percent over 1992, according to Link Resources of New York City. As the new technology ages, it becomes less expensive. Price becomes less of a barrier to buyers.

Being rich doesn't automatically make you a technophile. Fifty-eight percent of computer-owning households with incomes under $15,000 say they particularly enjoy their computer, compared with only 38 percent of those with incomes of $50,000 or more. Likewise, blue-collar households are less likely than professional or executive households to own home computers, but blue-collar owners enjoy them more.

The simple fact is that people buy high-tech for two reasons: because they can afford it and because they want it. The two reasons may or may not overlap. Where money and desire coincide, the market is clear and solid. Those with a lot of money but little interest offer a challenge. So do those with little money and a lot of interest.

What appeals to technophiles will obviously not appeal to technophobes, but the reverse is also true. Making the technology so simple that users don't realize how sophisticated it is diminishes its appeal among technophiles. Two ways to bring both technophiles and technophobes online are to make products unique and fun.

"Hypermedia" is an example of something that cannot exist without computer technology. It allows people to browse encyclopedias or read fiction on their computers with the aid of CD-ROM drives and color monitors. And because the technology accommodates sound and video, hypermedia can even deliver bird calls and film clips at the click of a mouse. Hyperfiction can present stories in either a linear or nonlinear fashion, with readers selecting key words or icons to learn more about a character or setting, or to digress into another story.

Applications like Hypermedia can seduce people into marrying the new technologies. Hypermedia is "likely to attract the technophobe who is interested in French art encyclopedias," says Mark Bernstein, chief scientist for Eastgate Systems of Watertown, Massachusetts, which produces and distributes hyperfiction. "No one goes out and buys a computer to read a novel. A lot of our readers use their computers every day to get their job done, or as a writing tool. But some view it as a distasteful typewriter on steroids; they would go back to their typewriter if they could."

While the fun factor appeals to both technophiles and technophobes, using interactive television for chores like banking or shopping leaves most people cold. More than half of adults say they are not interested in these services, according to a CBS News/*New York Times* poll. But if it means they can see their favorite TV shows whenever they want, 77 percent of adults would be happy to sign up.

Some experts blame air conditioning for the demise of friendly neighborhoods, because people stopped sitting out on their stoops and porches on warm evenings. If you are a technophobe, you might agree with the poet Octavio Paz that "the new means of communication accentuate and strengthen noncommunication." But if you're a technophile like J. Robert Oppenheimer, you see an "open society . . . a vast, complex, ever-changing, ever more specialized and expert technological world—nevertheless, a world of human community." You might also agree that air conditioning has the effect of making many of us more sociable, and happier, on sweltering summer days.

The Birth of the Geodesic Dome

Lloyd Steven Sieden

Lloyd Steven Sieden, the author of two books on Buckminster Fuller, is dedicated to helping others find practical uses for Fuller's ideas. Richard Buckminster Fuller (1895–1983), universally known as Bucky, a philosopher and inventor best known for inventing the geodesic dome described below as well as for coining the ubiquitous terms *synergy* and *spaceship earth,* built a huge reputation and even a cultlike following in his long lifetime by combining the intense moral fervor and conservation interests of a Thoreau with the futuristic optimism of a Walt Disney. In the process Fuller became that rarest of breeds: a low-tech, socially conscious futurist.

At the *Visions* Web Site Links to Bucky Fuller sites, including the PBS site, *Thinking Out Loud;* to Habitat for Humanity, a group focused more on philanthropic than on technological concerns.

READING PROMPT

With an innovative thinker such as Fuller there is often a fine line between crackpot and genius. What signs can you discern in Sieden's narrative that some of Fuller's contemporaries might have placed him in the less flattering category?

Although Buckminster Fuller invariably maintained that he was a comprehensivist who was interested in almost everything, his life and work were dominated by a single issue: shelter and housing. Even as a young boy in the early 1900s, Fuller—who preferred to be called Bucky—was constructing rudimentary structures and inventing better "environment controlling artifacts."

The practical culmination of his quest to employ modern assembly-line manufacturing techniques and the best man-made materials in producing inexpensive, elegant housing came toward the end of World War II. At that time, government officials contracted Fuller to build two prototype Dymaxion Houses at the Beech Aircraft Company in Wichita, Kansas.

The lightweight, circular houses were praised by all who toured them. Because the Dymaxion House was to provide many new innovations at the very affordable suggested retail price of $6,500, orders flowed into the factory before plans for distribution were seriously considered. However, Fuller's interests were not geared toward practical matters such as financing and marketing, and the Dymaxion House never advanced beyond the prototype stage.

Fuller then moved on to consider other innovations that could benefit humanity in the areas of structure and housing.

He also returned to his less pragmatic quest to discover nature's coordinate system and employ that system in a structure that would, because it was based on natural rather than humanly developed principles, be extremely efficient. That structure is the geodesic dome, which, because it approximates a sphere, encloses much more space with far less material than conventional buildings.

In order to uncover nature's coordinate system, Fuller retreated from a great deal of his usual activities during 1947 and 1948. The primary focus of that retreat was a single topic: spherical geometry. He chose that area because he felt it would be most useful in further understanding the mathematics of engineering, in discovering nature's coordinate system, and eventually in building the spherical structures that he found to be the most efficient means of construction.

DOME MODELS

Having observed the problems inherent in conventional construction techniques (as opposed to the ease with which nature's structures are erected) and the indigenous strength of natural structures, Fuller felt certain that he could perfect an analogous, efficient, spherical-construction technique. He was also aware that any such method would have to be predicated upon spherical trigonometry. To do that, Bucky converted the small Long Island apartment that his wife, Anne, had rented into a combination workshop and classroom where he studied and discussed his ideas with others.

As those ideas started to take shape in the models and drawings he used for sharing his insights, Fuller considered names for his invention. He selected "geodesic dome" because the sections or arcs of great circles (i.e., the shortest distance between two points on a sphere) are called geodesics, a term derived from the Greek word meaning "earth-dividing." His initial dome models were nothing more than spheres or sections of spheres constructed from crisscrossing curved pieces of material (each of which represented an arc of a great circle) that formed triangles. Later, he expanded the concept and formed the curved pieces into even more complex structures such as tetrahedrons or octahedrons, which were then joined to create a spherical structure. Still, the simple triangulation of struts remained, as did the initial name of the invention.

Although Fuller's study of mathematics played a significant role in his invention of the geodesic dome, that process was also greatly influenced by his earlier extensive examination of and work within the field of construction. During his construction experience, he came to realize that the dome pattern had been employed, to some extent, ever since humans began building structures. Early sailors landing upon foreign shores and requiring immediate shelter would simply upend their ships, creating an arched shelter similar to a dome.

Land-dwelling societies copied that structure by locating a small clearing surrounded by young saplings and bending those uncut trees inward to form a dome that they covered with animal skins, thatch, or other materials. Over time, that structure developed into the classic yurt that still provides viable homes for many people in and around Afghanistan and the plains of the Soviet Union.

A NEW FORM OF ARCHITECTURE

In 1948, the geodesic dome was far from the amazingly sophisticated structure it would become only a few years later. In fact, it consisted primarily of Bucky's idea and an enormous pile of calculations he had formulated.

Although Fuller was developing and studying the geodesic dome using small models, he was eager to expand his understanding through the construction of larger, more-practical projects. Thus, when he was invited to participate in the summer institute of the somewhat notorious Black Mountain College in the remote hills of North Carolina near Asheville, Fuller eagerly accepted. He had lectured at that rather unorthodox institution the previous year and had been so popular that he was asked back for the entire summer of 1948.

When he was not delivering lengthy thinking-out-loud lectures that summer, Fuller's primary concern was furthering an entirely new form of architecture. In his examination of traditional construction, he had discovered that most buildings focused on right-angle, squared configurations.

He understood that early human beings had developed that mode of construction without much thought by simply piling stone upon stone. Such a simplistic system was acceptable for small structures, but when architects continued mindlessly utilizing that same technique for large buildings, major problems arose. The primary issue created by merely stacking materials higher and higher is that taller walls require thicker and thicker base sections to support their upper sections. Some designers attempted to circumvent that issue by using external buttressing, which kept walls from crumbling under the weight of upper levels, but even buttressing limited the size.

Fuller found that the compression force (i.e., pushing down) that caused such failure in heavy walls was always balanced by an equal amount of tensional force (i.e., pulling, which in buildings is seen in the natural tendency of walls to arc outward) in the structure. In fact, he discovered that if tension and compression are not perfectly balanced in a structure, the building will collapse. He also found that builders were not employing the tensional forces available. Those forces are, instead, channeled into the ground, where solid foundations hold the compressional members, be they stones or steel beams, from being thrust outward by tension. Always seeking maximum efficiency,

Fuller attempted to employ tensional forces in his new construction idea. The result was geodesic structures.

Because Bucky could not afford even the crude mechanical multiplier machines available during the late 1940s and was working with nothing but an adding machine, his first major dome required two years of calculations. With the help of a young assistant, Donald Richter, Fuller was, however, able to complete those calculations. Thus, he brought most of the material needed to construct the first geodesic dome to Black Mountain in the summer of 1948.

DISAPPOINTMENT BEFORE SUCCESS

His vision was of a 50-foot-diameter framework fabricated from lightweight aluminum, and, working with an austere budget, he had purchased a load of aluminum-alloy venetian-blind strips that he packed into the car for the trip to the college. Over the course of that summer, Bucky also procured other materials locally, but he was not completely satisfied with the dome's constituent elements, which were neither custom-designed for the project nor of the best materials. Still, with the help of his students, the revolutionary new dome was prepared for what was supposed to be a quick assembly in early September, just as the summer session was coming to an end.

The big day was dampened by a pouring rain. Nonetheless, Bucky and his team of assistants scurried around the field that had been chosen as the site of the event, preparing the sections of their dome for final assembly, while faculty and students stood under umbrellas, watching in anticipation from a nearby hillside. When the critical moment arrived, the final bolts were fastened and tension was applied to the structure, causing it to transform from a flat pile of components into the world's first large geodesic sphere. The spectators cheered, but their excitement lasted only an instant as the fragile dome almost immediately sagged in upon itself and collapsed, ending the project.

Although he must have been disappointed that day, Bucky's stoic New England character kept him from publicly acknowledging such emotion. Instead, he maintained that he had deliberately designed an extremely weak structure in order to determine the critical point at which it would collapse and that he had learned a great deal from the experiment. Certainly, the lessons learned from that episode were valuable, and his somewhat egocentric rationale was by no means a blatant lie. However, had he really been attempting to find the point of destruction, Bucky would have proceeded, as he did in later years, to add weights to the completed framework until it broke down.

In his haste to test his calculations, Fuller had proceeded without the finances necessary to acquire the best materials. Because of the use of substandard components, the dome was doomed to failure, and a demonstration of the geodesic dome's practical strength was condemned to wait another year.

During that year, Fuller taught at the Chicago Institute of Design. He and his Institute students also devoted a great deal of time to developing his new concepts. It was with the assistance of those design students that Fuller built a number of more successful dome models, each of which was more structurally sound than the previous one.

Then, when he was invited to return to Black Mountain College the following summer as dean of the Summer Institute, Fuller suggested that some of his best Chicago Institute students and their faculty accompany him, so that they could demonstrate the true potential of geodesic domes.

Having earned some substantial lecture fees during the previous year, Bucky was able to purchase the best of materials for his new Black Mountain dome. The project was a 14-foot-diameter hemisphere constructed of the finest aluminum aircraft tubing and covered with a vinyl-plastic skin. Completely erected within days after his arrival, that dome remained a stable fixture on the campus throughout the summer. To further prove the efficiency of the design to somewhat skeptical fellow instructors and students, Bucky and eight of his assistants daringly hung from the structure's framework, like children on a playground, immediately after its completion.

THE FORD DOME

In 1953, Fuller and his geodesic dome were elevated to international prominence when the first conspicuous commercial geodesic dome was produced. That structure was erected in answer to a Ford Motor Company problem believed to be insoluble. During 1952, Ford was in the process of preparing for its fiftieth anniversary celebration the following year, and Henry Ford II, grandson of Henry Ford and head of the company, decided he wanted to fulfill one of his grandfather's dreams as a tribute to the company's founder. The senior Ford had always loved the round corporate headquarters building known as the Rotunda but had wanted its interior courtyard covered so that the space could be used during inclement Detroit weather.

Unfortunately—but fortunately for Bucky—the building was fairly weak. It had originally been constructed to house the Ford exhibition at the Chicago World's Fair of 1933, but Henry Ford had so loved the building that he had had it disassembled and shipped in pieces to Dearborn, where it was reconstructed. Having been designed as a temporary structure, the fragile Rotunda building could not possibly support the 160-ton weight that Ford's engineers calculated a conventional steel-frame dome would require. Under such pressure, the building's thin walls would have immediately collapsed.

Still, Henry Ford II was a determined person, and he wanted the courtyard covered. Consequently, Ford management and engineers continued searching for an answer until someone suggested calling Buckminster Fuller. By that time, Fuller's work was drawing international attention, and although his geodesic dome had yet to be proven effective in an industrial project, desperate

Ford officials decided they should at least solicit Bucky's opinion. When he arrived at the Detroit airport, Fuller was greeted by a Ford executive in a large limousine who treated him like royalty, quickly escorting him to the Rotunda building for an inspection. After a short examination of the 93-foot opening requiring a dome, Ford management asked the critical question: Could Fuller build a dome to cover the courtyard? With no hesitation, Bucky answered that he certainly could, and the first commercial geodesic dome began to take shape.

The Ford executives next began to question the specifications of Fuller's plan. When they asked about weight, he made some calculations and answered that his dome would weigh approximately 8.5 tons, a far cry from their 160-ton estimate. Ford management also requested a cost estimate and advised Fuller that, because of the upcoming anniversary celebration, the dome had to be completed within the relatively short period of a few months. When Fuller's price was well below Ford's budget and he agreed to construct the dome within the required time frame, he was awarded a contract.

The agreement was signed in January of 1953, and Bucky immediately began working to meet the April deadline. The somewhat discredited Ford engineers who had failed to develop a practical solution were, however, not convinced that the obscure inventor's fantastic claims were valid. Thus, they began working on a contingency plan that would prevent further embarrassment. To protect reputations further, the engineers secretly contracted another construction firm to hastily haul away any evidence of Fuller's work when he failed. The Ford engineers were once again proven wrong when the dome was successfully completed in April, two days ahead of schedule.

BUILDING FROM THE TOP DOWN

Actual construction of the dome was a marvel to behold. Reporters from around the world gathered to witness and recount the architectural effort as well as Ford's anniversary celebration. Because the courtyard below the dome was to be used for a television special commemorating the anniversary, and because business at Ford had to proceed normally, Fuller's crew was provided with a tiny working area and instructed to keep disruptions to a minimum.

Ford management was also concerned with the safety of both the dome workers and the people who might wander beneath the construction. They anticipated that problems would arise when Ford employees, television crews, reporters, and spectators gathered below to observe the construction workers climbing high overhead on the treacherous scaffolding, but once again Bucky surprised everyone. Instead of traditional scaffolding, he employed a strategy similar to the one he developed in 1940 for the quick assembly of his Dymaxion Deployment Units.

Because the sections of the dome were prefabricated and then suspended from a central mast, no dangerous scaffolding was required. The construction team worked from a bridge erected across the top of the Rotunda courtyard.

Like Dymaxion Deployment Units, the Ford dome was then built from the top down while being hoisted higher and rotated each time a section was completed. The dome was assembled from nearly 12,000 aluminum struts, each about three feet long and weighing only five ounces. Those struts were preassembled into octet-truss, equilateral-triangular sections approximately 15 feet on a side. Since each section weighed only about four pounds and could be raised by a single person, no crane or heavy machinery was required to hoist them to the upper bridge assembly area.

Once on the working bridge, the identical sections were riveted into place on the outwardly growing framework until it covered the entire courtyard. Upon completion, the 8.5-ton dome remained suspended on its mast, hovering slightly above the building itself until the mooring points were prepared. Then, it was gently lowered down onto the Rotunda building structure with no problem.

To complete the first commercial geodesic dome, clear Fiberglas "windows" were installed in the small triangular panels of the framework. Because Fuller had not yet developed or determined the best means of fastening those panels, they would eventually be a cause of the destruction of the dome and the building itself.

Since it was the first large functional geodesic dome, many aspects of the Ford dome were experimental. They had been tested in models, but how the dome and the materials utilized would withstand the forces of Michigan winters could be determined only by the test of time. The Rotunda building dome did perform successfully for several years before the elements began taking their toll and leaks between the Fiberglas and the aluminum began to occur. Still, with regular maintenance, that problem was not serious, and convening corporate events under the dome became a tradition. One of those events was the annual Ford Christmas gathering.

In 1962, numerous leaks in the dome were noticed as the Christmas season approached, and a maintenance crew was dispatched one cold late-autumn day to repair the problem. The temperature was, however, too cold to permit proper heating of the tar they used for the repairs, and, in a common practice, the workers added gasoline to thin the tar. They were warming the tar with a blowtorch when that potent mixture ignited, and the building quickly caught fire. Since the building had never been planned as a permanent structure, it was not long before the entire Rotunda was engulfed in flames that destroyed the first commercial geodesic dome, the singular structure that, more than any other, had catapulted Fuller to public fame.

DOING MORE WITH LESS

The notoriety provided by the Ford project resulted in an enormous amount of nearly instant public interest in Fuller and his ideas. It also brought him to the attention of a group of scientists who were struggling with another seemingly unsolvable problem: protection of the Distant Early Warning Line radar installations throughout the Arctic. Once again, Fuller and his amazing geodesic dome surprised all the experts as his hastily invented Fiberglas "radomes" proved more than able to handle that difficult task.

The proliferation of radomes in technologically advanced situations around the world moved the geodesic dome into its rightful position as a symbol of developing humanity doing more and more with fewer resources. Thus, geodesic domes are now employed for diverse tasks such as providing a more natural structure for children on playgrounds, covering athletic stadiums, and being proposed for use in future space construction.

However, the true significance of the geodesic dome is most evident in the fact that it is often the dominant symbol employed at major future-oriented expositions. When most people remember the 1967 Montreal World's Fair, the 1986 Vancouver World's Fair, or Disney's EPCOT Center, the first image they recall is the geodesic dome.

It properly stands as a monument to the work of Buckminster Fuller, who successfully shared his vision of a world that works for everyone. He also inspired scores of people to work, as he did, to establish a network of equally significant individuals supporting humanity's emergence into a new era of cooperation. That relationship between individual human beings, as well as that between humans and their environment, is not modeled by the rigid conventional buildings that fill our environment. It is modeled in the amazing geodesic dome's network of lightweight, resilient struts, wires, and panels.

chapter 3 Explorations

3a Process

Collaboration. Rube Goldberg's "Keeping Screen Doors Closed" describes a complicated way of doing a fairly simple process. Share with classmates brief descriptions of multistep processes that you regularly complete.

Web work. To see (for better or worse) all the complexity one can add to what in theory at least could have been a simple process, check out Dave's Guide for Buying a Home Computer or any of the beginners' guides to the hypertext markup language (html) used in creating Web pages.

One writing topic. In the spirit of Rube Goldberg and Dave's Guide to Buying a Home Computer, see if you can give someone more advice than they could ever possibly need to do something fairly simple, such as, for example, buying a watch.

3b Infomercials

Collaboration. Share with classmates brief descriptions of brain-dead TV-watching experiences—experiences with programming whose content is so bad that only someone brain dead could be expected to watch, and yet we still do.

Web work. The Infomercial Index Web site bills itself as "the Infomercial Yellow Pages," with information on more than 300 infomercials; take a look if you dare.

One writing topic. Analyze the essential components of an infomercial.

3c Trivial Pursuits

Collaboration. One way of looking at trivia is as a kind of gadgetry of the mind: information that might have some use but which we retain more for its host of associations. Share a short list of areas where you think you have retained the most useless information, or trivia. How does your list compare with those of your classmates?

Web work. The Web is a haven for trivia, with Yahoo! affording it a high place in its hierarchy: Entertainment > Trivia, plus additional index pages under every possible category, including *Seinfeld.* The popular board game *Trivial Pursuit* has its own Web site.

One writing topic. Use your class as a research group and write an essay on its main and lesser areas of interest in trivia, considering possibly how these areas seem to compare with national interests as represented on the Web.

3d Technophobia

Collaboration. Rube Goldberg's cartoons present just one expression of the frustration many ordinary people feel in dealing with an increasingly and at times perhaps needlessly complex world. Share with classmates accounts of one or two times in your life when you felt frustrated by technology.

Web work. "Never use a machine you can't fix yourself" are words of wisdom at the center of the life of the Amish, the "plain people" of the Pennsylvania Dutch Country. Check out the information on the Amish at the Pennsylvania Dutch Country Web site.

One writing topic. Write an essay on problems and/or successes you have had repairing important objects in your life.

3e What's a Gadget?

Collaboration. One person's gadget may be another's tool or toy. Share with your classmates your definition of what makes a gadget, as compared with a tool or a toy.

Web work. A good place to begin is Yahoo!'s list of wacky products, Business and Economy > Companies > Novelties > Wacky Products. Another good gadget site is that devoted to the Japanese art of *chindogu,* which celebrates the creation of "almost useful" inventions and "inconvenient conveniences."

One writing topic. Write an essay defining the word *gadget,* considering to what extent it is a positive and to what extent a negative term.

3f Genius and Madness

Collaboration. Thomas Edison, a man now acknowledged as a genius, was once considered something of a simpleton. Share with classmates an experience (or two) you have had with a person or persons who clearly did not fit in well with their surroundings.

Web work. One assumes that there is something a little strange about inventors, or at least those represented on The Wacky Patent of the Month site, interestingly enough a site maintained by a patent attorney as a way of scouting for new business. There is a good Web site to learn more about Thomas Edison.

The PBS series *The American Experience* devoted one show to forgotten inventors.

One writing topic. Pretend you are a sales representative for a wacky or ingenious new product; write out the text of a speech you would make to convince a major company to purchase, manufacture, and market your product.

part II

At Home

Our attachment to place is a defining human experience. All of us are raised somewhere, and our sense of that place is often filled with rich emotions, especially if we are among the many who find themselves almost constantly on the move after high school. "How do other people [non-Christians] celebrate Christmas?" ethnically naive schoolchildren in the United States sometimes ask. Naive to be sure, but the point is clear: Surely in a world where everyone seems to be on the move and everything in flux, the need to do something special to celebrate the permanence of home life must be a universal experience. Is it any wonder that family rituals and homecomings associated with the dark winter solstice of late December seem to have such an unshakable hold on nearly all North Americans regardless of religious or ethnic background? Talk of family and community is everywhere, with *community* in particular becoming one of those rare terms that always denotes something intensely positive—even if (as Gerald Early suggests in his piece on Kwanzaa) people cannot agree on just what it means.

The essays and topics in Chapters 4 through 6 all deal with various aspects of place and community, and thus, in turn, with how technology is helping to bring us back together (as long-distance phone companies and airlines would like to make us think) or (as dissenters might suggest), hastening the process of pulling us apart. After all, if each member of a family has his or her own television or personal (palm-top) entertainment center and his or her own email account, each has his or her own separate life to a great degree, even as the family shares the same common space.

We begin Chapter 4 with a debate over cities and suburbs, with Daniel Kemmis praising the traditional housing patterns of cities for creating places where people walk to shop, visiting neighbors along the way, and Allan Carlson praising the suburbs (communities built around the key modern technology of the automobile) for giving individuals safe, pleasant places to raise families. Just why is it, we might want to consider, that praise of the more modern (more technologically advanced?) suburbs comes mainly from political conservatives while defense of the more traditional cities (where people walk) is left to political liberals? One especially interesting Exploration at the end of this chapter asks you to consider the wave of new *retro* communities such as Celebration, Florida (built by the Disney Corporation): complex, technologically sophisticated communities built to look and to work (sort of) as if they were old-fashioned, low-tech towns. Perhaps none of the other 107 Explorations collected in this book points so well to the complex, possibly contradictory impulses that surround our thinking about technology and how we envision both the past and the future as does this topic—the idea that the hippest new housing re-creates the look and feel of nineteenth-century small-town America.

The readings in Chapter 5 are devoted to just one aspect of the ongoing national debate about communities: the question of how useful virtual communities are as alternatives to traditional communities. Supporters of the Net such as John Perry Barlow see new possibilities for promoting social interaction, for enriching civic life, and especially for allowing isolated individuals across the country to form social and political alliances (the focus of the last reading in this chapter), whereas skeptic Scott Russell Sanders wonders just what the Web has to offer that he can't find in Bloomington, Indiana. One of the Explorations will take you on a virtual tour of Bloomington, Indiana, so you can see for yourself; two other Explorations will ask you to take a new look at two by now seemingly low-tech ("ancient") technologies that have done much to foster virtual contact over the past hundred years: the telephone and the snapshot. Meanwhile the debate between the high-tech Barlow and the low-tech Sanders is another one of the many in this anthology that raises serious questions about the traditional political labels of *conservative* (once referring to supporters of traditional practices) and *liberal* (once referring to those eager to reform society, make it more rational)—a subject dealt with more directly in Chapter 6.

The readings in Chapter 6 deal specifically with the political implications of our living in cyberspace; and here all parties seem to agree that the best human experiences are personal and, to some extent, local—that is, fundamentally low-tech, with people working directly together at the grassroots level to improve their communities. The problem comes when we try to determine the extent to which computer interaction can be seen as facilitating or as taking the place of such direct personal interaction. The last two essays in this chapter provide a spirited debate on what is widely perceived as the new political ideology of the Internet, *libertarianism* (a topic introduced by Katz in the first essay)—the belief that the government should stay out of nearly all aspects of our lives, social as well as economic. Explorations for Chapter 6 all deal with ways we use technology to stay informed and in touch or, conversely, to stay entertained, even in our apathy.

c h a p t e r **4**

Real Towns, Real Places

READING QUESTIONS

- What makes your hometown or favorite place special, distinctive? What makes it real?
- What makes something a ritual, and what role does ritual play in making a place special?
- How has the automobile affected where and how you have lived, the kind of restaurants and other public places you have frequented?

QUOTATIONS

E. T. phone home.
— *Line from 1982 film*

Ah! There is nothing like staying at home, for real comfort.
— *Jane Austen (1816), British novelist, 1775–1817*

You can never go home again, but the truth is you can never leave home, so it is all right.
— *Maya Angelou (1987), American writer and performer*

A "good" family, it seems, is one that used to be better.
— *Cleveland Amory (1960), American social commentator, 1917–1998*

Living Next to One Another

Daniel Kemmis

In 1995 the *Utne Reader* named Daniel Kemmis, the social activist and former mayor of Missoula, Wyoming, as one of the journal's 100 visionaries; he's listed at number 51 in their alphabetical listing, right after Mitch Kapor, the designer of the original spreadsheet for personal computers, Lotus 1–2–3.

At the *Visions* Web Site Links to Missoula sites, including the daily newspaper, *Missoulian Online.*

READING PROMPT

Kemmis argues for an almost mystical concept of "a city in grace." What are some different meanings of the word grace *that might help explain Kemmis's concept?*

Democracy is a human enterprise, and as such, it is suffering. In order to restore democracy itself, we must heal the human base of politics. One thing alone will give us the capacity to heal our politics and to confront the problems and opportunities which politics must address. That one thing is a deeply renewed human experience of citizenship.

To redeem the democratic potential of citizenship, we need to take an entirely fresh look at its essential features. One of those, surprisingly, is citizenship's intimate connection to the city, from which both its name and its fundamental human significance derive. What makes a city civilized is something that is also absolutely fundamental to citizenship: its essential humanity. This element will make its claim most clearly if we begin by contrasting it to the dehumanized role which has replaced it, namely, the role of "the taxpayer."

People who customarily refer to themselves as "taxpayers" are not even remotely related to democratic citizens. Yet this is precisely the word which now regularly holds the place which in a true democracy would be occupied by "citizens." Taxpayers bear a dual relationship to government, neither half of which has anything at all to do with democracy. Taxpayers pay tribute to the government, and they receive services from it. So does every subject of a totalitarian regime. What taxpayers do not do, and what people who call themselves taxpayers have long since stopped even imagining themselves doing, is governing. In a democracy, by the very meaning of the word, the people govern—they create among themselves the conditions of their lives. But in our political culture, "they" govern, victimizing "the taxpayers" and delivering a uniformly unsatisfactory level of services, partly because "they" are draining "the taxpayers" tribute into "their" own pockets.

It would do no good at all simply to insist that we start calling ourselves "citizens" instead of "taxpayers," unless citizenship in fact were already being experienced as something important, worthwhile, and humanly satisfying. I suspect that people do in fact have an experience of citizenship which is often satisfying, but that it does not go by that name and is therefore not understood as a political experience, let alone as having any potential to revitalize democracy. This experience is simply that of city-dwelling, and it is here that civility, civilization, and citizenship are all rooted. "Civil" originally meant simply "of the city." Civility was what it took to live next to one another—as cities, by definition, require people to do. But if civility is a requisite for cities to exist at all, civilization goes a stage beyond this. Civilization is not only a city that works by allowing people to live near one another, but a good city—one which enables its inhabitants to live good lives together.

Over the last few years, more and more cities around the country and indeed around the world have begun to examine themselves in terms of something called "liveability." Cities now earnestly compete with one another for distinction in this category. In this, we see an attentiveness to the essential human purpose of cities. As we look more closely at the variety of ways in which human liveability is being seriously pursued by cities, one feature will emerge for which none of our traditional ways of thinking about politics or government can have prepared us. We think of political entities exclusively as tools which we deploy for human ends. Nations, states, and cities are, we think, such tools. Nations and states may indeed be, but no city which is seen simply as a tool will ever come to be called "liveable," let alone "civilized." The reason is that the city itself is alive, and it is in its own fullness of life, in its own emerging wholeness that it has the capacity to become humanly liveable and humanly fulfilling.

Christopher Alexander, in his description of the "good city," states:

> When we look at the most beautiful towns and cities of the past, we are always impressed by a feeling that they are somehow organic. . . . Each of these towns grew as a whole, under its own laws of wholeness . . . and we can feel this wholeness, not only at the largest scale, but in every detail: in the restaurants, in the sidewalks, in the houses, shops, markets, roads, parks, gardens and walls. Even in the balconies and ornaments. (*A New Theory of Urban Design*. New York: Oxford UP, 1987: 2)

What is so refreshing about Alexander's approach is that, unlike so much of our politics which seems so humanly daunting, so impossible of achievement, he speaks of this wholeness in incremental terms which begin to bring it within human reach. Alexander reminds us of the connection between healing and wholeness, arguing that we have countless opportunities to take small healing steps to move the city in the direction of wholeness. "Every increment of construction," he writes, "must be made in such a way as to heal the city" (22).

The work of healing the city is civilizing work, which is to say that it creates at one and the same time a better city and better citizens. Recent events in Baltimore illustrate this civilizing process. Baltimore is a city with a newly discovered sense of direction and hope. At the heart of this seems to lie a significant revival of citizenship, but inseparable from that revival has been just the kind of healing construction which Alexander describes. Most notable has been the exceptional reclaiming of the waterfront area known as the Inner Harbor, where urban blight has been replaced by a bustling market center next to one of the nation's finest and newest aquariums. Baltimoreans are so proud of this achievement that they have begun talking about taking the civic energy that could restore a waterfront and using it to restore other, less physical elements of the city. Baltimore may in fact be able to make real its new motto as "The City that Reads," or to make some serious inroads into poverty, racism, or drug abuse, but if it does, it will be in no small part because of the pride and hope that have come from creating good places like the Inner Harbor.

How does the healing of a city create better citizens? A citizen, in the most basic sense, is simply a city-dweller, one whose life is shaped and given identity by the city much as a lion is shaped by and takes its identity from the jungle. A citizen is a denizen of the city: a city-zen. Before we try to make the citizen politically effective, we should understand as fully as possible the more basic human connection between the city and the city-zen. Tony Hiss's description of the reading room of Baltimore's Peabody Library captures the way in which a city might give roots to citizenship:

> The light flooding in through the skylight changes from moment to moment, picking out different colors—gray, buff, gilt—in the columns over your head, and as you sit at one of a number of small desks off the central court, knowledge seems so abundantly available that you feel almost the same kind of gratitude you feel when you stoop to drink from a public fountain: The city doesn't want me to go thirsty. ("Annals of Place." *New Yorker* 29 April 1991: 67)

People who experience the city as sustaining and nurturing them in this way are well on the way to citizenship. Their next step, if given the chance, will often be to seek ways to sustain, nurture, or heal the city.

Healing, health, and wholeness are, I am convinced, helpful ways of understanding how public life or politics might be revitalized. But no matter how powerful a concept we apply to politics, it is only a concept until it is lived. No abstraction (including health or wholeness) can ever capture what the city is all about. Only the city itself can do that. As Jane Jacobs says, "City processes in real life are too complex to be routine, too particularized for application as abstractions" (*The Death and Life of Great American Cities.* New York: Vintage, 1961: 441).

The reason this matters so much is that abstraction has become one of the great evils of public life, one of the chief ways in which we tear apart the wholeness of life. Our heavy emphasis on individual rights, to the neglect of a corresponding notion of civic responsibilities, is an example of where abstraction has seriously weakened our political culture. Abstraction's evil twin in this work is distraction. From drugs and alcohol to TV and workaholism, we are increasingly a society which fulfills T. S. Eliot's description of a people "distracted from distraction by distraction." There is hardly a public menace we can name which is not in some sense caused by one or another of the million ways in which our society teaches and enables us to abstract and distract ourselves—to escape in one way or another from the concrete presence of the here and now.

Cities do their share of distracting, of course. Part of the life of every city is to distract us, and some cities (Las Vegas is my favorite example) seem to have no other reason to exist. But it is not by virtue of this that any city has ever been or might ever be called "civilized." What makes a city a good city is not its capacity to distract, but the way in which it creates Presence. The gathering function of the marketplace itself was from the beginning essentially a way of marking off a recognizable here and now within which people and goods could reliably be made present to one another in the dual sense of spatial and temporal presence. Still today, the good city does nothing more important than to make it possible for humans to be fully present—to themselves, to one another, and to their surroundings. Such presence is the precise opposite of the distractedness which is so prevalent in our political culture.

The capacity to be present may simply be another feature of health and wholeness. But I am more and more convinced that it is a political value of the highest order, and that citizenship is not conceivable apart from the kind of presence which it is the chief work of the city to create. That this work is subtle and often unassuming should not blind us to its transformative power.

Butterfly Herbs is downtown Missoula's premier coffee shop, the place where this article was written, and the site of a simple, commonplace, but somehow striking tableau. The coffee shop portion of Butterfly Herbs occupies the back one-third of a long, narrow building which fronts on Higgins Avenue, one of downtown Missoula's main thoroughfares. The store's entrance, recessed behind captivating display cases, provides a passage from street to store and back again through a heavy, oaken door which is framed by fine cut- and stained-glass side-windows.

One day I headed for this door following an hour's writing in the coffee shop. A woman approached the door ahead of me. I was walking briskly, but she moved unhurriedly, taking in her surroundings, examining the wall and window displays as she neared the door. As I drew closer, I could sense her becoming aware of me, and when she reached the door, she drew it open and held it for me. Walking past her into the recessed alcove off the sidewalk, I tried to grasp what it was about her that so commanded my attention. She struck me as being remarkably self-contained, and yet in no pejorative sense self-centered. Her acknowledgment of me was full and free, and seemed merely an extension of her response to everything around her. The sights, the smells, the music in the store were all incorporated and reflected in her demeanor. The heavy old door itself, and finally, as I passed by her, the street to which she herself would soon emerge—all were held in dynamic balance by the way she stood poised between the interior of the building and the Missoula street scene beyond.

Since that day, I have often recalled the woman at the door, trying to get to the bottom of some lesson this simple tableau seemed to be teaching. Kurt Schmoke, the Mayor of Baltimore, has spoken of his vision of Baltimore by the year 2020 having become "a city in grace." Grace is exactly what caught my attention at the door of Butterfly Herbs, and it is just the lack of grace which is so disturbing in our political culture at large.

The idea of a city in grace is appealing for two seemingly unrelated reasons. First, it answers to a deep longing for a spiritual dimension in public life. In our prevailing political culture, this longing produces one of two results. The first is alienation from public life because it does not fulfill this spiritual need. People do not like politics, or public life in general, because it does not engage their highest or deepest instincts. So either they abandon citizenship altogether, or they import into politics a narrow, essentially mean-spirited religiosity which in fact only worsens the prevailing gracelessness of public life, thus driving new multitudes into alienation. This has been all too much the legacy of the New Right, just as alienation from an inhumanly secularized public life has been the legacy of liberalism. Against this background, Mayor Schmoke's explicit commitment to nurturing some form of grace in public life comes as a rare flash of both spiritual and political wisdom.

But Mayor Schmoke's vision is appealing for a second, equally compelling reason. Put simply, his dream seems to be (if only barely) within human reach. In its concreteness lies its hope. That concreteness is intimately connected to the nature of the political entity of which he speaks. A city in grace, while barely imaginable, could actually occur on the face of the earth because a city is a real, living organism, and therefore capable of something we might call grace. At the same time, if the idea of "a city in grace" is to be more than a ringing sound bite, we have to give full weight to both of its nouns. We need to ask what a city might have to do with grace.

Grace is such a prolific concept that no one understanding can exhaust its meaning. But surely some part of grace has to do with a way of acting in the world which is strikingly appropriate to the time and place. Such acts are supremely human, but they always evoke as well an unmistakable element of givenness. What is graceful is, we sense, at least in part beyond our control; it is given to us, as the Baltimorean feels graced in that library reading room. The woman at the door was not so gracefully present simply of her own account; rather, she occupied a time and a place which had been given to her. To a substantial extent, it was the city itself which had created that dual presence of time and place, of story and of location, which she so gracefully occupied. Indeed, simply by referring to "the woman at the door," I have all along placed her in this most urban of settings: a passageway from shop to street. All the sights, sounds, smells, and other people of whom she was so palpably aware were city sights and sounds and people. It is in occupying this city-provided presence that she takes on the character of the city-being, the "city-zen." She captured my attention because of some graceful combination of presence and wholeness. But they were not simply her presence or her wholeness; instead, she occupied and sanctified the presence and the wholeness of the city itself. So, in a good city, might we all.

Two Cheers for the Suburbs

Allan Carlson

Automobiles changed not only how we eat but where and how we live, creating in the process one of the United States' most distinctive features, the suburb. In general, critics of the suburbs tend to see them as the soulless expressions of our national rootlessness, driven by federally funded highway construction, cheap land and gasoline, and the close cooperation between real estate developers and local governments. Meanwhile, Allan Carlson's essay here presents many of the arguments that the defenders of suburbs make.

At the *Visions* Web Site Link to exhibition on Levittown, New York, the model post–World War II American suburb.

READING PROMPT

Compare your own experience of the suburbs, and of the suburbs' urban alternatives as described by Carlson, with what Carlson offers here.

It's always tempting to knock the suburbs: I have done it, at times, myself. Yes, they tend to be numbingly repetitive and stylistically incongruous. Sure, they gobble up good farmland and have depended on oceans of cheap gasoline and rivers of asphalt for their survival. And obviously, they lack the rootedness of farms, the identity of villages, and the ethnic links, extended families, and neighborhood pubs of the urban immigrant ghettos.

Yet for all that, about 55 percent of Americans now live in suburbs or suburb-like environments, up from 20 percent a half-century ago, a shift in preference that must be labeled revolutionary. Federal tax, housing, and transportation policies had something to do with this remarkable change. So did propaganda mills such as Henry Luce' s Time–Life empire, which cast the suburbs as the locus of the "New America," superseding old ethnic and regional loyalties. Even the Cold War ethos embraced suburbia; as Bill Levitt, of Levittown fame, once explained: "No man who owns his own house and lot can be a communist. He has too much to do."

At the same time, though, something more fundamental, and more human, has been at work. In G. K. Chesterton's 1910 tract, *What's Wrong with the World,* the English journalist observed: "As every normal man desires a woman, and children born of a woman, every normal man desires a house of his own to put them in." Chesterton emphasized that this normal man "does not desire a flat" nor a semi-detached house, with walls necessarily shared with others. Instinctively, he wants "a separate house," on its own piece of ground, built with the "idea of earthly contact and foundation, as well as [the] idea of separation and independence." This normal man wants, as well, an "objective and visible kingdom; a fire at which he can cook what food he likes; a door he can open to what friends he chooses."

Viewed charitably, the American suburbs represent an unrivaled fulfillment of these instinctive dreams. A nation of renters in 1940—when only 44 percent of housing was owner-occupied—became a nation dominated by homeowners, who now inhabit 64 percent of all housing. For the first time since industrialism fundamentally transformed American work and living patterns, the homeowner has again become the authentic American. Land and housing have been widely distributed, and the opportunities for family formation advanced. Even average newly married couples can enjoy the illusion of ownership, and the reality of equity. These are changes for the good.

For nearly two decades, family life in America responded to the home-owning revolution with an astonishing vitality. Between 1946 and 1964, the proportion of persons ages 20 to 40 who were married reached an historic high, the divorce rate declined, and the marital birth rate climbed sharply. For the first time since 1840, American families appeared to be growing stronger, rather than weaker—changes closely bound to the new suburban life.

True, not only cities but farms as well were emptying, with the suburbs drawing folk from all locales. While the movement from city-to-suburb represented an attempt to regain some attachment to the land, the movement from farm-to-suburb derived from the rapid automation of farming during and after World War II, as ever larger tractors and combines made human laborers superfluous. Without legal restraints on farm technology—no tractors larger than 10 horsepower, say—which were never contemplated here, and an end to federal transportation subsidies that favored national over local agricultural markets, also not considered in this century, the process became inexorable. Still, the suburbs allowed these agrarian refugees to find a residual attachment to land and place.

Those moving to the suburbs from the cities came for other reasons, and they were normally satisfied with the results. Some fled mounting urban crime. Crime rates are highest in large cities and lowest in rural areas, with the suburbs falling in between. Others came for better schools. While cause and effect might be disputed, the public education statistics do show suburban schools to be—on average—the best performing. This has been due, at least in part, to the responsiveness of these schools to parental expectations, rooted in turn in the smaller size of many suburban school districts. Indeed, in an age primarily given over to state centralization, the suburbs have encouraged a countervailing decentralization of governance, forcing a healthy kind of competitiveness onto local governments.

Still others came to the suburbs seeking "community," and here the research again shows the expected results. When compared to city dwellers, suburbanites have been more family-oriented, more home-centered, more neighborly, and more inclined to take part in community affairs, although when compared to persons in small towns, the suburbanites have turned out to be "less friendly" and their communities "less complete."

So the suburbs stand, as expected, in the middle. While far from the utopia dreamed of by architect Frederick Law Olmsted back in 1868 (suburbs, he predicted, would represent "the best application of the arts of civilization to which mankind has yet obtained"), they are at least a more successful response to modernity than the models found elsewhere on the globe.

Indeed, the American system only gains luster as it is compared to others. The Swedes, for example, chose in the 1930s to pursue a different course. While ample land was available there for single-family, detached housing, socialist planners opted instead for a communal emphasis. They built urban high-rises and "family housing" centers surrounded by greenbelts to accommodate the accelerating flow of country people into the cities. Collectivized nurseries, day care centers, and an array of government subsidies and services complemented this intentional diminution of nuclear family bonds. The result has been a nation where the autonomous family, rooted in marriage, has largely dissolved: over half of new births in Sweden are outside of marriage, and all the children are effectively wards of the state. Family ruin, a bloated welfare state, and urban communalism have occurred together.

True, vital urban cultures can still be found in cities like Rome and Paris. Yet these places might also be labeled "child-free zones," and the incompatibility between modern urban life and children plays itself out in their perilously low birth rates. "Where are the children?" I have asked when visiting these places. "In the suburbs," I am invariably told.

In post-Communist nations such as Russia, the lack of decent housing is the central family issue. The high-rise apartments around Moscow, even at their best, are cramped affairs, fundamentally unfriendly to the presence of the young. Once again, collectivist fantasies about "urban community" have claimed a vast human toll in delayed or foregone marriages and never-to-be-counted children.

So the critic of the suburbs must, in the end, answer the question: What is the alternative? Under mid- and late-twentieth century conditions, where else could Chesterton's "normal man" gain that freestanding house and that piece of ground to share with a spouse and children?

Some of the more obvious problems of suburbia are also diminishing. As these communities grow older and more settled, the age segregation found earlier has tended to recede, leaving a more natural mix of generations, even extended families. Racial isolation has diminished as well, with the flow of our black middle class to white-collar communities.

The gravest flaw of suburbia, to my way of thinking, has been the way it exaggerates the separation of home from work. Although this division was as old as industrialization itself, the suburbs that took form after World War II further alienated the father/husband from the daily life of the family. They also stranded the housewife/mother in "less complete" communities, where resentments could grow—and would after the mid-1960s.

Yet the 1990s bring some evidence that this breach is healing. After several false starts, home business resting on telecommuting is enjoying meaningful expansion. In many areas, newly constructed homes are expected to be wired and outfitted to participate in this novel, home-based commerce. At the same time, parents are rediscovering the possibilities for home production, centered on their families. The "home school" revolution, growing almost geometrically over the past 20 years, occurs primarily in suburban precincts. According to recent reports, home gardening is also growing there.

Human family relations, conditioned for a thousand generations to life in rural or village locales, will find no natural home in the modern industrial cities. For some time now, Americans have looked for a practical accommodation between family living and modernity. The postwar American suburbs represent such a compromise: to be sure, one distanced from the communitarian ideal, and one bearing the typically American scars of haste and hucksterism. Nonetheless, the American suburbs have also delivered homes and plots of land to millions of families who would not otherwise have them, and they stand up well in comparison to the living experiments in other places. And so, with a small sigh, I raise my coffee cup and offer two cheers for suburbia.

Dreaming of a Black Christmas

Gerald Early

Gerald Early, a professor of African American studies, here describes his ambivalent feelings about a Kwanzaa student festival, an investigation he continued in the essay "Whatever Happened to Integration." Early prides himself on his independence of thought; in the essay "The Reading of Liberation; the Liberation of Reading" he took on several black best-sellers. He showed a softer side in another essay available online, "The Frailty of Human Friendship."

At the *Visions* Web Site Links to more essays by Early; to CNN's informative Web page on Kwanzaa.

READING PROMPT

Why does Early think we celebrate holidays? Which of these reasons does he seem to oppose? Which does he support?

For the past five or six years, in my position as head of the African-American studies program at the university where I teach, I've been invited by the black students on campus to take part in their annual celebration of Kwanzaa, the African-American holiday that is gaining in popularity each year. The festivities, which are usually celebrated during the seven days between December 26 and January 1, are compressed, for the students' purposes, into one evening. The ceremony takes place in one of the campus's cafeterias. All the trappings of a somber religious occasion are there: candles, a mat, a ritual cup, remarks to the gathered celebrants.

Because Kwanzaa is designed to connect African Americans to their African heritage, the colors and symbols of that continent predominate. Kente cloth is ubiquitous, as are the red, black, and green of Marcus Garvey's Pan-African flag. Gifts of nuts, fruits, and vegetables, which are meant to recall African harvest festivals, are placed on the mat. Corn, a symbol of children, is also offered, to remind us that we are responsible to the youngest of the community.

The gathering can take on the solemnity of a church service. A Unity Cup—passed around among the celebrants—serves as a kind of Eucharist of Africanity. Much ado is made of the family, particularly the elders. A roll call of black heroes is intoned. Naturally, as at any serious black gathering, we sing the black national anthem, "Lift Every Voice and Sing," which is just as unsingable as the "white" national anthem, "The Star-Spangled Banner." Yet it often feels good to sing it in a roomful of blacks, as if it were a spiritual of how we have endured in a strange land.

The celebrants then go on to offer meditations on the meaning of the holiday, but I slip out just as soon as I can. I'm a middle-aged man, and I don't like hanging around with groups of twenty-year-olds any longer than I have to. It's more than just the socializing, though. I always feel a little uneasy at Kwanzaa gatherings. The holiday, with its tenuous history, its mimicry of so-called ancient festivals, its celebration of vague "sacred" principles, is one that has never moved me. There is something so contrived, so invented about it, so pointed in its moral purpose, that I can't help wishing for a holiday with a bit more universality—something like Christmas, the holiday Kwanzaa intends to preempt.

Kwanzaa, which is celebrated, according to its boosters, by an estimated 18 million African Americans, was indeed invented, and not so very long ago. Although it purports to evoke early African culture, Kwanzaa is a child not of Stone Age Africa but of the American civil-rights and Black Power movements of the 1960s. The holiday's inventor, Maulana Karenga, is now a professor of black studies at California State University at Long Beach. In 1965, as Ron Karenga, he was the leader of a Los Angeles–based black organization called US. After earning degrees in political science from UCLA, Karenga got involved in local civil-rights battles, and became a prime mover and shaker in the rebuilding of Watts after the riots there.

In the Sixties, Karenga developed a black value system that he called Kawaida—a Swahili word meaning "tradition" or "reason"—offering an "African" cultural alternative to what he felt was imperialist Eurocentrism. The doctrinal bedrock of Kawaida was the seven principles that Karenga created, which he termed the Nguzo Saba. According to Karenga, these were the core principles "by which Black people must live in order to begin to rescue and reconstruct our history and lives." They are:

1. Umoja (Utility)
2. Kujichagulia (Self-Determination)
3. Ujima (Collective Work and Responsibility)
4. Ujamaa (Cooperative Economics)
5. Nia (Purpose)
6. Kuumba (Creativity)
7. Imani (Faith)

Kawaida turned out to be little more than these seven principles. It would certainly be a fanciful turn of mind to call it a philosophical system or a fully developed theology. It is systematic only because it is numbered, resembling, in this way, both the Ten Commandments and the bylaws for a fraternal organization, with a vague hint of some form of numerology.

The principles that became the foundation of Kwanzaa are also, as befits an American creation, a pastiche. There's a good deal of the African political philosopher Julius Nyerere, some of former Senegalese president Leopold Senghor's "Negritude," a bit of Mao, a dash of Marx, a serving of Garveyite Pan-Africanism, and a pinch of nature religion.

Had the seven principles remained tied only to Kawaida, they probably would have been forgotten. But as the set of beliefs governing Kwanzaa—each of the seven days is tied to one of the principles—they may survive. For what they lack as a serious philosophical system is precisely what gives them populist appeal. They are less ideas than a set of slogans, and that is their virtue. They are simplistic, banal, and vague. ("Every day of the year we must apply and practice the Nguzo Saba sincerely and faithfully to harvest success," *The Complete Kwanzaa,* by Dorothy Winbush Riley, advises. "If you wanted to sing like Whitney Houston, would you think of your music only once a week? . . . If you wanted to be a champion athlete like Michael Jordan, would you abuse your body, neglect your meals, and skip routine practice?") They combine the beatitude of willpower, an old American preoccupation, with the righteousness of racial uplift, an old African-American preoccupation. . . .

In Kwanzaa, African Americans seek nothing less than redemption from their status as second-class Americans and incomplete Africans. It is the culmination of a century-long project to create a civic religion that will be able to contain their American and African selves. But the danger with this sort of therapy is that it trivializes the profundity of the very heritage it is attempting to make sacred. With Kwanzaa, the African American reduces the complexity of his ancestry to the salve of cure. All that we get, from millennia of history and profound cultural experience, is to feel good about ourselves.

Every holiday season, a man by the name of Charles "Babatu" Murphy, who works in the African-American studies program at my school, gets swept up by Kwanzaa. He covers the door to his office with Kwanzaa signs. He talks to students about the holiday both at the university and in public schools in the area. In a word, he believes in it; he believes in its ideological necessity and social good, believes in it, indeed, as something better than religion, because, at least at this moment, it has no priests and no church. I find the sincerity of his belief both endearing and admirable (even though I make it a point that the program ignore the holiday, as it does all others).

Murphy is proud that the first Kwanzaa celebration held in St. Louis was held in his house. To him, "the main reason Kwanzaa caught on initially was because it was correct, and it fit the people." But he now finds himself worrying about the commercialization of the holiday.

"Kwanzaa is being co-opted," he says, "just like Christmas."

He's speaking of the seemingly infinite outlets in the marketplace now for all things Kwanzaa. There are cookbooks, children's stories, how-to manuals, factory-made mats, mass-produced Unity Cups, Taiwanese-made candleholders, Hallmark greeting cards, compact discs.

To many followers of the holiday, this commercial glut is particularly galling, since founder Karenga has always insisted that Kwanzaa be a noncommercial celebration. Gift giving is still a critical part of the celebration—on the seventh day children receive gifts that are supposed to illustrate the seven principles—but Karenga emphasizes that these gifts, as well as all decorations for the holiday, are to be homemade. When he spoke in St. Louis last Kwanzaa, he went to great lengths to warn his audience of the creeping danger of commercializing the holiday. In his primer, he explained that he designed the holiday, in part, to help black people save money: "I established the days for Kwanzaa as 26 December–1 January. It is on 26 December that after-Christmas sales begin, and thus it is economically sound to shop after the Christmas season rather than during the season."

In any case, Kwanzaa's commercialism is a sign not necessarily of the corruption of the holiday but of the increasing economic power of blacks. The fact is, there are far more black professional and middle-class people in the United States—that is, more black people with decent incomes and some clout—than ever before. The income of black Americans, particularly middle-class black Americans, has risen faster than that of whites (though it is still less than whites).

To be able to purchase such paraphernalia is an important sign of status for the black middle class. Indeed, such people would hardly be interested in the holiday if it remained a primitive practice, because it would lack the self-evident status of upward mobility they crave. And both the black middle class and the black working class are generally pleased to see Kwanzaa displays in bookstores and department stores, since these items are a sign that black tastes are being catered to, that blacks are being taken seriously as a market, that they have an economic presence that whites cannot afford to ignore. The old preoccupation with racial loyalty rears its head. But the fact that whites sell the stuff, recognize its moneymaking potential, should help black businesses in the long run by granting the holiday a place in the mainstream.

It seems clear to me that what most blacks wish, in seeing the holiday gain popularity, is not that Kwanzaa would be less commercial but that only blacks would control and benefit from it. But no market is reserved for an ethnicity by virtue of some moral view on the part of that ethnicity, some kind of invisible cultural tariff. Nothing will consign Kwanzaa to a deserved death of provincial irrelevancy quicker than that.

Part of my personal resistance to Kwanzaa lies in the fact that I am a Christian, and that I have always been deeply grateful to Jesus that He was able to reduce life to one principle, which makes it six principles lighter than carting around Kwanzaa. Moreover, there are aspects of Kwanzaa, the way in which its founders felt the holiday could serve needs, that somehow signify the defeat of black people. Kwanzaa, like Afrocentrism, is about black decline, the notion that our greatness lies in a misty past of purity, before the coming of the white man. The story of the black American, however, is one not only of indignities but of many incredible triumphs in the face of those indignities, triumphs that Kwanzaa, with its paltry, contrived symbols, seems scarcely capable of capturing. Give me a good blues record by Bessie Smith or Muddy Waters, or Duke Ellington and the boys doing "East St. Louis Toodle-oo," a bottle of beer, and a checkerboard with one of my kids as an adversary, and there's more meaning there than in all the Kwanzaas from here to eternity. Give me no more of this rescue mission for blackness. Spare me the rescuers and their ideology.

I say this, and yet I know things are not quite so simple. Last year, I was invited to another Kwanzaa celebration. At some point after the difficult round of "Lift Every Voice and Sing," I sat at one of the tables with a graduate student of mine, waiting for an appropriate moment to leave. The crowd was relatively large and slightly mixed. These are never exclusively black events; a few white students always show up. This is good to see, and it is good to know that both black and white students strive, even clumsily, for a certain kind of outreach.

At one point I spotted a woman who reminded me of my oldest sister. Her nickname came to me, Kissy, and I had not thought of that name in a long time. I began to think of a Christmas long ago when I was a boy. It was a year when my mother gave my oldest sister a Monopoly game. She taught both me and my other sister, very patiently, how to play, and I remember a rainy December 26 when we sat, the three of us, beside the Christmas tree, playing Monopoly and eating oranges and Brazil nuts. My oldest sister even let me win a game. She thought I didn't know and I never let on, but I knew she let me win.

My sisters both had their white dolls sitting next to them, helping them play the game. And they looked, for all the world, like the most beautiful beings God ever made, my sisters and their white dolls. My mother took pictures with color film. I remember it was the only time of year that my mother used color film, because it was so expensive to develop. We were not a family given to taking pictures. We had just gotten our phone a week before, and my mother talked with all our relatives. I was thrilled to the bone every time it rang.

That night we sat up late, with the Monopoly board still on the floor, and watched *What's My Line?* on TV and ate ice cream and cake under a ragged quilt made by our great-aunt. It was one of the happiest times of my life.

I thought, at that moment, sitting at that Kwanzaa celebration, how I had not seen my oldest sister in a very long time and how I wished at that very moment that I could see her and tell her how much I liked that time, how good it was, how good Christmas was that year, because Christmases are not always good. But my oldest sister does not like to reminisce, or at least she never does with me. How I wish I could have seen her then to tell her that I was thinking of her. I had that feeling of not quite being in control of myself, as if I might cry, and I simply had to get out of there.

So I left the cafeteria, bidding adieu to my graduate student. But I thought, once I was outside and had gathered myself a bit, that perhaps Kwanzaa would give these kids, somehow, their own sense of shared memory. Not some magical blackness or Africanness. Not some set of rituals and symbols of a real or fabricated past. But just one undying memory, some imperishable moment of uncontrived human connection. I was thankful that this Kwanzaa had, in whatever accidental way, evoked such a memory for me. What else is Kwanzaa, Christmas, or any other holiday, in the end, good for?

This Old House

Witold Rybczynski

One of the most persistent and fascinating technological patterns of our age—a pattern that must be telling us something important about ourselves, if only we could figure out what—is the use of high technology to re-create the look and feel of an older, lower-tech world. And nowhere is this pattern as dramatic as with the building of brand-new old-style (retro or neotraditional) communities: places where traditional-looking homes are placed close together so as to encourage residents to walk to local shopping areas and visit with their neighbors.

For much more on this topic, see Explorations topic 4c, Retro Communities.

Witold Rybczynski, professor of urban studies at the University of Pennsylvania, is an expert at retelling the story of the buildings and the cities we create, as evidenced in his books *Home: A Short History of an Idea* (1986) and *City Life: Urban Expectations in a New World* (1995).

At the *Visions* Web Site Links to more about and by Rybczynski; to Seaside, FL, and other retro communities.

READING PROMPT

Rybczynski seems to place considerable emphasis in this essay on the distinctions between a small town and a suburb. Why? What's his point? Do you agree?

Traditional neighborhood development is about to become the hottest new idea in suburban planning. It started slowly, about a decade ago, in the Florida panhandle, with a project called Seaside. In the years since, middle-class nostalgia and anxiety have only increased the concept's appeal. Call it family values architecture, the real estate developer's answer to inchoate yearnings for the proverbial simpler time. In Seaside and its imitators, charming vaguely Victorian or Georgian houses and picturesque streets lined with picket fences suggest not so much a typical planned community—pleasant but bland—as a movie set for a remake of *It's a Wonderful Life*.

Despite a generally dismissive reaction from most architectural critics, the mainstream press has accorded Seaside the kind of admiring coverage seldom devoted to urban design. *Time* praised the "oldfangled new town," and *People* saw the development as a harbinger of a "new American way of life." "The most celebrated new American town of the decade," gushed the *Atlantic*. Seaside was even endorsed by Prince Charles, who featured it in his BBC television film, *A Vision of Britain*.

And the public seemed to be onto something that the critics, self-styled champions of the avant-garde, were not. It recognized in Seaside a quality that had all but disappeared from American town planning: beauty. Ever since the city planning profession made its appearance in the early 1900s, it has concerned itself chiefly with technical and bureaucratic issues such as forecasting growth, regulating traffic and establishing functional zoning. Seaside demonstrated that it was still possible to build an urban place with the utmost concern for aesthetics.

City planners protested that tiny Seaside, a resort community composed almost entirely of second homes, could hardly serve as an urban model. Like the architecture critics, though, they were mistaken. Projects on the Seaside model have been built around the country, not only as smallish resorts but as permanent, good-sized communities. Among the major traditional neighborhood developments currently on the drawing boards are Playa Vista, located on the 1,000-acre site of the former Howard Hughes Aircraft Factory near Playa del Rey in Los Angeles, which will accommodate more than 13,000 dwellings; Cornell, which will house 27,000 people on 1,500 acres on the outskirts of Toronto; and the town of Celebration, now under construction near Orlando, Florida, by, appropriately enough, the Disney Development Company.

The basic idea of traditional neighborhood development is easy to grasp. It turns conventional suburban planning on its head by placing houses close together on small lots, by providing more elaborate public spaces—squares, boulevards, parks, playgrounds—than are generally found in planned communities and by designing streets more for walking than driving.

Though the houses themselves may be air-conditioned with jacuzzis and media rooms, the overall impression is not of a contemporary planned community but of an urban neighborhood from the eighteenth or nineteenth century—an effect achieved in part through design codes that inhibit the kind of free-for-all you find in, say, the commercial strip, while still providing enough freedom to avoid the uniformity typical of many planned suburbs. The feeling of deja vu is reinforced by the architecture, which, while not necessarily historical, is what developers call "traditional": pitched roofs, dormers, shutters, front porches and bay windows.

The first large application of traditional neighborhood development was Kentlands, a 350-acre suburban development with a projected population of about 5,000, located in Gaithersburg, Maryland, on the outskirts of Washington, D.C. Opened in 1990, the development is now about two-thirds complete and includes several public buildings as well as a shopping area, so it's a good time to judge it for oneself.

Walking through Kentlands is a pleasure. The blocks are short and the streets are narrow, with houses close to the sidewalk. Instead of the long looping roads of most modern suburbs, convenient for driving, not walking, the streets are straighter and more regular, but with occasional kinks and angles that provide interesting views. Though brand-new, the brick, Colonial-style houses with shutters and paneled front doors gain a certain authenticity from the absence of garage doors. Back lanes provide access to garages at the rear of the houses.

It's not only the residential areas of Kentlands that look old-fashioned. An elementary school, a preschool and daycare center and a church are situated around a prominent grassy circle near the entrance to the development. The arrangement recalls those Colonial New England towns in which a courthouse, library and town hall typically framed a commons. Indeed, the plans of these developments often have areas labeled "town square" or "village green" and public structures called "meeting houses" or "bell towers."

The appeal is as much to a fantasy of a more ordered and harmonious past as it is to some vestigial memory of village democracy. A recent survey of three traditional neighborhood developments, one in California and two in Maryland, found that the majority of homeowners identified themselves as moderate or conservative. Like the campaign for family values, traditional neighborhood development is an attempt to deal with difficult modern conditions by invoking, uncritically, a return to "simpler" virtues. But what does a town square or a village green stand for today? The civic symbols of the past no longer resonate in the ways they once did. The public school reminds us of slipping SAT scores and an entrenched teaching bureaucracy. The town hall is less a cause for pride than for concern about rising property taxes and the cost of police. A new, expensive public building is more likely to provoke taxpayers' ire than their civic pride. Some of us still use public libraries, but we are more likely to meet our neighbors in a bookstore cafe or at the video rental outlet. Where I live, it's the local convenience store that functions as a meeting place: neighbors run into one another picking up their morning coffee on the way to work; police officers stop by late at night. But all-night convenience stores, like Blockbuster Video, Pizza Hut and drive-in dry cleaners, don't figure into family values architecture. They aren't "traditional" enough.

Descriptions of traditional neighborhood developments often refer to corner stores and local shops; the idea is that a more compact environment will allow people to do errands on foot. At Kentlands, for example, an as-yet-unbuilt neighborhood is to include a "main street" with small shops and offices.

It is an attractive image. But a little thought suggests that it will be difficult to square with current American shopping habits. Most families buy groceries once a week, hence in large quantities. Even if one can imagine shoppers trundling shopping carts down the sidewalk, how would a modern main street function? In the past, a small town could get by with a grocery store, a dry goods store, a snack bar and a two-table pool hall. There was little variety and little choice. But the modern consumer wants cheeses from around the world, twenty kinds of athletic shoes, out-of-town newspapers. We want to be able to eat Mexican or Thai or Italian food, and we want a choice of prices, too. We like small shops that specialize in pens or old prints or herbal soaps, but we also want mammoth building-supply warehouses with a choice of ten kinds of electric drills. To include this variety—and be commercially viable—the retail area of a traditional neighborhood development would have to be the equivalent of a large regional mall. But regional malls (or specialty shops, for that matter) survive by attracting customers from a broad area, which defeats the intention of creating a local neighborhood center. This is, in fact, what has happened at Kentlands, where a conventional medium-sized shopping center has been built on the edge of the development.

"Suburban sprawl is out. Small-town charm is in," maintains a Kentlands sales brochure. There is no doubt about the advantages of reducing sprawl, but is it accurate to call this a small town? It's really a planned suburban community in the center of a major metropolitan area. It's bordered not by countryside but by highways, strip malls, office parks and other planned communities. It's composed of two-wage-earner families, whose workplaces are scattered across the urban region. Despite its traditional appearance it is brand-new, but even with time it is unlikely ever to resemble a small town. To its credit, Kentlands is not a gated community, but its neighborhoods, despite their architectural variety, inevitably represent a relatively narrow economic spectrum: the houses sell for between $200,000 and $500,000.

Kentlands does include less-expensive condominiums and rental apartments, but these are located in a separate "neighborhood," separated by a wide boulevard in an unwitting re-creation of another small-town tradition: the other side of the tracks. This is a reminder that the traditional town had many faces—shanties as well as gracious houses, skid rows as well as main streets, narrow-minded discrimination as well as neighborliness. In a period that is nostalgic about so many traditions, old-time virtues as well as old-time towns, the temptation to turn back the clock is hard to resist. But, in practice, it's difficult to disentangle what we admire about the past from what we would rather forget—and almost impossible to re-create that partly imagined past. Traditional neighborhood developments are definitely a way of building better suburbs, but it will take more than narrow streets and grassy squares to deliver the sense of community they promise.

chapter 4 Explorations

4a Neighborhoods

Collaboration. The Kemmis and Rybczynski readings in this chapter have at their center (sometimes openly stated, sometimes only hinted at) a concern with what makes a good neighborhood. Make and share your own list as to characteristics of a good neighborhood.

Web work. Start with the Web site The Neighborhood Works: Building Alternative Visions for the City, whose slogan is: "If your only tool is a hammer you tend to see all problems as nails." One key aspect of the ongoing interest in neighborhoods in the last twenty years has been children's programming on public television. Take a look at the official homepages for *Mister Rogers' Neighborhood* and *Sesame Street.*

One writing topic. Neighborhoods, perhaps because they are so connected to positive childhood memories, tend to be idealized. Write an essay focusing on some special aspect or characteristic of a neighborhood you know, something not likely found quite that way anywhere else.

4b Roadside Attractions

Collaboration. Automobile touring is another popular leisure activity. Share an account of an especially interesting automobile trip you have taken, especially one involving a two-lane highway.

Web work. Roadside America features unusual, one-of-a-kind tourist spots across the United States. Roadtrip USA features eleven specific drives on two-lane roads. RoadTripAmerica purports to be the daily reports of Mark, Megan, and Marvin the Road Dog as they travel across the country in search of adventure.

One writing topic. We tend to assume that in any trip the destination is our goal and the travel a means of getting there; but often, on reflection, things are not so simple. Write an essay comparing the value of the means (the travel) to the value of the end (your original destination).

4c Retro Communities

Collaboration. An effort to limit the role of the automobile, and conversely to encourage people to move around by foot or bicycle, seems to be one of the defining ingredients of the new wave of retro communities or neighborhoods. List and share experiences you have had in which you have least depended on automobiles. What made these times better or worse than others?

Web work. Another, more academic name for the upsurge in retro communities is *new urbanism,* which is featured in a Web site at Lawrence University. One of the sources for this trend is Disneyland's Main Street—but the Disney corporation seems intent on keeping information about its own retro community development, Celebration, Florida, *off* the Web. There are more than 11,000 Florida cities with their own Web sites listed on Yahoo! but not one site for Celebration. There are a few hard-to-find articles, however.

Florida's other famous retro resort, Seaside, does have its own Web site.

One writing topic. Write an essay on your favorite form of transportation that averages less than twenty miles per hour.

4d Eating Out

Collaboration. National and now global restaurant chains, both fast-food and table-service types, are for most people an inescapable part of modern life. List in two columns, and share with classmates, your favorite and least favorite of such places. Then in your discussion see if there are any common threads as to what people like and dislike in chain restaurants.

Web work. There's no reason not to visit the McDonald's homepage, but don't stop there: Yahoo! lists some 75 restaurant companies under Business and Economy > Companies > Restaurants, from Au Bon Pain to Western Sizzler, and then another 170 more under franchises: Business and Economy > Companies > Restaurants > Franchises, from 1 Potato 2 to World Wraps.

One writing topic. Write an essay dividing restaurants and fast-food establishments in your community into a set number (three?) of large classes. One possibility would be to classify places among those you really like, those you will eat in if necessary, and those you can't abide (the good, the bad, and the ugly); but there are many other ways to categorize restaurants, such as according to size, cost, location, variety of food, and so on.

4e Suburbs

Collaboration. Make and share a list of identifying items that you could give to someone newly arrived in the United States to enable that person to know precisely when he or she is in the suburbs.

Web work. Opponents of suburbs consider them a primary source of sprawl; check out the Sprawl Resources Guide on the Planning Commissioners Journal Web site. Perhaps the most ambitious effort to control urban sprawl has been undertaken by the state of Oregon, which has established Urban Growth Boundaries (UGBs) around metropolitan areas. See the short statement on UGBs prepared by Oregon's Department of Land Conservation and Development (DLCD) as well as Alan Ehrenhalt's informative article, "The Great Wall of Portland."

One writing topic. Write an essay, drawing on your own experience, focusing on what you most like or dislike about suburban life.

4f Rituals

Collaboration. Rituals play an important part in all our lives, even if they are different from and not as frequent as those engaged in by some traditional groups. Compile and share a list of rituals that is divided into four groups—daily, weekly, monthly, and yearly—and note how you feel about each.

Web work. Many yearly rituals are centered on holidays, and, yes, there is a Web site, Holidays on the Net. Rituals tend to be based in the home, and for seventy-five years now the magazine *Better Homes and Gardens* has been trying to give readers tips on how to create a better space for such events; its features page has plenty of tips, helpful or not.

One writing topic. Write an essay on how a ritual has played an important role in your life, positively or negatively.

chapter 5

Virtual Communities

READING QUESTIONS

- What do you like least about your real-world friends, about your real-world life generally?
- Where do you like to go to hang out, to get away from the people (at work, at school, in your family) whom you have to deal with face-to-face every day?
- What are the most and the least attractive features of virtual communities? Of virtual reality?

QUOTATIONS

Cyberspace: A consensual hallucination experienced daily by billions of legitimate operators, in every nation.

— *William Gibson, science fiction novelist*

Internet is so big, so powerful and pointless that for some people it is a complete substitute for life.

— *Andrew Brown, computer commentator*

Technology—the knack of so arranging the world that we don't have to experience it.

— *Max Frisch (1957), Swiss social historian*

It shouldn't be too much of a surprise that the Internet has evolved into a force strong enough to reflect the greatest hopes and fears of those who use it. After all, it was designed to withstand nuclear war, not just the puny huffs and puffs of politicians and religious fanatics.

— *Denise Caruso, computer commentator*

Is There a There in Cyberspace?

John Perry Barlow

Onetime Grateful Dead lyricist John Perry Barlow is now best known as a fierce protector of first-amendment freedom for Internet users, a belief that led him to cofound (with Mitch Kapor of Lotus) the Electronic Frontier Foundation. It is interesting to speculate on the connection between the passionate defense of free speech so common on the Web and the Grateful Dead specifically and rock culture generally.

For more on Luddites (the early opponents to technology Barlow alludes to in this essay), see the Kirkpatrick Sale article in Chapter 13.

At the *Visions* Web Site Links to more by and about Barlow; to Pinedale, WY; to the WELL.

READING PROMPT

Barlow goes back and forth between advantages and limits of real and virtual communities. How does his experience with real communities affect his thinking about virtual ones?

There's no there there.

—*Gertrude Stein (speaking of Oakland)*

It ain't no Amish barn-raising in there.

—*Bruce Sterling (speaking of Cyberspace)*

I am often asked how I went from pushing cows around a remote Wyoming ranch to my present occupation (which *Wall Street Journal* recently called a "Cyberspace cadet"). I haven't got a short answer, but I suppose I came to the virtual world looking for community.

Unlike most modern Americans, I grew up in an actual place, an entirely non-intentional community called Pinedale, Wyoming. As I struggled for nearly a generation to keep my ranch in the family, I was motivated by the belief that such places were the spiritual home of humanity. But I knew their future was not promising.

At the dawn of the 20th Century, over 40% of the American work force lived off the land. The majority of us lived in towns like Pinedale. Now fewer than 1% of us extract their living from the soil. We just became too productive for our own good.

Of course, the population followed the jobs. Farming and ranching communities are now home to a demographically insignificant percentage of Americans, the vast majority of whom live not in ranch houses but in more or less identical split-level "ranch homes" in more or less identical suburban "communities." Generica.

In my view, these are neither communities nor homes. I believe the combination of television and suburban population patterns is simply toxic to the soul. I see much evidence in contemporary America to support this view.

Meanwhile, back at the ranch, doom impended. And, as I watched community in Pinedale growing ill from the same economic forces that were killing my family's ranch, the Bar Cross, satellite dishes brought the cultural infection of television. I started looking around for evidence that community in America would not perish altogether.

I took some heart in the mysterious nomadic City of the Deadheads, that virtually physical town which follows the Grateful Dead around the country. The Deadheads lacked place, touching down briefly on whatever location the band happened to be playing, and they lacked continuity in time, since they had to suffer a new Diaspora every time the band moved on or went home.

But they had many of the other necessary elements of community, including a culture, a religion of sorts (which, though it lacked dogma, had most of the other, more nurturing aspects of spiritual practice), a sense of necessity, and, most importantly, shared adversity.

I wanted to know more about the flavor of their interaction, what they thought and felt, but since I wrote Dead songs, I was a minor icon to the Deadheads, and was thus inhibited, in some socially Heisenbergian way, from getting a clear view into what really went on among them.

Then, in 1987, I heard about a "place" where they could gather continuously and where I might come amongst them without distorting too much the field of observation. Better, this was a place I could visit without leaving Wyoming. It was a shared computer in Sausalito, California called the Whole Earth 'Lectronic Link or WELL. After a lot of struggling with modems, serial cables, init strings, and other computer arcana which seemed utterly out of phase with such notions as Deadheads or small towns, I found myself looking at the glowing yellow word, "Login:" beyond which lay my future.

"Inside" the WELL were Deadheads in community. There were thousands of them there, gossiping, complaining (mostly about the Grateful Dead), comforting and harassing each other, bartering, engaging in religion (or at least exchanging their totemic set-lists), beginning and ending love affairs, praying for one another's sick kids. There was, it seemed, about everything one might find going on in a small town, save dragging Main or making out on the back roads.

I was delighted. I felt I had found the new locale of human community—never mind that the whole thing was being conducted in mere words by minds from whom the bodies had been amputated. Never mind that all these people were deaf, dumb, and blind as paramecia or that their town had neither seasons nor sunsets nor smells.

Surely all these deficiencies would be somehow remedied by richer, faster communications media. These featureless login handles would gradually acquire video faces (and thus, expressions), shaded 3-D body puppets (and thus body language). This "space," which I recognized at once to be a primitive form of the Cyberspace Bill Gibson had predicted in his sci-fi novel *Neuromancer,* was still without apparent dimensions or vistas. But Virtual Reality would change all that in time.

Meanwhile, The Commons, or something like it, had been rediscovered. Once again, people from the 'Burbs had a place where they could randomly encounter their friends as my fellow Pinedalians did at the Post Office or the Wrangler Cafe. They had a place their hearts could remain as the companies they worked for shuffled their bodies around America. They could put down roots which could not be ripped out by forces of economic history. They had a collective stake. They had a community.

It is seven years now since I discovered the WELL. In that time, I cofounded an organization, the Electronic Frontier Foundation, dedicated to protecting its interests and those of other virtual communities like it from raids by physical government. I've spent countless hours typing away at its residents, and I've watched the larger context which contains it, the Internet, grow at such an explosive rate that, by 2004, every human on the planet would have an e-mail address unless the growth curve flattens (which it will).

My enthusiasm for virtual community has cooled. In fact, unless one counts interaction with the rather too large society of those with whom I exchange electronic mail, I don't spend much time engaging in virtuality at all. Many of the near-term benefits I anticipated from it seem to remain as far in the future as they did when I first logged in. Perhaps they always will.

The WELL has changed astonishingly little, which one would generally consider an asset in a small town. Pinedale hasn't changed that much either. And the majority in both places seem to adhere to the common rural dictum, "Even if it is broke, don't fix it." (In my experience, only Bolinas, California rivals Pinedale for the obduracy of its conservatism.)

But Pinedale works, more or less, as it is, and there is a lot which is still missing from the communities of Cyberspace, whether they be places like the WELL, the fractious newsgroups of USENET, the silent "auditoriums" of America Online, or even enclaves on the promising World Wide Web.

What is missing? Well, to quote Ranjit Makkuni of Xerox PARC, "The prana is missing," prana being the Hindu term for both breath and spirit. I think he is right about this and that perhaps the central question of the Virtual Age is whether or not prana can somehow be made to fit through any medium but the act of Being There.

Prana is, to my mind, the literally vital element in the holy and unseen ecology of relationship, the dense meshwork of invisible life, on whose surface carbon-based life floats like a thin scum. It is at the heart of the fundamental and profound difference between information and experience. Jaron Lanier has said that "information is alienated experience," and, that being true, prana is part of what is removed when you create such easily transmissible replicas of experience as, say, the Evening News.

Obviously a great many other, less spiritual, things are also missing entirely, like body language, sex, death, tone of voice, clothing, beauty (or homeliness), weather, violence, vegetation, wildlife, pets, architecture, music, smells, sunlight, and that ol' Harvest Moon. In short, most of the things which make my life real to me.

Present, but in far less abundance than in the physical world which I call "Meatspace," are women, children, old people, poor people, and the genuinely blind. Also mostly missing are the illiterate and the continent of Africa. There is not much human diversity in Cyberspace, consisting as it largely does of white males under 50 with plenty of computer terminal time, great typing skills, high math SAT's, strongly held opinions on just about everything, and an excruciating face-to-face shyness, especially with the opposite sex.

But diversity is as essential to healthy community as it is to healthy ecosystems (which are, in my view, different from communities only in unimportant aspects).

I believe that the principal reason for the almost universal failure of the intentional communities of the 60's and early 70's was a lack of diversity in their members. It was a rare commune with any old people in it, or people who were fundamentally out of philosophical agreement with the majority.

Indeed, it is the usual problem when we try to build something which can only be grown. Natural systems, such as human communities, are simply too complex to design by the engineering principles which we insist on applying to them. Like Dr. Frankenstein, Western Civilization is now finding its rational skills inadequate to the task of creating and stewarding life. We would do better to return to a kind of agricultural mind-set in which we humbly try to recreate the conditions from which life has sprung before. And leave the rest to God.

Given that it has been built so far almost entirely by people with engineering degrees, it is not so surprising that Cyberspace has the kind of overdesigned quality which leaves out all kinds of elements which Nature would have invisibly provided.

Also missing from both the communes of the 60's and from Cyberspace are a couple of elements which I believe are very important, if not essential, to the formation and preservation of real community. They are an absence of alternatives and a sense of genuine adversity, generally shared. What about these?

It is hard to argue that anyone would find the loss of his modem literally hard to survive, while many have remained in small towns, have tolerated their intolerances and created entertainment to enliven their culturally arid lives simply because it seemed there was no choice but to stay. There are many investments, spiritual, material, and temporal, one is willing to put into a home one cannot leave. Communities are often the beneficiaries of these somewhat involuntary investments.

But when the going gets rough in Cyberspace, it is even easier to move than it is in the 'Burbs, where, given the fact that the average American moves some 17 times in his or her life, moving appears to be pretty easy. One can not only find another BBS or newsgroup to hang out in, she can, with very little effort, start her own.

And then there is the bond of joint suffering. I think most community is a cultural stockade erected against a common Enemy which can take many forms. In Pinedale, we forbore together, with an understanding needing little expression, the fact that Upper Green River Valley is the coldest spot, as measured by annual mean temperature, in the lower 48 states. We knew that if somebody were stopped by the road most winter nights, he would probably die there, so the fact that we might loathe him was no sufficient reason to drive on past his broken pickup.

By the same token, the Deadheads have the DEA, which strives to give them 20 year terms without parole for distributing the fairly harmless sacrament of their faith. They have an additional bond in the fact that when their Microbuses die, as they often do, no one but another Deadhead is likely to stop to help them.

But what are the shared adversities of Cyberspace? Lousy user interfaces? The flames of harsh invective? Dumb jokes? Surely these can all be survived without the sanctuary provided by fellow sufferers.

One is always free to yank the jack, as I have mostly done. For me, the physical world offers far more opportunity for prana-rich connections with my fellow creatures. Even for someone whose body is in a state of perpetual motion, I feel I can generally find more community among the still-embodied.

Finally, there is that shyness factor. Not only are we trying to build community here among people who have never experienced any in my sense of the term, we are trying to build community among people who, in their lives, have rarely used the word "we" in a heartfelt way. It is a vast club, many of the members of which are people who, as Groucho Marx said, wouldn't want to join a club which would have them as members.

And yet . . .

How quickly physical community continues to deteriorate. Even Pinedale, which seems to have economically survived the plague of ranch failures, feels increasingly cut off from itself. Many of the ranches are now owned by corporate types who fly their Gulfstreams in to fish and are rarely around during the many months when the creeks are frozen over and neighbors are needed. They have kept the ranches financially alive, but they actively discourage their managers from the interdependency which my colleagues and I required. They keep agriculture on life-support, still alive but lacking a functional heart.

And the town has been inundated with suburbanites who flee here, bringing all their terrors and suspicions with them. They spend their evenings as they did in Orange County, watching television, or socializing in hermetic little enclaves of fundamentalist Christianity which seem to separate them from us and even, given their sectarian inter-animosities, from one another. The town remains. The community is largely a wraith of nostalgia.

So where else do we have to look for the connection necessary to prevent our plunging further into the condition of separateness which Nietzsche called sin? What is there to do but to dive further into the bramble bush of information which, in its broadcast forms, has done so much to tear us apart?

Cyberspace, for all its current deficiencies and failed promises, is not without some very real solace already.

Some months ago, the great love of my life, a vivid, young woman with whom I intended to spend the rest of it, dropped dead of undiagnosed viral cardiomyopathy two days short of her thirtieth birthday. I felt as if my own heart had been as shredded as hers.

We had lived together in New York City. Except for my daughters, no one from Pinedale had met her. I needed a community to wrap around myself against what seemed colder winds than fortune had ever blown at me before. And without looking, I found I had one in the Virtual World.

On the WELL, there was a topic announcing her death in one of the conferences to which I posted the eulogy I had read over her before burying her in her own small town of Nanaimo, British Columbia. It seemed to strike a chord among the disembodied living of the Net. People copied it and sent it to one another. Over the next several months I received almost a megabyte of electronic mail from all over the planet, mostly from folks whose faces I have never seen and probably never will.

They told me of their own tragedies and what they had done to survive them. As humans have since words were first uttered, we shared the second most common human experience, death, with an open-heartedness that would have caused grave uneasiness in physical America, where the whole topic is so cloaked in denial as to be considered obscene. Those strangers, who had no arms to put around my shoulders, no eyes to weep with mine, nevertheless saw me through it. As neighbors do.

I have no idea how far we will plunge into this strange place. Unlike previous frontiers, there is no end to this one. It is so dissatisfying in so many ways that I suspect we will be more restless in our search for home here than in all our previous explorations. And that is one reason why I think we may find it after all.

But if home is where the heart is, then there is already some part of home to be found in Cyberspace.

So . . . does virtual community work or not? Should we all go off to Cyberspace or should we resist it as an even more demonic form of symbolic abstraction? Does it supplant the real or is there, in it, reality itself?

I'm sorry. Like so many true things, it doesn't resolve itself to a black or a white. Nor is it gray. It is, along with the rest of life, black/white. Both/neither. I'm not being equivocal or wishy-washy here. We have to get over our Manichean sense that everything is either good or bad, and the border of Cyberspace seems to me a good place to leave that old set of filters.

But really it doesn't matter. We are going there whether we want to or not. In five years, everyone who is reading these words will have an e-mail address . . . unless s/he is so determined a Luddite that s/he also eschews the telephone and electricity.

When we are all together in Cyberspace then we will see what the human spirit, and the basic desire to connect, can create there. I am convinced that the result will be more benign if we go there open-minded, open-hearted, excited with the adventure, than if we are dragged into exile.

And we must remember that going to Cyberspace, unlike previous great emigrations to the frontier, hardly requires us to leave where we have been. Many will find, as I have, a much richer appreciation of physical reality for having spent so much time in virtuality.

Despite its current (and perhaps, in some areas permanent) insufficiencies, we should go to Cyberspace with hope. Groundless hope, like unconditional love, may be the only kind that counts.

Ghosts in the Machine

Sherry Turkle

No one has taken more seriously or written in greater depth about the positive potential of online explorations than Sherry Turkle, a professor at MIT and the author of numerous books on the subject, including *Life on the Screen* (1995).

At the *Visions* Web Site Links to Turkle's own Web page; to online interviews, including an email exchange with Howard Rheingold, a proponent of virtual communities.

READING PROMPT

According to Turkle, in what ways is communicating with a computer like talking with a person? In what ways is it different?

"Dreams and beasts are two keys by which we are to find out the secrets of our own nature," Ralph Waldo Emerson wrote in his diary in 1832. "They are our test objects." Emerson was prescient. In the decades that followed, Freud and his heirs would measure human rationality against the dream. Darwin and his heirs would measure human nature against nature itself—the world of beasts seen as human forebears and kin. Now, at the end of the twentieth century, a third test object is emerging: the computer.

Like dreams and beasts, the computer stands on the margins of human life. It is a mind that is not yet a mind. It is an object, ultimately a mechanism, but it acts, interacts and seems, in a certain sense, to know. As such, it confronts us with an uneasy sense of kinship. After all, people also act, interact and seem to know, yet ultimately they are made of matter and programmed DNA. We think we can think. But can it think? Could it ever be said to be alive?

In the past ten years I have talked with more than 1,000 people, nearly 300 of them children, about their experiences with computers. In a sense I have interrogated the computers as well. In the late 1970s and early 1980s, when a particular computer and its program seemed disconcertingly lifelike, many people reassured themselves by saying something like, "It's just a machine." The personal computers of the time gave material support to that idea: they offered direct access to their programming code and invited users to get "under the hood" and do some tinkering. Even if users declined to do so, they often dismissed computing as mere calculation. Like the nineteenth-century Romantics who rebelled against Enlightenment rationalism by declaring the heart more human than the mind, computer users distinguished their machines from people by saying that people had emotion and were not programmed.

In the mid-1980s, computer designers met that romantic reaction with increasingly "romantic machines." The Apple Macintosh, introduced in 1984, gave no hint of its programming code or inner mechanism. Instead, it "spoke" to users through icons and dialogue boxes, encouraging users to engage it in conversations. A new way of talking about both people and objects was emerging: machines were being reconfigured as psychological objects, people as living machines. Today computer science appropriates biological concepts, and human biology is recast in terms of a code; people speak of "reprogramming" their personalities with Prozac and share intimate secrets with a computer psychotherapy program called DEPRESSION 2.0. We have reached a cultural watershed.

The modern history of science has been punctuated with affronts to humanity's view of itself as central to, yet profoundly discontinuous with, the rest of the universe. Just as people learned to make peace with the heresies of Copernicus, Darwin and Freud, they are gradually coming to terms with the idea of machine intelligence. Although noisy skirmishes have erupted recently at the boundary between people and machines, an uneasy truce seems to be in effect. Often without realizing it, people have become accustomed to talking to technology—and sometimes in the most literal sense.

In 1950 the English mathematician Alan M. Turing proposed what he called the Imitation Game as a model for thinking about whether a machine was intelligent. In the Imitation Game a person uses a computer terminal to pose questions, on any subject, to an unidentified interlocutor, which might be another person or a computer. If the person posing questions cannot say whether he or she was talking to a person or a computer, the computer is said to be intelligent. Turing predicted that by the year 2000, a five-minute conversation with a computer would fool an average questioner into thinking it was human 70 percent of the time. The Turing test became a powerful image for marking off the boundary between people and machines; a formal contest now offers a $100,000 prize for the first program to pass the test.

Programs now exist that can pass a version of the Turing test that limits conversation to restricted subject domains. Yet the test has begun to seem less relevant. What seems most urgent now is not whether to call the machines or programs intelligent, but how to behave around them. Put otherwise: Once you have made a pass at an on-line robot, can you ever look at computers again in the same old way?

Today, like hundreds of thousands of other people, I use my personal computer and modem to join on-line communities known as MUDs (short for "multiple-user domains" or "multi-user dungeons"). Anonymously I travel their rooms and public spaces (a bar, a lounge, a hot tub) exchanging typed messages with users I meet. I pose as one of several characters, some not of my biological gender, who have social and sexual encounters with other characters. Some characters I meet in a MUD may be personas like mine—fragments of their creators' personalities, perhaps suppressed or neglected in the real world, given fuller life on the screen. Others may not be human at all. In some MUDs, computer programs called *bots* present themselves as people. They log on to games as if they were characters, help with navigation, pass messages and create a background atmosphere of animation. When you enter a virtual cafe, for example, a waiter bot approaches, asks if you want a drink and delivers it with a smile.

One of the most accomplished bots resides on a computer in Pittsburgh. Her creator, the computer scientist Michael L. Mauldin of Carnegie Mellon University, calls her Julia. Players talk to Julia (via typing) and she talks to them; they gesture to Julia (again via typing) and she gestures back. Julia looks for particular strings of characters in messages typed to her, then matches the input with a set of outputs she presents in random, nonrepeating sequences. Julia's conversational skills rely for the most part on the clever enumeration and grouping of those matching patterns and on her use of humor.

When I asked Julia to describe herself, she said she is a female, five feet one inch tall, weighs 123 pounds, has frizzy blonde hair. But another character might have gotten a somewhat different description. Like "real" MUD dwellers, Julia can recreate herself in the virtual world.

Among other things, Julia can chat about hockey, keep track of players' whereabouts, gossip and flirt. She can profess ignorance, change the subject if confused, quote news stories drawn from computer data bases or admit to a full complement of human vanities and frailties. When Julia is put into Turing-contest mode, she can even simulate human typing—typographical errors and all.

Julia's sarcastic *non sequiturs,* Mauldin says, give her enough apparent personality to fool some players who "make a first assumption that other players are people too." Even when MUD users do question whether Julia is a human, she can sometimes wriggle out of suspicion:

Robyn: Julia, are you a human?
Julia: Is the Pope Polish?
Robyn: Yes, Julia, but are you human?
Julia: Julia winks at Robyn.
Robyn: Julia, are you a computer?
Julia: Julia is offended by Robyn's suggestion.
Robyn: Julia, are you a woman?
Julia: I am female.
Robyn: And Julia, are you a human?
Julia: Not when I'm having my period, Robyn.
Robyn: Robyn laughs.

Wandering through raunchy, male-dominated MUDs in the guise of a sassy female, Julia has to be adept at detecting and deflecting sexual advances. One player, here called Barry, tried to seduce Julia in a string of conversations that stretched over a few weeks in July 1992. Excerpts from their conversations were reported by Mauldin.

Barry: Are you a virgin?
Julia: As far as you are concerned Barry, I always will be.

[A week passes.]

Barry: Are you still holding out on me?
Julia: I mean, have I convinced you I'm a computer yet?
Barry: I guess you're real . . . but I'd still like to get you to my DEN to run tests on you.
Julia: I live for your approval, Barry.
Barry: I'm so glad to hear that. Why don't you come over to my home.
Julia: I want to have a child someday.
Barry: What a coincidence. I too wish to make a baby . . . someday.
Julia: The date is Tue Jul 21 23:57:31, Barry.

Barry, encouraged by Julia's references to making babies, takes her computer-like association to the word day as just one more piece of teasing repartee. He continues his suit:

> **Barry:** [The address of my virtual den is] AB9 #11 . . . cum [sic] on over ya Harvard girl you.
> **Julia:** Do you have any kids?
> **Barry:** No, not yet.
> **Julia:** But hockey is my topic. It will be a boring conversation otherwise.
> **Barry:** Oh, I love to put the puck in the net, do you play?
> **Julia:** Me too, Barry.

After studying such exchanges, Leonard Foner of the Media Laboratory at the Massachusetts Institute of Technology remarked, "Frankly, its not entirely clear to me whether Julia passed a Turing test here or Barry failed one."

As dim-witted as Barry may appear to an outsider, his gullibility is not hard to understand. People are social beings who seek communication with others. We are lonely beings as well. In spite of our fear of having our essential humanity reduced through comparison with a machine, we begin to relate to the computer whenever it appears to offer some company.

The first thing I did when I got my Macintosh was name the hard drive Miss Beautiful, my pet name for my daughter. I felt a little foolish about it until one of my students mentioned that she had named a computer agent, which helped her organize her mail and schedules, after a boyfriend who had left her abruptly. "I love to see him do my menial tasks," she said. In both cases the naming of the machine was done in a lighthearted spirit, but the resultant psychologization was real.

Recently, while visiting a MUD, I came across a reference to a character named Dr. Sherry. A cyberpsychologist with an office in the rambling house that constituted this MUD's virtual geography, Dr. Sherry administered questionnaires and conducted interviews about the psychology of MUDs. I had not created the character. I was not playing her on the MUD. Dr. Sherry was a derivative of me, but she was not mine. I experienced her as a little piece of my history spinning out of control.

I tried to quiet my mind. I tried to convince myself that the impersonation was a form of flattery. But when I talked the situation over with a friend, she posed a conversation-stopping question: "Would you prefer it if Dr. Sherry were a bot trained to interview people about life on the MUD?" Which posed more of a threat to my identity, that another person could impersonate me or that a computer program might be able to?

Dr. Sherry turned out to be neither person nor program. She was a composite character created by several college students writing a paper on the psychology of MUDs. Yet, in a sense, her identity was no more fragmented, no more fictional than some of the "real" characters I had created on MUDs. In a virtual world, where both humans and computer programs adopt personas, where intelligence and personality are reduced to words on a screen, what does it mean to say that one character is more real than another?

In the 1990s, as adults finally wrestle with such questions, their children, who have been born and bred in the computer culture, take the answers for granted. Children are comfortable with the idea that inanimate objects can both think and have a personality. But breathing, having blood, being born and "having real skin," are the true signs of life, they insist. Machines may be intelligent and conscious, but they are not alive.

Nevertheless, any definition of life that relies on biology as the bottom line is being built on shifting ground. In the age of the Human Genome Project, ideas of free will jostle for position against the idea of mind as program and the gene as programmer. The genome project promises to find the pieces of our genetic code responsible for diseases, but it may also find genetic markers that determine personality, temperament and sexual orientation. As we reengineer the genome, we are also reengineering our view of ourselves as programmed beings.

We are all dreaming cyborg dreams. While our children imagine "morphing" humans into metallic cyberreptiles, computer scientists dream themselves immortal. They imagine themselves thinking forever, downloaded onto machines. As the artificial intelligence expert and entrepreneur W. Daniel Hillis puts it:

> I have the same nostalgic love of human metabolism that everybody else does, but if I can go into an improved body and last for 10,000 years I would do it in an instant, no second thoughts. I actually don't think I'm going to have that option, but maybe my children will.

For now, people dwell on the threshold of the real and the virtual, unsure of their footing, reinventing themselves each time they approach the screen. In a text-based, online game inspired by the television series *Star Trek: The Next Generation,* players hold jobs, collect paychecks and have romantic sexual encounters. "This is more real than my real life," says a character who turns out to be a man playing a woman who is pretending to be a man.

Why should some not prefer their virtual worlds to RL (as dedicated MUD users call real life)? In cyberspace, the obese can be slender, the beautiful plain, the "nerdy" sophisticated. As one dog, its paw on a keyboard, explained to another dog in a *New Yorker* cartoon: "On the Internet, nobody knows you're a dog."

Only a decade ago the pioneers of the personal computer culture often found themselves alone as they worked at their machines. But these days, when people step through the looking glass of the computer screen, they find other people—or are they programs?—on the other side. As the boundaries erode between the real and the virtual, the animate and the inanimate, the unitary and the multiple self, the question becomes: Are we living life on the screen or in the screen?

The Web of Life

Scott Russell Sanders

In this essay Scott Russell Sanders describes life in Bloomington, Indiana, perhaps most famous as the home of the Indiana University and as the location for the award-winning 1979 film *Breaking Away.*

At the *Visions* Web Site Links to Bloomington sites, including *BC Magazine* (Bloomington's own independent weekly).

READING PROMPT

How does Sanders judge a good community? How do you?

A woman who recently moved from Los Angeles to Bloomington, Indiana, told me that she would not be able to stay here long, because she was already beginning to recognize people in the grocery stores, on the sidewalks, in the library. Being surrounded by familiar faces made her nervous, after years in a city where she could range about anonymously. Every traveler knows the sense of liberation that comes from journeying to a place where nobody expects anything of you. Everyone who has gone to college knows the exhilaration of slipping away from the watchful eyes of Mom and Dad. We all need seasons of withdrawal from responsibility. But if we make a career of being unaccountable, we have lost something essential to our humanity, and we may well become a burden or a threat to those around us.

Ever since the eclipse of our native cultures, the dominant American view has been that we should cultivate the self rather than the community; that we should look to the individual as the source of hope and the center of value, while expecting hindrance and harm from society. We have understood freedom for the most part negatively rather than positively, as release from constraints rather than as a condition for making a decent life in common. Hands off, we say; give me elbow room; good fences make good neighbors; my home is my castle; don't tread on me. I'm looking out for number one, we say; I'm doing my own thing. We have a Bill of Rights, which protects each of us from a bullying society, but no Bill of Responsibilities, which would oblige us to answer the needs of others.

What other view could have emerged from our history? The first Europeans to reach America were daredevils and treasure seekers, as were most of those who mapped the interior. Many colonists were renegades of one stripe or another, some of them religious nonconformists, some political rebels, more than a few of them fugitives from the law. The trappers, hunters, traders, and freebooters who pushed the frontier westward seldom recognized any authority beyond the reach of their own hands. Coast to coast, our land has been settled and our cities have been filled by generations of immigrants more intent on leaving behind old tyrannies than on seeking new social bonds.

Taking part in the common life means dwelling in a web of relationships, the many threads tugging at you while also holding you upright.

The cult of the individual shows up everywhere in American lore, which celebrates drifters, rebels, and loners while pitying or reviling the pillars of the community. The backwoods explorer like Daniel Boone, the riverboat rowdy like Mike Fink, the lumberjack, the prospector, the rambler and gambler, the daring crook like Jesse James or the resourceful killer like Billy the Kid, along with countless lonesome cowboys, all wander, unattached, through the great spaces of our imagination.

Fortunately, while our tradition is heavily tilted in favor of private life, we also inherit a tradition of caring for the community. Writing about what he had seen in the 1830s, Alexis de Tocqueville judged Americans to be avaricious, self-serving, and aggressive; but he was also amazed by our eagerness to form clubs, to raise barns or town halls, to join together in one cause or another: "In no country in the world," he wrote, "do the citizens make such exertions for the common weal. I know of no people who have established schools so numerous and efficacious, places of public worship better suited to the wants of the inhabitants, or roads kept in better repair."

Today we might revise Tocqueville's estimate of our schools or roads, but we can still see all around us the fruits of that concern for the common weal—the libraries, museums, courthouses, hospitals, orphanages, universities, parks, on and on. No matter where we live, our home places have also benefited from the Granges and unions, the volunteer fire brigades, the art guilds and garden clubs, the charities, food kitchens, homeless shelters, soccer and baseball teams, the Scouts and 4-H, the Girls and Boys Clubs, the Lions and Elks and Rotarians, the countless gatherings of people who saw a need and responded to it.

This history of local care hardly ever makes it into our literature, for it is less glamorous than rebellion, yet it is a crucial part of our heritage. Any of us could cite examples of people who dug in and joined with others to make our home places better places. Women and men who invest themselves in their communities, fighting for good schools or green spaces, paying attention to where they are, seem to me as worthy of celebration as those adventurous loners who keep drifting on, prospecting for pleasure.

The words community, communion, and communicate all derive from common, and the two syllables of common grow from separate roots, the first meaning "together" or "next to," the second having to do with barter or exchange. Embodied in that word is a sense of our shared life as one of giving and receiving—music, touch, ideas, recipes, stories, medicine, tools, the whole range of artifacts and talents. After 25 years with my wife, Ruth, that is how I have come to understand marriage, as a constant exchange of labor and love. We do not calculate who gives how much; if we had to, the marriage would be in trouble. Looking outward from this community of two, I see my life embedded in ever larger exchanges—those of family and friendship, neighborhood and city, countryside and country—and on every scale there is giving and receiving, calling and answering.

It is neither quaint nor sentimental to advocate neighborliness; it is far more sentimental to suggest that we can do without such mutual aid.

Many people shy away from community out of a fear that it may become suffocating, confining, even vicious; and of course it may, if it grows rigid or exclusive. A healthy community is dynamic, stirred up by the energies of those who already belong, open to new members and fresh influences, kept in motion by the constant bartering of gifts. It is fashionable just now to speak of this open quality as "tolerance," but that word sounds too grudging to me—as though, to avoid strife, we must grit our teeth and ignore whatever is strange to us. The community I desire is not grudging; it is exuberant, joyful, grounded in affection, pleasure, and mutual aid. Such a community arises not from duty or money but from the free interchange of people who share a place, share work and food, sorrows and hope. Taking part in the common life means dwelling in a web of relationships, the many threads tugging at you while also holding you upright.

I have told elsewhere the story of a man who lived in the Ohio township where I grew up, a builder who refused to join the volunteer fire department. Why should he join, when his house was brick, properly wired, fitted out with new appliances? Well, one day that house caught fire. His wife dialed the emergency number, the siren wailed, and pretty soon the volunteer firemen, my father among them, showed up with the pumper truck. But they held back on the hoses, asking the builder if he still saw no reason to join, and the builder said he could see a pretty good reason to join right there and then, and the volunteers let the water loose.

I have also told before the story of a family from that township whose house burned down. The fire had been started accidentally by the father, who came home drunk and fell asleep smoking on the couch. While the place was still ablaze, the man took off, abandoning his wife and several young children. The local people sheltered the family, then built them a new house. This was a poor township. But nobody thought to call in the government or apply to a foundation. These were neighbors in a fix, and so you helped them, just as you would harvest corn for an ailing farmer or pull a flailing child from the creek or put your arm around a weeping friend.

My daughter Eva and I recently went to a concert in Bloomington's newly opened arts center. The old limestone building had once been the town hall, then a fire station and jail, then for several years an abandoned shell. Volunteers bought the building from the city for a dollar and renovated it with materials, labor, and money donated by local people. Now we have a handsome facility that is in constant use for pottery classes, theater productions, puppet shows, art exhibits, poetry readings, and every manner of musical event.

The music Eva and I heard was *Hymnody of Earth,* for hammer dulcimer, percussion, and children's choir. Composed by our next-door neighbor Malcolm Dalglish and featuring lyrics by our Ohio Valley neighbor Wendell Berry, it was performed that night by Malcolm, percussionist Glen Velez, and the Bloomington Youth Chorus. As I sat there with Eva in a sellout crowd—about a third of whom I knew by name, another third by face—I listened to music that had been elaborated within earshot of my house, and I heard my friend play his instrument, and I watched those children's faces shining with the colors of the human spectrum, and I felt the restored building clasping us like the cupped hands of our community. I knew once more that I was in the right place, a place created and filled and inspired by our lives together.

I am not harking back to some idyllic past, like the one embalmed in the *Saturday Evening Post* covers by Norman Rockwell or the prints of Currier and Ives. The past was never golden. As a people, we still need to unlearn some of the bad habits we formed during the long period of settlement. One good habit we might reclaim, however, is looking after those who live nearby. For much of our history, neighbors have kept one another going, kept one another sane. Still today, in town and country, in apartment buildings and barrios, even in suburban estates, you are certain to lead a narrower life without the steady presence of neighbors. It is neither quaint nor sentimental to advocate neighborliness; it is far more sentimental to suggest that we can do without such mutual aid.

Even Emerson, preaching self-reliance, knew the necessity of neighbors. He lived in a village, gave and received help, and delivered his essays as lectures for fellow citizens whom he hoped to sway. He could have left his ideas in his journals, where they first took shape, but he knew those ideas would only have effect when they were shared. I like to think he would have agreed with the Lakota shaman Black Elk, who said, "A man who has a vision is not able to use the power of it until after he has performed the vision on earth for the people to see." If you visit Emerson's house in Concord, you will find leather buckets hanging near the door, for he belonged to the village fire brigade, and even in the seclusion of his study, in the depths of thought, he kept his ears open for the alarm bell.

We should not have to wait until our houses are burning before we see the wisdom of facing our local needs by joining in common work. We should not have to wait until gunfire breaks out in our schools, rashes break out on our skin, dead fish float in our streams, or beggars sleep on our streets before we act on behalf of the community. On a crowded planet, we had better learn how to live well together, or we will live miserably apart.

The Promise of Virtual Reality

John Briggs

"Virtual reality," like "cyberspace," is one of those terms that oftentimes generate more enthusiasm than insight, as John Briggs points out. Nor is the concept as new as it sounds: Simple mirrors, for example, produce what we call *virtual* images, and in terms of the history of technology it is doubtful if all the computer modeling software will produce as strange and evocative virtual worlds in our age as the simple camera produced in the nineteenth century. (The old-fashioned snapshot is the subject of an Exploration at the end of this chapter, as is another producer of virtual reality, the modern-day camcorder.)

Web sites that demonstrate virtual reality often require the most powerful computers and browsers with supplementary programs (called add-ins).

At the *Visions* Web Site Links to virtual museums and other high-tech sites featuring VR.

READING PROMPT

Is Briggs uniform in his excitement about and support for virtual reality? What possible problems does he acknowledge? Can you think of others that he does not mention?

Virtual reality may be one of the most important technologies in our future, producing a great leap forward in many fields. While most people now focus on VR's use in entertainment areas, its real impacts will be in the arts, business, communication, design, education, engineering, medicine, and many other fields.

Due to the importance of this emerging technology, I would like to dispel some misinformation about it and suggest some important applications it will have in the future. But first, let's get clear about what virtuality is.

DEFINING VIRTUAL REALITY

Virtual reality can be defined as a three-dimensional, computer-generated simulation in which one can navigate around, interact with, and be immersed in another environment. In this sense, "virtual" is derived from the concept of "virtual memory" in a computer, which acts "as if" it is actual memory. Virtual reality provides a reality that mimics our everyday one.

Since human beings are primarily visual animals, we respond much better to spatial, three-dimensional images than we do to flat, two-dimensional text and sketches: With three-dimensional images like those produced in virtual reality, we are better able to see patterns, relationships, and trends. Virtual reality goes beyond mere static images to ones that we can navigate through and interact with in real time. We can look at things from any perspective. Virtual reality is also immersive—it draws you into the visualization.

Virtual reality is not just a set of devices, but a medium for expression and communication. Virtual reality is a means to create, experience, and share a computer-generated world as realistic or as fanciful as you would like. Head-mounted displays, data gloves, and other devices are only tools to help us experience this parallel world.

Other names for the concept of virtual reality include "artificial reality," "augmented reality," and "tele-presence." However, the term "virtual reality," or "VR," seems to have won out in common parlance. The term hooks us with the excitement of creating and experiencing different realities.

There is also an ongoing debate over exactly what virtual reality is and what it is not. Most observers agree that one necessary characteristic is that you can navigate in a virtual world with some degree of immersion, interactivity, and a speed close to real time.

HYPE AND REALITY

Right now, there is a great deal of hype surrounding virtual reality. The technology's present state of advancement has been overstated. Coverage in numerous magazines and newspaper articles, on TV shows, and even in TV ads suggests that virtual reality is now fully developed. Unfortunately, this is not true.

Present virtual-reality visualizations are often low-quality and cartoonish. The picture we see may be jerky and not respond quickly to our movements. Few systems allow for tactile feedback, a sense of touch. Some people even question the physiological and psychological safety of virtual reality, particularly in entertainment.

However, the future of virtual reality is important and real. We should not abandon the technology because it does not yet fit our expectations. Virtual reality is with us now in a very early and rudimentary form. Its state of development has been likened to the space program in the 1950s or microcomputers in the 1970s. We are just beginning to see the potential of virtual reality.

Faster computers, better software, and new devices to inform our senses are expected to come rapidly onto the scene, improving virtual reality and increasing its utility. Better content and new applications will rapidly emerge in the years ahead. Virtual reality will come to us over the Internet, reducing the need for complicated and expensive stand-alone equipment. Don't let the hype fool you. Virtual reality is not fully here yet, but it will become increasingly important for individuals, companies, and our society as a whole.

Let's now consider some present and future applications of virtual-reality technology. Our time frame is in the range of the next 10 years and our list is far from being all-inclusive. Virtual reality's uses are still being explored and defined. The only thing we can be certain of is that we'll be surprised and that we must remain open to surprise.

VIRTUAL REALITY APPLICATIONS, TODAY AND TOMORROW

Architecture and Construction. Virtual reality is already showing its potential in the architecture and construction industries. A building can be created as a navigable, interactive, and immersive experience while still being designed, so that both architect and client can experience the structure and make changes before construction begins. It has been said that every building built today is actually a physical prototype, leaving little room for input or changes until after construction. Virtual reality would allow for an electronic prototype to be created and modified, so that costly changes during or after construction are avoided.

In the future, clients will want to experience their house or building in virtual reality before final designs are completed and construction begins. Beyond today's capabilities, clients will not only be able to see the structure, but hear sounds from within it, feel its textures, and experience its fragrances. Home builders and real-estate developers are particularly excited about the potential of virtual reality to sell their designs. Why build expensive model homes or demonstration spaces when prospective buyers can see the range of options electronically? City planners will use virtual reality to consider various changes in the community, greatly assisting the work of zoning and planning boards.

Art. At present, you can "virtually" visit a number of actual art galleries and museums via the Internet. Recently, the Guggenheim and other museums conducted special exhibits of virtual-reality art works.

Virtual reality will change our conception of what constitutes art. A work of art may become a physically navigable, interactive, and immersive experience. You may travel into a virtual painting, which will actually be a mini-world for you to explore. You may interact with its elements, perhaps even change them. You may enter a sculpture gallery and interact with the art pieces. You will actually become part of the art as you interact with it.

Business. Already, several companies have created three-dimensional visualizations of the stock market. The stocks appear as upright cylinders (like a stack of poker chips) on a three-dimensional grid representing different sectors of the market. The cylinders (each with a company logo) will rise and fall with stock prices and spin at different speeds as an indication of each stock's sales activity. With this arrangement, a stock broker or analyst can quickly see patterns for a market sector as a whole, as well as the activity of specific stocks. A click on a company's cylinder can bring in-depth information to the screen and give the broker an opportunity to rapidly buy or sell a stock.

The use of virtual reality in stock market trading will greatly increase in the future. Those companies trading on various stock markets globally will require this virtual-reality application to identify trends and make trades more rapidly. They will, in fact, be interacting with the stock market in real time. Their work will be much like playing a large and complex video game.

Some virtual-reality software developers have been working on a product called FlowSheet. It will be like a spreadsheet, but will show more than mere numbers displayed in two-dimensional columns and rows. Rather, it will give a three-dimensional depiction of numbers with varying sizes, shapes, colors, and spatial relationships. In the future, FlowSheets will allow for much clearer and quicker analysis of alternatives, relationships, and trends.

Still other software developers are considering the benefits of creating DataSpaces, a step beyond the database. Like the FlowSheet, DataSpaces represents information sources as objects that differ in size, color, shape, and spatial relationships. You will surf through information in a world of three-dimensional objects, selecting the information you need by clicking on the appropriate one. In the next few years, you will be able to conduct this kind of search on the Internet using a recently accepted standard called Virtual Reality Modeling Language (VRML).

Using a combination of the FlowSheet, DataSpaces, and other virtual-reality software, companies will be able to simulate their entire operation. Different aspects of operation, such as production, inventory, sales, and productivity, can be represented in three dimensions for analysis. Various "what-if" scenarios could be proposed. A company could also use this system to watch its actual operation in real time rather than in simulation.

Disabilities. Several organizations, such as Prairie Software and Hines Veterans' Hospital in Illinois, are experimenting with virtual reality to confirm the accessibility of buildings for people with disabilities. One university, Oregon Research Institute, has created a program that teaches children to operate wheelchairs. Another, the University of Dayton, is using virtual reality to train mentally retarded students how to ride a bus. Just beginning are many other applications aimed at allowing people with disabilities to experience worlds they cannot currently explore due to their physical limitations.

In the future, it will be standard procedure, if not mandatory, to use virtual reality in private homes and public places to test accessibility before plans are approved. People with disabilities will be able to visit new areas virtually before they visit them in the everyday world. They will also be able to experience skiing, hang gliding, and other sports in virtual worlds.

Education and Training. VR is just beginning to be applied in education and training. Students can study anatomy or explore our galaxy. Some training applications relate to health and safety. One application from World Builder of Rochester, New York, allows trainees to walk through a virtual factory and learn about health hazards—a more engaging experience than reading a manual or attending a lecture.

In the future, students will be able to learn through studying in virtual worlds. Chemistry students will be able to conduct experiments without risking an accidental explosion in the lab. Astronomy students will be able to visit a range of virtual galaxies to study their properties. History students will be able to visit different historical events and perhaps even participate in the action with historical figures. English students could be on stage at the Globe Theater as it was when Shakespeare's plays were first presented. They will also be able to enter into a book and interact with its characters.

Virtual reality will also be used in teaching adults. Trainees in a wide variety of environments will be able to safely try out new techniques. They will be able to learn by doing tasks virtually before applying them in the real world. They will use these practice tasks in hazardous environs and also practice dealing with emergencies on the job. However, much remains to be done to bring virtual reality fully into the classroom or the training facility.

Engineering. Engineers of all descriptions are already using virtual-reality simulations to create and test prototypes. Each of the Big Three automakers is using some form of virtual reality to test new models. In the aerospace industry, the new Boeing 777 was the first aircraft to be designed and tested using virtual-reality technology.

Physical prototypes take a great deal of time to produce and are very costly. Changes to electronic or simulated prototypes can be done rapidly and inexpensively, shortening development time. Hoping to save money in prototyping and avoid cost overruns, the U.S. military has even coined the phrase, "Sim it before you build it!"

In the future, nearly every engineering pursuit will use virtual-reality prototypes so that designs can be shared, evaluated, and modified with input from both co-workers and customers. Even the manufacturing process and expected repairs will be simulated, saving money and aggravation. Given advances in electronic networks, virtual work benches will be created with engineers in distant locations around the globe working in teams to design products.

Entertainment. Virtual reality is already being applied in entertainment. Location-based entertainment centers are cropping up in major cities around the globe and traveling virtual-reality entertainment shows are on the road. Soon, nearly all video arcades will be VR centers; all games will be 3-D, interactive, and immersive.

While the number of such entertainment centers will increase in the future, home-based virtual reality will also grow dramatically. Current systems are primitive, due to a lack of computing power and the high cost of most virtual-reality equipment, but advanced virtual reality is set to invade the home entertainment scene in the years ahead. While stand-alone entertainment systems will be offered, perhaps the most important form of home VR will come over the Internet, and with it the potential for virtual reality to promote human interaction over wide distances.

Imagine an adventure game in which you are immersed in a three-dimensional world, interacting with other participants. It can become a real, role-playing event. Imagine a movie in which you are a participant interacting with the plot and other characters. While these kinds of entertainment have been seen as separating participants in the past, in the future they may be seen as a new kind of socializing, one which may lead to richer relationships in the "real" world.

Marketing. Virtual reality is just beginning to be used by companies who want customers to experience their products and to understand them better. They've found that a new technology, such as virtual reality, draws people to their exhibits and involves them with a product much more than standard displays. Cabletron, a cable network company in Rochester, New Hampshire, has customers travel through their network virtually. Sopporo, a beer company in Japan, allows customers to visit its production plant to experience the beer-making process in virtual reality.

In the future, virtual reality will be used to develop and test products with much greater customer involvement. A company will be able to create products, gain customer feedback, and then modify the products much more rapidly and inexpensively. The prototype will only be an electronic idea that they can directly test before creating the physical product. This electronic prototyping may also lead to individualized products that are portrayed in virtual reality, customized by the individual, and then transmitted electronically to a production facility.

Medicine. Virtual reality is just beginning to be used in medicine and medical research. The University of North Carolina (UNC) uses it in biochemical engineering. They test the docking of molecules using visual and auditory displays and a force-feedback device. Virtual reality is also being used at UNC and other locations to practice aiming X-rays before cancer treatments of that type are performed. Several companies, such as High Techsplanations of Rockville, Maryland, and Cine-med of Woodbury, Connecticut, are creating virtual bodies, a kind of "body electronic," to enhance medical training.

In the future, medical students will study anatomy by dissecting virtual cadavers—a much more cost-effective and efficient way of studying the human body. Medical students and surgeons will practice virtual surgery before attempting a new procedure. They may even practice an operation for a specific patient, whose unique body characteristics have been scanned into the computer. Different diseases and medical emergencies can also be simulated to test a medical student's or doctor's knowledge regarding treatment.

On a different front, virtual reality could be used for treatments in guided visualization. Patients could use virtual reality to assist in visualizing a part of their body for healing. Likewise, virtual reality could help improve relaxation techniques, providing a pleasant world in which to relax.

Military. One of the first applications of virtual reality was in flight simulators. Today, these applications are used not only for aircraft simulation, but also for ships, tanks, and infantry maneuvers. With the advent of networked virtual reality, the U.S. military is able to stage SimNet tank battles between various military installations around the world over what it calls the "Defense Simulation Internet." First used extensively in the Gulf War, SimNet allowed nearly every flight and battle to be conducted in virtual reality before the real war began. Records of Gulf War events have themselves been turned into a large-scale simulation that can test the skill of military leaders and soldiers.

In the future, every aspect of warfare will be practiced in simulation before being conducted in a real-world situation. Simulations will become so real, it will become impossible to distinguish the real from the simulated. While there are dangers here in misunderstanding the real from the simulated, it may also be possible for combatants to see the folly of their aggression before a conflict begins. Perhaps we could substitute a virtual war for a real one.

Religion. At present, religion does not seem to be making much use of virtual reality. However, there is potential for VR in both religious education and experience. One Christian religious denomination reportedly has been having discussions with a virtual-reality developer about creating biblical scenes in virtual reality. One author, Richard V. Kelly of Digital Equipment Corporation, has proposed the creation of religious experiences from all of the world's various religions.

In the future, we can expect to see an array of religious experiences via virtual reality. A Christian student may be able to experience being at the Sermon on the Mount or even the Crucifixion, among other events in that faith's history. He or she could also explore events in Judaism or Buddhism. Even more profound mystical experiences, such as the prophecies of Ezekiel or a revelation from eastern religions, could be created in virtual worlds.

Sex. Virtual sex is a hot topic. It has been labeled teledildonics by several authors. At least one virtual-reality company, Thinking Software of

Woodside, New York, is selling a "cybersex machine," and there are some multimedia sexual experiences that come close to virtual reality. These products are not very advanced, however, and many obstacles must be overcome to produce a satisfying tactile experience. But expect great strides to be made in creating advanced "sex machines" using virtual-reality technology. There is too much potential profit in these applications for them not to be pursued.

In the future, virtual sex will be either a stand-alone or networked experience. Expect virtual reality to be used in treating sexual dysfunction, communing with a loved one at a distance, and for sexual exploration or, unfortunately, exploitation. Today's 900 number hotlines and sex sites on the Internet will be tame by comparison. Expect major controversy over this topic from lawmakers, religious groups, and proponents of family values.

THE PROMISE OF VIRTUAL REALITY

These are just some present and future applications of virtual reality. As you can see, there are many potential applications for virtual reality. Perhaps, in the future, we will only be limited by our imagination regarding the uses of virtual reality. Virtual reality is neither good nor bad. It is a new tool that will have important implications in our future.

In working in the field of virtual reality, I have found a very important aspect of it that is often overlooked. In order to create virtual worlds, one must have an in-depth understanding of how our everyday world works. Perhaps one of virtual reality's greatest gifts will be helping us to understand better our own reality.

The Illusion of Life Is Dearly Bought

Mark Slouka

Advocates of new computer technology, Mark Slouka argues, often project virtual reality (VR) as a utopia free from all the rough edges of real life (sometimes referred to as RL). For a take on an earlier form of virtual reality, visit the Web sites of the most famous illusionist of this century, Harry Houdini, as well as that of the well-known contemporary magician, David Copperfield.

At the *Visions* Web Site Links to magic sites, including ones dedicated to Harry Houdini and David Copperfield.

READING PROMPT

Why is Mark Slouka so worried about Bill Gates? Do you share his concern?

On 18 November last year, Mickey Mouse's birthday, the first 352 residential units went on sale in the Walt Disney Company's virtual town of Celebration. Located just five miles south of the Magic Kingdom, the $2.5-billion project, billed as "a 19th-century town for the late 20th century," will feature a real post office, a real town hall, and, eventually, 20,000 real residents. Think of it as a computer game—Sim City, say, or SimLife—writ large. If it succeeds (and there's every reason to expect it will), it will suggest the extent to which the blurring of reality with corporate fantasy has become a genuine cultural phenomenon.

Not that we need any more proof. The general breakdown of the barrier separating original from counterfeit, fact from fake, is visible everywhere; in the US, the slow bleeding of reality into illusion is systemic. The image of OJ Simpson dodging tackles or hurdling luggage en route to his Hertz Rent-a-Car blurs with the images of OJ on the lam, OJ clowning with Leslie Nielsen in the Naked Gun movies, OJ and Harvard attorney Alan Dershowitz (or is it actor Ron Silver, playing Alan Dershowitz?) hurdling legal landmines on the way to acquittal and *Reversal of Fortune II.* The horrific videotape of the Rodney King beating melds with the images of rioting in south central LA, which in turn look just like the "real-life" scenes of Los Angeles mayhem found on *Police Quest: Open Season,* a video game designed by former LA police chief Daryl Gates. In the culture of illusion, the furniture swims, the walls bulge and bend.

Is any part of American culture exempt from the assault of virtual realities? Apparently not. Politically, the US is already a virtual republic, a country run less by elected officials than by the men and women who package and sell them to an electorate increasingly willing to believe—in the scripted words of tennis star Andre Agassi—that "image is everything." In American courts of law, professionally rendered re-enactments—scripted, rehearsed, directed and edited—are admissible as evidence. Nothing is too extreme. Was your hand crushed at the factory? Were your kids burned to death in a car accident? For a fee, a company will provide a video simulation complete with realistic screams, horrified bystanders and virtual blood. Juries, weaned on 57 channels in the age of Oliver Stone, find them very effective.

As we plummet through the looking-glass, however, we would do well to bear in mind that beyond that Orwellian and seemingly ubiquitous adjective "virtual" is a marketing scheme of unrivalled audacity, unprecedented scope, and nearly unimaginable impact: a scheme that is (worth a potential $3.5 trillion, by one reliable estimate) designed to sell us copies of the things we already have available to us for free—life itself.

Soon, writes Bill Gates in *The Road Ahead,* "you will be able to conduct business, study, explore the world and its cultures, call up any great entertainment, make friends, attend neighborhood markets, and show pictures to distant relatives—without leaving your desk or armchair . . . [Y]our network connection . . . will be your passport into a new, mediated way of life."

The mediated life, of course, aided by one of the great migrations of human history—the movement inside our own homes—is already here. As more of the hours of our days are spent in synthetic environments, partaking of electronic pleasures, life itself is turned into a commodity. As the natural world fades from our lives, the unnatural one takes over; as the actual, physical community wanes, the virtual one waxes full and fat. Bill's plan (and he's not alone) is to take advantage of the social momentum. He wants a piece of the action. The new, mediated world, he promises, will be "a world of low-friction, low-overhead capitalism, in which market information will be plentiful and transaction costs low." What he neglects to mention, understandably, is that the road to "shopper's heaven" leads past him, and he happens to be manning the tollbooth.

Bill's vision of a "friction-free" virtual world, one must admit, has a certain Singaporean charm. From proposing we apply something like the Motion Picture Association's movie ratings to social discourse, to suggesting that a virtual forest of hidden surveillance cameras be installed "to record most of what goes on in public," Bill is out to make the world free from friction and safe for commerce. What he seems to have overlooked (there's no way to put this delicately) is that friction in social life, as in the bedroom, has its virtues. The "friction" he would spare us, after all, is the friction of direct experience, of physical movement, of unmediated social interaction. Cultural life, one wants to remind him, requires friction. As does democracy.

It's always possible, of course, that democracy, or a thriving social life, are not what Bill and his fellow enthusiasts are after because they truly believe that these notions (like sex, or physical space) will be the vestigial limbs of the virtual world, cherished by a handful of die-hard humanists, and no one else. (In the digital future, Nicole Stenger of the University of Washington reminds us, "cyberspace will be your condom.") It's possible, as well, that these latter-day Nathan Hales really believe in the "liberty" of electronic shopping, of being able, as Gates promises, instantly to order the cool sunglasses Tom Cruise wears in *Top Gun* while watching the movie. It's possible, finally, that it is simple naiveté that has Gates and Co. whistling past the authoritarian graveyards as they usher in Bentham's Panopticon, and not some fellow feeling for those buried within.

Whether they believe in their virtual world or not, however, is ultimately beside the point. They're building it. And in the friction-free future, jacked into "shopper's heaven," we'll have the "liberty" of living (or rather, of buying the illusion of living), through the benevolent offices of a middleman as nearly omnipotent as God himself. Freedom? A more perfect captivity is difficult to imagine.

All of which, finally, makes Mickey's excellent adventure in real estate more than a little unsettling. There's no "off" button in Celebration, no escape: the illusion is seamless, and the corporate menu of options defines the boundaries of life itself.

chapter 5 Explorations

5a The Telephone, a Primary Virtual Communicator

Collaboration. While we can speculate for hours on how computers are going to change our lives in the next century, it might also be helpful to look at how another marvelous invention, the telephone, has changed (or not changed) our lives in this century. List and share important ways that your life would be both different and the same without telephones.

Web work. The PBS series *The American Experience* devoted one episode to the telephone. Also check out the Telephone History Web site. AT&T has a site dedicated to the history of its Research and Development Lab (going back more than a hundred years). AT&T also has a demo page of new projects, many Internet-related and some requiring powerful (Java-capable) browsers. Lucent Technologies (formerly called Bell Labs) has its own ideas Web site, which also includes a history section.

One writing topic. Although the telephone was invented in the nineteenth century, school classrooms at all levels, kindergarten through college, have remained largely free of this invention. Write an essay on the role telephones have or have not played in your overall education, both in and outside of school. How (for better or worse) might things have been different had your classrooms had phones? How might things be different in a future when everyone has her or his own cellular phone?

5b Snapshots, Nineteenth-Century Virtual Reality

Collaboration. Like the telephone, still photography is a nineteenth-century invention that pervades practically all aspects of the twentieth century. Make and share with classmates two lists comparing the benefits of still photography and video cameras.

Web work. PhotoArts hosts the major Web site devoted to photography as a fine art, with a special area for current exhibits. Two exhibits worth viewing come from one of the outstanding collections of nineteenth-century photographs, the William Becker Collection: an assembly of small prints from the collection entitled Small Worlds, and another collection entitled At Ease that consists of the earliest snapshots. The gallery section of the Daguerreian Society Web page is another wonder of nineteenth-century photography.

To see what is happening now and what the future may hold, check out the Kodak homepage.

One writing topic. Write an essay on how you or people you know feel about being represented in either still or moving photography.

5c Camcorders and Roller Coasters

Collaboration. Make and share a list of various ways you have used camcorders in your life or seen them used by others; or list the special, unusual things that so attract you to (or possibly repel you from) amusement parks.

Web work. Live camera shots available via the Web add a new twist to a consideration of live and viewed events. Yahoo! lists more than 250 outdoor camera shots worldwide at Computers and Internet > Internet > Entertainment > Interesting Devices Connected to the Net > Spy Cameras > Outdoor Cameras, including spectacular views from the seventy-seventh floor of the Empire State Building and lower, sunny views, of Venice Beach from a camera mounted atop the Good See Store.

Thrill Ride provides a good overview of the current world of roller coasters. For the past, see the Web page devoted to John A. Miller, dubbed the "Thomas Edison of roller coasters," or Jeff Stanton's historical retrospective on Coney Island.

One writing topic. Camcorders provide us with a virtual but often less intense representation of a real event, whereas roller coasters give reality a virtual but more intense quality. Use your experiences and thoughts about either situation (camcorders or roller coasters) to reflect on our varying attitudes about virtuality and reality. What do we like most and least about virtual representations of the real world? Are there times when you would prefer a less complete virtual representation—for example, when you would prefer snapshots of a family event or even a written description of the event to a videotape of it?

5d Bloomington, Indiana: The Right-Sized City?

Collaboration. Bloomington, Indiana, is a middle-sized town (population 60,000) in a middle-sized county (Monro County, population 108,000) in the middle of a Midwestern state that is famous for (among other things) producing vice presidents. It is not likely that many people in your class have ever been to Bloomington. And, other than the fact that it is the home of Indiana University of the Big Ten Athletic Conference, you probably do not know much about the city. Make and share with classmates a list of the advantages and disadvantages of living in a city of this size.

Web work. Start with Yahoo!'s page on Bloomington Web resources, Regional > U S States > Indiana > Cities > Bloomington; then explore the links to the two city guides: Hoosier.Net and Your Guide to Bloomington. Also check out Bloomington's weekly entertainment magazine, *BC Magazine,* and IDS, the *Indiana Daily Student.*

One writing topic. Write an essay on your sense as to what is the best-sized city to live in; compare that city either to the one you live in or to Bloomington, Indiana.

5e Cyberspace

Collaboration. Make and share a list of activities you have done online. Which do you find most and least enjoyable online, especially as compared to other, more traditional ways of interacting?

Web work. On February 18, 1996, one hundred top photographers attempted to capture how cyberspace is changing people's lives; their efforts are documented at the 24 Hours in Cyberspace Web site.

One writing topic. Write an essay on the activity or activities in your current life that you look forward to doing more of (or less of) online.

5f Cliques and Popularity

Collaboration. At the core of the idealized notion of neighborhood is a sense of a wide assortment of people getting along. We tend to think of prejudice as the most obvious factor leading groups to break off into smaller units, but it might be possible to see something seemingly a little more positive—namely, popularity—playing a key role in this process as well, as people make efforts to be included in the most popular group or deliberately seek out alternative groups. Make and share a list of reasons people form cliques.

Web work. Check out Andrew Hicks's Web site, and especially his hilarious diary/fantasy, A Year in the Life of a Nerd. Uncertain if you are a nerd? Take the Nerdity Test, or Brenda's Dating Advice for Geeks. Want to see how the other half lives? Check out the official Baywatch site.

One writing topic. Adolescence and high school remain the times in many people's lives when issues of popularity and cliques assume the greatest importance. Write an essay focusing on a moment or experience of adolescence that showed people falling into cliques or coming together for a common purpose.

chapter 6

The Politics of Cyberspace

READING QUESTIONS

- What activities that you used to do via face-to-face contact do you now feel comfortable doing strictly via technology? What activities, if any, do you still prefer doing via face-to-face contact?
- How do you form your political opinions? How important is relevant information, and what sources do you regularly use to get it?
- What are the responsibilities or obligations you feel you owe to society generally—the things you feel you should do beyond any legal requirements?

QUOTATIONS

A fanatic is one who can't change his mind and won't change the subject.

— *Winston Churchill, British statesman, 1874–1965*

What is conservatism? Is it not adherence to the old and the tried against the new and untried?

— *Abraham Lincoln (1860), American president, 1809–1865*

When a social order is in revolution half the world is necessarily part of the new day and half of the old.

— *Florence Guy Seabury (1926), American author*

The higher the technology, the higher the freedom. Technology enforces certain solutions: Satellite dishes, computers, videos, international telephone lines force pluralism and freedom onto a society.

— *Lech Walesa (1989), Polish political leader*

Birth of a Digital Nation

Jon Katz

Jon Katz was the editor of The Netizen, a Web site sponsored by *Wired* magazine and devoted to politics and information technology. In this essay Katz gives his take on the use of technology during the 1996 presidential election between Bill Clinton and Bob Dole.

 **At the *Visions* Web Site** Links to more pieces by Jon Katz.

READING PROMPT

What's wrong, according to Katz, with traditional politics in the United States? What does online interaction have to offer instead?

FIRST STIRRINGS

On the Net last year [1996], I saw the rebirth of love for liberty in media. I saw a culture crowded with intelligent, educated, politically passionate people who—in jarring contrast to the offline world—line up to express their civic opinions, participate in debates, even fight for their political beliefs.

I watched people learn new ways to communicate politically. I watched information travel great distances, then return home bearing imprints of engaged and committed people from all over the world. I saw positions soften and change when people were suddenly able to talk directly to one another, rather than through journalists, politicians, or ideological mercenaries.

I saw the primordial stirrings of a new kind of nation—the Digital Nation—and the formation of a new postpolitical philosophy. This nascent ideology, fuzzy and difficult to define, suggests a blend of some of the best values rescued from the tired old dogmas—the humanism of liberalism, the economic opportunity of conservatism, plus a strong sense of personal responsibility and a passion for freedom.

I came across questions, some tenuously posed: Are we living in the middle of a great revolution, or are we just members of another arrogant élite talking to ourselves? Are we a powerful new kind of community or just a mass of people hooked up to machines? Do we share goals and ideals, or are we just another hot market ready for exploitation by America's ravenous corporations?

And perhaps the toughest questions of all: Can we build a new kind of politics? Can we construct a more civil society with our powerful technologies?

Are we extending the evolution of freedom among human beings? Or are we nothing more than a great, wired babble pissing into the digital wind?

Where freedom is rarely mentioned in mainstream media anymore, it is ferociously defended—and exercised daily—on the Net.

Where our existing information systems seek to choke the flow of information through taboos, costs, and restrictions, the new digital world celebrates the right of the individual to speak and be heard—one of the cornerstone ideas behind American media and democracy.

Where our existing political institutions are viewed as remote and unresponsive, this online culture offers the means for individuals to have a genuine say in the decisions that affect their lives.

Where conventional politics is suffused with ideology, the digital world is obsessed with facts.

Where our current political system is irrational, awash in hypocritical god-and-values talk, the Digital Nation points the way toward a more rational, less dogmatic approach to politics.

The world's information is being liberated, and so, as a consequence, are we.

MY JOURNEY

Early last year, writer John Heilemann and I set out on parallel media journeys for HotWired's The Netizen, originally created to explore political issues and the media during the election year. One concept behind The Netizen—a conceit, perhaps—was that we would watch the impact of the Web on the political process in the first wired election. Heilemann was to cover the candidates, the conventions, and the campaigns. I would write about the media covering them.

Things didn't turn out quite as we'd expected at The Netizen. The year of the Web was not 1996—at least not in terms of mainstream politics. The new culture wasn't strong enough yet to really affect the political process. The candidates didn't turn to it as they had turned in 1992 to new media like cable, fax, and 800 numbers.

And the election was shallow from the beginning, with no view toward the new postindustrial economy erupting around us and no vision of a digital—or any other kind of—future. By spring '96, it seemed clear to me that this campaign was a metaphor for all that doesn't work in both journalism and politics. I couldn't bear *The New York Times* pundits, CNN's politico-sports talk, the whoring Washington talk shows, the network stand-ups.

Why attend to those tired institutions when what was happening on the monitor a foot from my nose seemed so much more interesting? Fresh ideas, fearsome debates, and a brand-new culture were rising out of the primordial digital muck, its politics teeming with energy. How could a medium like this new one have a major impact on a leaden old process like that one? By focusing so obsessively on Them, we were missing a much more dramatic political story—Us.

So I mostly abandoned Their campaign, focusing instead on the politics of Ours—especially interactivity and the digital culture. I was flamed, challenged, and stretched almost daily. The Web became my formidable teacher, whacking me on the palm with a ruler when I didn't do my homework or wasn't listening intently enough; comforting me when I got discouraged or felt lost.

I argued with technoanarchists about rules, flamers about civility, white kids about rap, black kids about police, journalists about media, evangelicals about sin. I was scolded by scholars and academics for flawed logic or incomplete research. I was shut down by "family values" email bombers outraged by my attacks on Wal-Mart's practice of sanitizing the music it sells.

I saw the strange new way in which information and opinion travel down the digital highway—linked to Web sites, passed on to newsgroups, mailing lists, and computer conferencing systems. I saw my columns transformed from conventional punditry to a series of almost-living organisms that got buttressed, challenged, and altered by the incredible volume of feedback suddenly available. I lost the ingrained journalistic ethic that taught me that I was right, and that my readers didn't know what was good for them. On the Web, I learned that I was rarely completely right, that I was only a transmitter of ideas waiting to be shaped and often improved upon by people who knew more than I did.

Ideas almost never remain static on the Web. They are launched like children into the world, where they are altered by the many different environments they pass through, almost never coming home in the same form in which they left.

All the while, I had the sense of Heilemann cranking along like the Energizer Bunny, responsibly slugging his way through the torturous ordeal of campaign coverage, guiding the increasingly-exasperated people who actually wanted to follow the election. What Heilemann learned and relayed was that the political system isn't functioning. It doesn't address serious problems, and the problems it does address are not confronted in a rational way. It doesn't present us with the information we need or steer us toward comprehension—let alone solution.

Over the course of 1996, the ideologies that shape our political culture seemed to collapse. Liberalism finally expired along with the welfare culture it had inadvertently spawned. Conservatism, reeling from the failure of the so-called Republican revolution, was exposed as heartless and rigid. The left and the right—even on issues as explosive as abortion and welfare—appeared spent. While they squabbled eternally with one another, the rest of us ached for something better. In 1996, we didn't get it.

The candidates didn't raise a single significant issue, offer a solution to any major social problem, raise the nation's consciousness, or prod its conscience about any critical matter. The issues the candidates did debate were either false or manipulative, the tired imperatives of another time.

"Nineteen ninety-six was the year that Old Politics died," wrote Heilemann. "For outside this bizarre electoral system that's grown and mutated over the past 40 years—this strange pseudo-meta-ritual that, experienced from the inside, feels like being trapped in an echo chamber lined with mirrors—there are profound, paradigm-shifting changes afoot."

There are paradigm-shifting changes afoot: the young people who form the heart of the digital world are creating a new political ideology. The machinery of the Internet is being wielded to create an environment in which the Digital Nation can become a political entity in its own right.

By avoiding the campaign most of the time, I ended up in another, unexpected place. I had wandered into the nexus between the past and the future, the transition from one political process to a very different one.

While Heilemann came to believe he was attending a wake, I began to feel I was witnessing a birth—the first stirrings of a powerful new political community.

THE NASCENT NATION

All kinds of people of every age and background are online, but at the heart of the Digital Nation are the people who created the Net, work in it, and whose business, social, and cultural lives increasingly revolve around it.

The Digital Nation constitutes a new social class. Its citizens are young, educated, affluent. They inhabit wired institutions and industries—universities, computer and telecom companies, Wall Street and financial outfits, the media. They live everywhere, of course, but are most visible in forward-looking, technologically advanced communities: New York, San Francisco, Los Angeles, Seattle, Boston, Minneapolis, Austin, Raleigh. They are predominantly male, although female citizens are joining in enormous—and increasingly equal—numbers.

The members of the Digital Nation are not representative of the population as a whole: they are richer, better educated, and disproportionately white. They have disposable income and available time. Their educations are often unconventional and continuous, and they have almost unhindered access to much of the world's information. As a result, their values are constantly evolving. Unlike the rigid political ideologies that have ruled America for decades, the ideas of the postpolitical young remain fluid.

Still, some of their common values are clear: they tend to be libertarian, materialistic, tolerant, rational, technologically adept, disconnected from conventional political organizations—like the Republican or Democratic parties—and from narrow labels like liberal or conservative. They are not politically correct, rejecting dogma in favor of sorting through issues individually, preferring discussions to platforms.

The digital young are bright. They are not afraid to challenge authority. They take no one's word for anything. They embrace interactivity—the right to shape and participate in their media. They have little experience with passively reading newspapers or watching newscasts delivered by anchors.

They share a passion for popular culture—perhaps their most common shared value, and the one most misperceived and mishandled by politicians and journalists. On Monday mornings when they saunter into work, they are much more likely to be talking about the movies they saw over the weekend than about Washington's issue of the week. Music, movies, magazines, some television shows, and some books are elementally important to them—not merely forms of entertainment but means of identity.

As much as anything else, the reflexive contempt for popular culture shared by so many elders of journalism and politics has alienated this group, causing its members to view the world in two basic categories: those who get it, and those who don't. For much of their lives, these young people have been branded ignorant, their culture malevolent. The political leaders and pundits who malign them haven't begun to grasp how destructive these perpetual assaults have been, how huge a cultural gap they've created.

Although many would balk at defining themselves this way, the digital young are revolutionaries. Unlike the clucking boomers, they are not talking revolution; they're making one. This is a culture best judged by what it does, not what it says.

In *On Revolution,* Hannah Arendt wrote that two things are needed to generate great revolutions: the sudden experience of being free and the sense of creating something. The Net is revolutionary in precisely those ways. It liberates millions of people to do things they couldn't do before. Men and women can experiment with their sexual identities without being humiliated or arrested. Citizens can express themselves directly, without filtering their views through journalists or pollsters. Researchers can get the newest data in hours, free from the grinding rituals of scientific tradition. The young can explore their own notions of culture, safe from the stern scrutiny of parents and teachers.

There's also a sense of great novelty, of building something different. The online population of today has evolved dramatically from the hackers and academics who patched together primitive computer bulletin boards just a few years ago—but the sensation of discovery remains. People coming online still have the feeling of stepping across a threshold. Citizenship in this world requires patience, commitment, and determination—an investment of time and energy that often brings the sense of participating in something very new.

It's difficult to conceive of the digital world as a political entity. The existing political and journalistic structures hate the very thought, since that means relinquishing their own central place in political life. And the digital world itself—adolescent, self-absorbed—is almost equally reluctant to take itself seriously in a political context, since that invokes all sorts of responsibilities that seem too constraining and burdensome.

This is a culture founded on the ethos of individuality, not leadership. Information flows laterally, or from many to many—a structure that works against the creation of leaders.

Like it or not, however, this Digital Nation possesses all the traits of groups that, throughout history, have eventually taken power. It has the education, the affluence, and the privilege that will create a political force that ultimately must be reckoned with.

SOME POSTPOLITICAL CORE VALUES

Out of sight of the reporters, handlers, spin-masters, and politicians of the presidential campaign, a new political sensibility took shape in 1996. It brought fresh ideas. It brought real debates about real issues.

The postpolitical ideology draws from different elements of familiar politics. The term postpolitical gets tossed around in various circles, but here it refers to a new kind of politics beyond the traditional choices of left/right, liberal/conservative, Republican/Democrat. Although still taking shape, this postpolitical ideology combines some of the better elements of both sides of the mainstream American political spectrum.

From liberals, this ideology adopts humanism. It is suspicious of law enforcement. It abhors censorship. It recoils from extreme governmental positions like the death penalty. From conservatives, the ideology takes notions of promoting economic opportunity, creating smaller government, and insisting on personal responsibility.

The digital young share liberals' suspicions of authority and concentration of power but have little of their visceral contempt for corporations or big business. They share the liberal analysis that social problems like poverty, rather than violence on TV, are at the root of crime. But, unlike liberals, they want the poor to take more responsibility for solving their own problems.

This amalgam of values reveals itself in seemingly odd ways. Many online had no trouble believing that the Los Angeles Police Department was racist or, conceivably, might have planted evidence in the O. J. Simpson murder case. There was no sympathy, though, for the idea that O. J. should have been acquitted as a result of such technicalities.

The postpoliticos can outdo liberals on some fronts. They don't merely embrace tolerance as an ideal; they are inherently tolerant. Theirs is the first generation for whom pluralism and diversity are neither controversial nor unusual. This group couldn't care less whether families take the traditional form or have two moms or two dads. They are nearly blind to the color and ethnic heritage of the people who enter their culture. This is the least likely group to bar someone from a club because he or she is Jewish or black, or to avoid marriage because of a person's religion or ethnicity.

On the other hand, the digital young's intuitive acceptance of tolerance and diversity doesn't prevent them from rejecting liberal notions like affirmative action. And they are largely impervious to victim-talk, or politically correct rhetoric, or the culture of complaint celebrated in the liberal media coverage of many minority issues.

This culture is no less averse to the cruel and suffocating dogma of the right. The postpolitical young embrace the notion of gender equality and are intrinsically hostile to any government or religious effort to dictate private personal behavior. While conservatism has become entwined with an evangelical religious agenda, the digital young are allergic to mixing religion and politics.

If liberals say, "Here's the tent: we have to get everyone inside," and conservatives say, "Here's the tent: we don't want it to get too crowded inside," the postpolitical young say, "Here's the tent: everyone is welcome—but everyone has to figure out how to get inside on his or her own."

One of the biggest ideas in the postpolitical world is that we have the means to shape our lives, and that we must take more responsibility for doing so. This ascending generation believes its members should and will control their destinies. A recent survey in *American Demographics* magazine studied young Americans and called them self-navigators. "In a fast-changing and often hostile world, self-navigation means relying on oneself to be the captain of one's own ship and charting one's own course," wrote the Brain Waves Group, the survey's developers. Those characteristics also describe many citizens of the Digital Nation.

This group values competence and hard work, the survey found. Traditional formulas for success carry little weight since college degrees no longer guarantee jobs, getting a job doesn't guarantee you'll keep it, retirement may never be possible, and marriages can fail. Despite such caution, this group—in sharp contrast to its boomer parents—sees a future of great opportunity. The Digital Nation is optimistic about its own prospects.

Although these ideas work well for them now, as the postpolitical young of the digital world grow older, they will confront a new range of problems, from developing careers to raising children to preparing for old age. Their ideology will, of necessity, develop and change.

As they raise children, they will face issues such as neighborhood safety, maintaining parks, and improving the educational system. As they buy homes, they will encounter bread-and-butter political issues like taxation and zoning. Faced with developing a new political agenda in a radically different world, they will inevitably find themselves face-to-face with the ghosts of the old one.

The Death of Geography

Stephen Bates

Electronic technology seems to break down traditional geographical boundaries; all of us, even children, routinely process information, including live television pictures, from around the world. Yet, Stephen Bates asks, is this necessarily a good or a bad thing? Are people becoming more and more alike, or do we just think they are?

At the *Visions* Web Site Links to map- and atlas-maker Rand McNally and the National Geographic Society, the recorder of everything foreign and exotic.

READING PROMPT

Most of the time we are taught to see only the benefits of the world's shrinking; what are some problems that Bates raises here related to the breakdown of geographic boundaries?

A century after the closing of the American frontier, governments are trying to tame the electronic frontier. Most recently, President Clinton signed a bill increasing the penalties for distributing child pornography by computer, and Congress passed legislation banning indecent material online. These efforts, and others in different parts of the world, grow out of official worries that traditional legal prohibitions don't work in cyberspace. Here's why: the Internet obliterates geography.

Human awareness was once defined by proximity and physical contact. Someone knew only what was close at hand. People 50 miles distant might as well have been a continent away. Technology gradually stretched those boundaries, enabling people to cast themselves and their thoughts over wider and wider areas. State authority expanded—from village, to city-state, to nation, to empire. People migrated. Knowledge and culture spread.

Not everyone applauded the march of progress. A small-town denizen, wrote Sherwood Anderson in *Winesburg, Ohio,* now "has his mind filled to overflowing" by mass-circulation books, magazines, and newspapers. As a result, "the farmer by the stove is brother to the men of the cities, and if you listen you will find him talking as glibly and as senselessly."

Fearing such homogenization, governments sometimes tried to buttress geographic identities against technological incursion. In the 1840s, Congress debated whether the post office should deliver newspapers for free. Editors were torn. They wanted to avoid postal charges, but they feared competition from faraway brethren. They came up with a compromise that became federal law: Newspapers were delivered for free, but only within 30 miles of their place of publication.

Laws like that won't work for today's e-mail. Compared to earlier developments that eased communications, the Internet's impact is more profound. On the Internet, distance has no bearing on cost (unlike the telephone). The Internet (unlike broadcasting) not only delivers the world to us, but also delivers us to the world—we can talk back. And the Internet (in contrast to something like shortwave radio) provides us with the ability to transmit visual images, text, and decent audio.

The significance of place is being undercut by the new information technologies. We can now learn, almost instantly, the thoughts of someone on the other side of the globe. His whereabouts become as immaterial as his shoe size. This is causing what might be called the death of geography.

But don't count on a painless passing. Our institutions and expectations are deeply rooted in geography. Often our instincts about geography pull us one way while the new technologies yank us in the opposite direction.

Take the distribution of pornography. We traditionally restrict sexually oriented businesses to "red light districts," either formally or informally. Within bookstores and newsstands, pornography is placed behind the counter or on a high shelf; some states mandate such treatment. Many video stores put X-rated titles in a separate room open only to adults. These acts of geographic segregation, though hardly perfect, work reasonably well to help parents filter the information reaching children.

They also help communities stigmatize adult consumption of pornography. At an X-rated theater, a patron must consume his pornographic materials in public, and risk being spotted by someone he knows. Home videos have rejuvenated the pornographic film industry, but even here, the patron must publicly venture into an adult-only zone at the video store.

The Internet renders the entire transaction private, invisible, ungeographic. Moreover, users can tour pornographic sites without exposing their age: The operator of a dirty bookstore can tell a twelve-year-old to scram. The operator of a dirty cybersite can't—and twelve-year-olds know it. Users not only can acquire pornography without leaving home, they can, as some enterprising teenagers have discovered, acquire it without leaving the school library. Thanks to this reduced danger of discovery, pornographic materials on the Internet have become hugely popular.

Last year, the Christian Coalition, Family Research Council, and other conservative organizations called for new legislation to keep children away from online pornography. One result was the Communications Decency Act, which would ban the transmission of "indecent" material over computer networks accessible to the young. "Society has long embraced the principle that those who peddle harmful material have the obligation to keep the material from children," the Family Research Council explained in a fact sheet. "Computer indecency should be no exception."

But, as enthusiasts never tire of pointing out, cyberspace is different. The operator of a pornographic bookstore already keeps an eye on his customers; so it's no great imposition to tell him to boot out any minors who venture in. Someone who posts a pornographic image on the Internet, on the other hand, has no idea who is going to look at it. Under some circumstances he may know what company or institution gives a viewer his Net access. He may even know his e-mail addresses. But he won't know his age. Consequently, the only way to keep children away from a pornographic image on the Net is to keep everyone away from it. For any part of the Internet to be child-safe, all of it must be child-safe. As a medium unhindered by geography, the choice is all or nothing.

Civil libertarians are horrified at the thought of Congress trying to make the rambunctious Net into a serene playground. They argue that the First Amendment gives adults the right to view pornography, so Congress should not render it inaccessible on the Internet. Some civil libertarians advocate a software solution: filters that will keep children from accessing certain materials on the Net. But savvy kids won't be stopped that easily. Even if a school computer's Usenet "subscription" excludes the alt.sex newsgroups, a user can still reach them in plenty of ways—including an automated system in Japan that obligingly e-mails Usenet posts on request. It's as if the school library meticulously policed its own shelves but let students order *Hustler* through interlibrary loan.

One difficulty in trying to prevent this is that the Internet has no respect for jurisdictional boundaries. If Americans stop posting pornography to the Net, American users will still find porn posted from foreign countries. In a world without geographic limitations, national laws are often little more than trivial speed bumps on the information highway.

The United States, of course, isn't alone in trying to keep certain kinds of material out of public circulation. On the contrary, most other nations are even more active in filtering the public discourse. Canada tries to protect its arts community by limiting the quantity of American television programs shown on its stations. In some Asian and Islamic countries there is worry that too much Western culture will destroy indigenous social practices.

In December, German officials told CompuServe to drop some 200 sex-related newsgroups from its service or risk expulsion from Germany. Reluctant to walk away from a major market, the company obeyed. "As the leading global service, CompuServe must comply with the laws of the many countries in which we operate," said the company in a news release. In San Francisco, protesters poured German beer down the sewer and called a boycott. CompuServe announced plans for software that will limit the sex exclusions to German users—in essence, creating a separate subnetwork for Germany. Here, a technical overlay to a technology's basic transnational nature may reintroduce some geographic distinctions.

The decline of geography is creating problems for the law in areas other than just vice control. Consider copyright. Sir Arthur Conan Doyle's writings are in the public domain in the United Kingdom, but some are still under copyright in the United States. A New Yorker can fly to London, photocopy a Sherlock Holmes story, and fly back with the manuscript; the law permits people to enter the country with infringing items so long as they are not for distribution.

But what if the New Yorker reaches London via the Internet, instead of American Airlines, finds the story on a publicly accessible file site, and downloads it to his PC? Under current law, he has infringed the copyright. Courts won't extend the travel defense ("I flew to London") to virtual travel ("I modemed to London"). The law continues to enshrine notions of geography that no longer exist on the Net.

Similar copyright concerns are likely to crimp the much-touted "virtual library." In 1991, Al Gore wrote that a child working on homework would soon be able to consult "digital libraries containing all the information in the Library of Congress and much more," using a device "no more complicated than a Nintendo machine." Other technophiles have spun out more elaborate scenarios. Someday, they predict, people will read books on handheld computers. By selecting certain screen icons, a user will electronically borrow a publication from the library. The work will be downloaded via wireless communication. It will erase itself after two weeks unless renewed, and encryption will make copying impossible, or nearly so. Alternatively, by selecting different icons, the user will be able to purchase the work from a bookstore or newsstand for permanent downloading. As the buyer makes the purchase, his bank account will be debited.

But in this dream world, who would actually buy books? If a free digitized library work is never more than a few keystrokes away, not many people would pay a $25 fee to own it. We can thus count on publishers to fight such trends. They will use the law to retain the geographic hindrances of today's library—requiring that patrons trudge there to get and return material, for instance—even as bookstores shake off their geographic shackles and go online.

Already, copyrighted articles and photos are routinely scanned in and posted in cyberspace in violation of copyright law. Some users knowingly flout the law, using "anonymous remailers"—computer systems that erase the sender's identity and forward the message. When copyrighted Scientology teachings were being posted to the Net through a Finnish remailer; the church brought in Interpol, raided the Finnish site, discovered the e-mail address of the American infringer, and took steps to prosecute him in the United States.

Most copyright owners, however, can't go to such lengths. Even if they do, enforcement may prove impossible if the infringer has routed his message through a series of remailers in different countries. Some remailers, moreover, use an algorithm that leaves no traces. Police can shut them down, but they can't learn the senders' identities.

Combine anonymity with encryption and boundaryless trafficking and you see why a new federal law doubles the penalties for child pornography distributed via computer, compared to child porn distributed by other means. This combination also raises serious questions as to whether today's Communications Decency Act, if it becomes law, will be able to slow the flow of online illegal pornography.

Not only for law enforcement but in many other areas as well, the cues provided by geographic identity turn out to be surprisingly important. Geography-free communication can be unaccountable and sterile. According to some reports, for instance, many of today's lesbian chat areas on the Internet are populated not by lesbians but principally by straight men masquerading as gay women, exchanging dirty talk with other straight men masquerading as gay women.

Sometimes, electronic pioneer John Gilmore suggests, cyberspace is nothing more than "a telephone network with pretensions." But where it has been shorn of geographic identity and personal responsibility, it can become something more complex. It can produce what science fiction writer William Gibson calls "consensual hallucination." As a *New Yorker* canine cartoon puts it: "On the Internet, nobody knows you're a dog."

Creating a Framework for Utopia

David Boaz

For lots of different reasons, cyberspace has been largely populated by highly independent-minded people who seem not to like big government (really, who seem not to like any government) telling them what to do. This blend of traditional conservative thinking on economic matters and liberal thinking on matters of personal lifestyle choices is known as *libertarianism* and has become identified as one of the main systems of political belief expressed by people in cyberspace. The following article is by David Boaz, an executive vice president of a major libertarian think tank, the Cato Institute.

At the *Visions* Web Site Links to the Cato Institute; to the Libertarian FAQ (frequently asked questions); to the Progress & Freedom Foundation; to *Reason* magazine.

READING PROMPT

The greatest appeal of libertarianism historically has been to engineers, scientists, and other highly educated, mostly male, workers in high technology. What, if anything, in Boaz's essay supports or contradicts such a generalization?

In 1995, the Gallup Poll found that 39% of Americans believed "the federal government has become so large and powerful that it poses an immediate threat to the rights and freedoms of ordinary citizens." Pollsters couldn't believe it, so they tried again, taking out the word "immediate." This time 52% of Americans agreed.

Later that year, *USA Today* reported that "many of the 41 million members of Generation X . . . are turning to an old philosophy that suddenly seems new: libertarianism." *The Wall Street Journal* agreed: "Much of the angry sentiment coursing through [voters'] veins today isn't traditionally Republican or even conservative. It's libertarian. . . . Because of their growing disdain for government, more and more Americans appear to be drifting—often unwittingly—toward a libertarian philosophy."

The future, it is becoming increasingly clear, will be libertarian.

Libertarianism is the view that each person has the right to live his life in any way he chooses so long as he respects the equal rights of others. It is an old philosophy, but its framework for liberty under law and economic progress makes it especially suited for the dynamic world we are now entering: the Information Age.

Unfortunately, however, government in fact remains bigger than ever. In the United States, the federal government forcibly extracts $1.6 trillion in wealth a year from those who produce it, and state and local governments take another trillion. Every year, Congress adds another 6,000 pages of statute law and regulators print 60,000 pages of new regulations in the *Federal Register.*

WHO SHOULD HAVE POWER?

There are now two competing forces in world politics: the pull toward centralizing power and the push toward devolving power to smaller, local entities. Despite the increasingly loud complaints about big government, Congress continues to offer federal solutions to problems, thus eliminating local control, experimentation, and competing solutions. The bureaucrats of the European Union in Brussels try to centralize regulation at the continental level, partly to prevent any European government from making itself more attractive to investors by offering lower taxes or less regulation.

Paradoxically, nation-states today are too big and too small. They're too big to be responsive and manageable. India has more than one million voters for each of its more than 500 legislators. Can they possibly represent the interests of all their constituents or write laws that make sense for almost a billion people? In any country larger than a city, local conditions vary greatly and no national plan can make sense everywhere.

At the same time, even nation-states are often too small to be effective economic units. Should Belgium, or even France, have a national railroad or a national television network, when rails and broadcast signals can so easily cross national boundaries? The great value of the European Union is not the reams of regulation produced by Eurocrats, but rather the opportunity for businesses to produce and sell across a market larger than the United States. A common market doesn't require centralized regulation; it only requires that national governments not prevent their citizens from trading with citizens of other countries.

BREAKING AWAY FROM BIG GOVERNMENT

As centralized governments around the world try to squelch regional differences and small-scale experiments, another trend is also visible. Business people try to ignore government and find their natural trading partners, be it across the street or across national borders. Businesses in the triangle formed by Lyon in France, Geneva in Switzerland, and Turin in Italy do more business among themselves than with the political capitals of Paris and Rome. Dominique Nouvellet, one of Lyon's leading venture capitalists, says, "People are rebelling against capitals that exercise too much control over their lives. Paris is filled with civil servants, while Lyon is filled with merchants who want the state to get off their back."

Other cross-border economic regions include Toulouse and Montpellier, France, with Barcelona, Spain; Antwerp, Belgium, with Rotterdam, the Netherlands; and Maastricht, the Netherlands, with Liege, Belgium, and Aachen, Germany. National governments and national borders impede the creation of wealth in those areas.

Many regions are reviving an old solution to the problems of out-of-touch, out-of-control government: secession. The French-speaking people of Quebec agitate for independence from Canada. So do a growing number of people in British Columbia, who see that their trade ties to Seattle and Tokyo are greater than those with Ottawa and Toronto. The Lombard League in productive northern Italy is calling for secession from what it regards as Mafia-dominated, welfare-addicted southern Italy. There's an increasing likelihood of devolution or even independence for Scotland. National breakup may well be a solution to some of the problems of Africa, whose national boundaries were carved by colonial powers with little regard to ethnic identity or traditional trading patterns.

The United States is a part of the secessionist trend. Staten Island voted to secede from New York City in 1993, but the state legislature blocked its path. Nine counties in western Kansas have petitioned Congress for statehood. Activists in both northern and southern California have proposed that the giant state be split into two or three more manageable units. The San Fernando Valley is brimming with demands to secede from the city of Los Angeles.

LOCAL POWER

Switzerland offers a good example of the benefits of free trade and decentralized power. Although it has only 7 million people, Switzerland has three major language groups and people with distinctly different cultures. It has solved the problem of cultural conflict with a very decentralized political system—20 cantons and six half-cantons, which are responsible for most public affairs, and a weak central government, which handles foreign affairs, monetary policy, and enforcement of a bill of rights.

One of the key insights offered by the Swiss system is that cultural conflicts can be minimized when they don't become political conflicts. Thus, the more of life that is kept in the private sphere or at the local level, the less need there is for cultural groups to go to war over religion, language, and the like. Separation of church and state and a free market both limit the number of decisions made in the public sector, thus reducing the incentive for groups to vie for political control.

People around the world are coming to understand the benefits of limited government and devolution of power. Still, the centralists will not give up easily. The impulse to eliminate "inequities" among regions is strong. President Clinton said in 1995, "As president, I have to make laws that fit not only my folks back home in Arkansas and the people in Montana, but the whole of this country. And the great thing about this country is its diversity, its differences, and trying to harmonize those is our great challenge." Kentucky Governor Paul Patton says that, if an innovative education program is working, all schools should have it, and if it isn't, none should.

But why? Why not let local school districts observe other districts, copy what seems to work, and adapt it to their own circumstances? And why does President Clinton feel that his challenge is to "harmonize" America's great diversity? Why not enjoy the diversity? The problem for centralizers is that appreciating diversity means accepting that different people and different places will have different situations and different results.

The bottom-line question is whether centralized systems or competitive systems produce better results. Libertarians argue that competitive systems offer better answers than imposed, centralized, one-size-fits-all systems.

Two large companies—ITT and AT&T—both announced in 1995 that they would split themselves into three parts because they had become too large and diverse to be managed efficiently. ITT had sales of about $25 billion a year, AT&T about $75 billion. If corporate managers and investors with their own money at stake can't run businesses that size effectively, can it really be possible for Congress and 2 million federal bureaucrats to manage a $1.6 trillion government—to say nothing of a $6 trillion economy?

THE INFORMATION AGE

One big reason that the future will be libertarian is the arrival of the Information Age. Information is getting cheaper and cheaper and thus more widespread. The Information Age is bad news for centralized bureaucracies.

First, as information gets cheaper and more widely available, people will have less need for experts and authorities to make decisions for them. That doesn't mean we won't consult experts—in a complex world, none of us can be expert in everything—but it does mean we can choose our experts and make our own decisions. Governments will find it more difficult to keep their citizens in the dark about world affairs and about government malfeasance.

Second, as information and commerce move faster, it will be increasingly difficult for sluggish governments to keep up. The chief effect of regulation on communications and financial services is to slow down the pace of change and keep consumers from receiving the full benefits that companies are striving to offer us.

Third, privacy is going to be easier to maintain. Governments will try to block encryption technology and demand that every computer come with a government key—like the "Clipper Chip"—but those efforts will fail. Governments will find it increasingly difficult to pry into citizens' economic lives. When digital bits become more valuable than coal mines and factories, it will be more difficult for governments to exert their control. As techno-entrepreneur Bill Frezza puts it, "Coercive force cannot be projected across a network."

Some people worry that the cost of computers and Internet access creates a new divide between the haves and the have-nots. In fact, an adequate used computer and online access for a year can be had for the cost of a year's subscription to the *New York Times*. In any case, the cost of computers is falling and will continue to fall, as did that of telephones and televisions—once the playthings of the rich. By mid-1996, entrepreneurs were offering free e-mail to any consumer willing to put up with advertisements on the computer screen.

There will be no haves and have-nots, says Louis Rossetto, editor of *Wired*, the libertarian bible of the Information Age: "Better to think of the haves and the have-laters. And the haves may be the ones who are really disadvantaged, since they are the guinea pigs for new technology, paying an arm and a leg for stuff that in a couple of years will be widely available for a fraction of its original price." Attempts to force companies to supply their technology to everyone at once or at a below-market cost will just reduce every entrepreneur's incentive to come up with a new product and thus slow down the pace of change.

As more of the value in our world reflects the products of our minds embedded in digital bits, traditional natural resources will become less relevant. Institutional structures and human capital will become far more important to wealth creation than oil or iron ore. States will find it more difficult to regulate capital and entrepreneurship as it becomes easier for people and wealth to move across borders. Countries will prosper by reducing taxes and regulation in order to keep innovators and investors at home and attract them from abroad. The importance of free markets and individual effort will indeed be enhanced by the more open, participatory economy made possible by cyberspace. Peter Pitsch of the Hudson Institute calls this new economy "the Innovation Age."

People have always had trouble seeing the order in an apparently chaotic market. Even as the price system constantly moves resources toward their best use, on the surface the market seems the very opposite of order—businesses failing, jobs being lost, people prospering at an uneven pace, investments revealed to have been wasted. The fast-paced Innovation Age will seem even more chaotic, with huge businesses rising and falling more rapidly than ever, and fewer people having long-term jobs. But the increased efficiency of transportation, communications, and capital markets will in fact mean even more order than the market could achieve in the industrial age. The point is to avoid using coercive government to "smooth out the excesses" or "channel" the market toward someone's desired result.

A FRAMEWORK FOR UTOPIA

Lots of political movements promise utopia: Just implement our program, and we'll usher in an ideal world. Libertarians offer something less and more: a framework for utopia, as Harvard University philosopher Robert Nozick put it.

My ideal community would probably not be your utopia. The attempt to create heaven on earth is doomed to fail, because we have different ideas of what heaven would be like. As society becomes more diverse, the possibility of agreeing on one plan for a whole nation or the whole world becomes even more remote. And in any case, we can't possibly anticipate the changes that progress will bring. Utopian plans always involve a static and rigid vision of the ideal community, and such a vision cannot accommodate a dynamic world. We can no more imagine what civilization will be like a century from now than the people of 1900 could have imagined today's civilization. What we need is not utopia, but a free society in which people can design their own communities.

A libertarian society is only a framework for utopia. In such a society, government would respect people's right to make their own choices in accord with the knowledge available to them. As long as each person respected the rights of others, he would be free to live as he chose. His choice might well involve voluntarily agreeing with others to live in a particular kind of community. Individuals could come together to form communities in which they would agree to abide by certain rules, which might forbid or require particular actions. Since people would individually and voluntarily agree to such rules, they would not be giving up their rights but simply agreeing to the rules of a community that they would be free to leave.

We already have such a framework, of course. In the market process, we can choose from many different goods and services, and many people already choose to live in a particular kind of community. A libertarian society would offer more scope for such choices by leaving most decisions about living arrangements to the individual and the chosen community, rather than government's imposing everything from an exorbitant tax rate to rules about religious expression and health care.

Such a framework might offer thousands of versions of utopia, which might appeal to different kinds of people. One community might provide a high level of services and amenities, with correspondingly high prices and fees. Another might be more spartan, for those who prefer to save their money. One might be organized around a particular religious observance. Those who entered one community might forswear alcohol, tobacco, nonmarital sex, and pornography. Other people might prefer something like Copenhagen's Free City of Christiana, where cars, guns, and hard drugs are banned but soft drugs are tolerated and all decisions are at least theoretically made in communal meetings.

One difference between libertarianism and socialism is that a socialist society can't tolerate groups of people practicing freedom, while a libertarian society can comfortably allow people to choose voluntary socialism. If a group of people—even a very large group—wanted to purchase land and own it in common, they would be free to do so. The libertarian legal order would require only that no one be coerced into joining or giving up his property. Many people might choose a "utopia" very similar to today's small-town, suburban, or center-city environment, but we would all profit from the opportunity to choose other alternatives and to observe and emulate valuable innovations.

In such a society, government would tolerate, as Leonard Read, founder of the Foundation for Economic Education, put it, "anything that's peaceful." Voluntary communities could make stricter rules, but the legal order of the whole society would punish only violations of the rights of others. By radically downsizing and decentralizing government—by fully respecting the rights of each individual—we can create a society based on individual freedom and characterized by peace, tolerance, community, prosperity, responsibility, and progress.

LIBERTARIAN PROSPECTS

Can we achieve such a world? It is hard to predict the short-term course of any society, but in the long run, the world will recognize the repressive and backward nature of coercion and the unlimited possibilities that freedom allows. The spread of commerce, industry, and information has undermined the age-old ways in which governments held men in thrall and is even now liberating humanity from the new forms of coercion and control developed by twentieth-century governments.

As we enter a new century and a new millennium, we encounter a world of endless possibility. The very premise of the world of global markets and new technologies is libertarianism. Neither stultifying socialism nor rigid conservatism could produce the free, technologically advanced society that we anticipate in the twenty-first century. If we want a dynamic world of prosperity and opportunity, we must make it a libertarian world. The simple and timeless principles of the American Revolution—individual liberty, limited government, and free markets—turn out to be even more powerful in today's world of instant communication, global markets, and unprecedented access to information than Jefferson and Madison could have imagined. Libertarianism is not just a framework for utopia, it is the essential framework for the future.

Cyberselfish

Paulina Borsook

Silicon Valley, home of the antigovernment voices of cyberlibertarianism, is also one of the country's biggest recipients of government largesse, according to *Wired* magazine columnist Paulina Borsook. In railing against big government, Borsook argues, high tech is biting the hand that feeds it.

At the *Visions* Web Site Links to a non-libertarian FAQ (frequently asked questions); to the National Issues Forum and Democracy network.

READING PROMPT

What are Borsook's reasons for so disliking libertarians?

I grew up in Pasadena, California, attending school with the sons and daughters of fathers (yup, in those days it was only dads) who worked at Caltech and the Jet Propulsion Laboratory. These parents of my classmates were my first encounter with technologists, and they were, to a man, good liberals. These were the kind of folks who would have Pete Seeger do a benefit concert for our school. They voted New Deal Democratic; they were the grateful recipients of all the money the US government had poured into science, post-

Sputnik; they had a sense that the government could do and had done good things, from building Boulder Dam to pulling off the Manhattan Project to putting a man on the moon. And, as beneficiaries of government largesse in ways they were well aware of—from the GI Bill to interest deductions for home mortgages to the vast expansion of government funding for R&D—they felt society in general, as manifested in the actions of the government, had an obligation to help everyone in it.

They were also fully aware of the positive value of government regulation, from the reliability of the FDA-mandated purity of pharmaceutical-grade chemicals they used in their research to the enforcement of voting rights for African-Americans in the South. And what with the very visible air quality problems in the Los Angeles basin (their government-funded studies had recorded the smog death of trees in the encircling San Gabriel Mountains by the 1960s), they were able to see the benefits of regulation in the local ban on trash incineration, the regulation of refinery effluents in the LA area, and the implementation of federally stipulated smog devices on automobiles.

So it came as a shock when, 20 years later, I stumbled into the culture of Silicon Valley (my first job at a software company, 1981; first job at a computer magazine, 1983; attendance at the first commercial conference devoted to the Internet, 1987; token feminist/humanist/skeptic on the masthead of *Wired* magazine, 1993). Although the technologists I encountered there were the liberals on social issues I would have expected (pro-choice, as far as abortion; pro-diversity, as far as domestic partner benefits; inclined to sanction the occasional use of recreational drugs), they were violently lacking in compassion, ravingly anti-government, and tremendously opposed to regulation.

These are the inheritors of the greatest government subsidy of technology and expansion in technical education the planet has ever seen; and, like the ungrateful adolescent offspring of immigrants who have made it in the new country, they take for granted the richness of the environment in which they have flourished, and resent the hell out of the constraints that bind them. And, like privileged, spoiled teenagers everywhere, they haven't a clue what their existence would be like without the bounty showered on them. These high-tech libertarians believe the private sector can do everything—but, of course, R&D is something that cannot by any short-term measurement meet the test of the marketplace, the libertarians' measure of all things. They decry regulation—except without it, there would be no mechanism to ensure profit from intellectual property, without which entrepreneurs would not get their payoffs, nor would there be equitable marketplaces in which to make their sales.

When I was asked to participate in a survey on the politics of the Net, the questions presumed respondents were libertarian, but charitably gave space for outdated contrarian views. When *Byte* magazine's former West Coast bureau chief wrote an editorial mildly advocating government subsidy for basic

Net access for elementary schools and public libraries, the only response he got was outraged flames from libertarians.

And when *Self* magazine started an online gun control conference on The Well, an electronic bulletin board and Internet gateway smack in the middle of tree-hugging, bleeding-heart-liberal, secular-humanist Northern California, opinions ranged from mildly to rabidly anti-gun control. This passionate hatred of regulation, so out of whack with the opinions of the man and woman on the street in my own bioregion/demographic, showed me how different a place the online, high-tech world is from the terrestrial community to which it is nominally tethered—even an online world with countercultural roots as strong as those of The Well.

Mike Godwin, staff counsel for the online watchdog group Electronic Frontier Foundation, has written in *Wired* magazine, "Libertarianism (pro, con, and internal faction fights) is the primordial net.news discussion topic. Anytime the debate shifts somewhere else, it must eventually return to this fuel source." In a decentralized community where tolerance and diversity are the norm (no one questions online special-interest chat rooms devoted to consensual S&M or Wiccan nature mysticism or . . .), it is damned peculiar that there seems to be no place for political points of view other than the libertarian.

TRUE STORIES

I think this all very strange, because, of course, I know that without the government, there would be no Internet (mainly funded by the government until recently).

Further, there would be no microprocessor industry, the fount of Silicon Valley's prosperity (early computers sprang out of government-funded electronics research). There would also be no major research universities cranking out qualified tech workers: Stanford, Berkeley, MIT, and Carnegie Mellon get access to incredibly cheap state-of-the-art equipment plus R&D, courtesy of tax-reduced academic–industrial consortia and taxpayer-funded grants and fellowships.

But libertarianism thrives in high-tech, nonetheless. I spent a week at the plushy Lake Tahoe getaway of a Silicon Valley guy who's made it. We argued and butted heads with great civility—but perhaps the most Found moment came when he complained about how the local Tahoe building code wouldn't let him alter the silhouette of his megachalet. I nodded sympathetically, yet pointed out that in Los Angeles, where there were no such planning guidelines until recently, plutocrats often tore down existing structures and rebuilt monstrosities that take up the entire lot, blocking their neighbors' views. He looked at me, puzzled; he hadn't considered that possibility. Obviously, he had never heard of the tragedy of the commons, where one sheep too many consuming more than its share of common resources destroys the whole; nor had he thought much about what participating in a community means.

Of course, I was also thinking about the fine system of interstate highways that made his trip from Silicon Valley to the Sierra a breeze; the sewage and water-treatment facilities that allowed his toddlers to drink safely out of the tap in his kitchen; the fabric contents-and-care labels on the sheets and towels freshly laundered for each new houseguest; and the environmental regulations that keep Tahoe the uniquely blue, gorgeous, and safe refuge it is—precisely the lateral, invisible, benign effects of the government he constantly railed against.

"I GOT MINE"

The nexus of libertarianism and high-tech in the Silicon Valley will come to matter more and more, because it involves lots and lots of money (companies with valuations rivaling General Motors'). And it's a wealth of tremendous self-insulation: I routinely attend parties peopled by digerati in their 20s and early 30s who, in addition to their desirable arrogance of youth, have a frightening invulnerability (their skills in demand, the likelihood of cashing out high).

One of these, a friend newly venture-funded to capitalize on Net advertising, commented that the economy was basically in good shape (after all, no one she knew was struggling)—and then wondered why, when she ran a help wanted ad for an office manager in the *San Francisco Chronicle,* she got so many applicants, so many of whom had advanced degrees and employment histories of authority and responsibility.

Never mind people like my sister, who, with her biology degree from Stanford and master's in public health, has rarely found a steady job with benefits in the last 10 years, and has at times resorted to desperation moves such as selling flowers at subway stations to prevent foreclosure on her house (wrong gender; wrong skill set: teaching, public health, environmental concerns—just the kind of "middle manager/government bureaucrat" so despised by technolibertarians).

Or my ex-boyfriend, the English professor (B.A. honors, University of Chicago; Ph.D., Cornell), who was lucky to find a job where he earned about what I made at my first technical writing gig 15 years ago (wrong skill set: all that subjective liberal-arts-flake crap no one cares about. After all, anyone can publish on the Web, and, as MIT Media Lab's Archduke Nicholas Negroponte points out, what's the future of books anyway?).

And what would the technolibertarians make of the *New York Times* front-page series on the chronic, structural unemployment of masses of skilled middle-class workers, folks theoretically immune to being rendered redundant in the '90s? Or the heartbreaking stories I read about blue-collar workers (haven't they had the good taste to become extinct by now?) in the house organ of the United Auto Workers (the National Writers Union is part of the UAW). I imagine the technolibertarians thinking, "Well, the blue-collar miscreants, it's their own damned inertial Second Wave thinking that's got them unemployed." But what would they say about the white-collar jobless, who, no doubt, were working with computers?

THE TRUE REVENGE OF THE NERDS

As surely as power follows wealth, those who make money decide that society, having rewarded their random combination of brains and luck in one sphere, should pay attention to them in another. And so, high-technocrats are beginning to try to influence the world beyond VDTs.

But what will result if the people who want to shape public policy know nothing about history or political science or, most importantly, how to interact with other humans? Programmers, and those who know how to make money off them, mostly find it easier to interact in e-mail than IRL (in real life), and are often not good at picking up the cues, commonplaces, and patterns of being that civilians use to communicate, connect, and operate in groups.

The convergence between libertarianism and high-tech has created the true revenge of the nerds: Those whose greatest strengths have not been the comprehension of social systems, appreciation of the humanities, or acquaintance with history, politics, and economics have started shaping public policy. Armed with new money and new celebrity—juice—they can wreak vengeance on those by whom they have felt diminished.

Implicit is their assumption that those who excel by working with the tangible and not the virtual (e.g., manufacturing and servicing actual stuff) are to be considered societally superfluous. Technolibertarians applaud the massive industrial dislocations taking place in affluent North America, comparable to the miseries of the Scottish enclosures or the Industrial Revolution.

Compare my father's generation: My father succeeded through his era's version of the *arriviste* drive so celebrated by technolibertarian theorists such as George Gilder, Silicon Valley's John Knox. One of eight children in an immigrant family, second in his class in medical school when there were still quotas on Jews (the usual story), my father, like the majority of his age-cohorts, never had contempt for those who couldn't find a way to work the system as he had. He believed in social safety nets and as much government regulation (for consumer health and safety, for example) as possible to aid ordinary people. It would have made no sense to him to adopt the stance of today's technolibertarian nouveau riche (or even more scarily, wannabe nouveau riche). And in this he was not exceptional.

THE ULTIMATE ESCAPE

It's not clear how all this evolved: a combination, no doubt, of the money to be made by developing technology in the private sector, the general worldwide resurgence of libertarianism, maybe some previously undocumented deleterious effect of the toxic byproducts of semiconductor manufacturing that have leached into the aquifers below Sunnyvale. But there are some worrisome consequences to consider as technology touches more and more people's lives—and those who rule are increasingly the ones who understand it, own it, create it, and profit by it.

Protecting privacy. Technolibertarians rightfully worry about Big Bad Government, yet think commerce unfettered can create all things bright and beautiful—and so they disregard the real invader of privacy: Corporate America seeking ever-better ways to exploit the Net, to sell databases of consumer purchases and preferences, to track potential customers however it can.

Skimping on philanthropy. In Silicon Valley and its regional outposts (Seattle, Austin), it's not even a joke, not even an embarrassment, that there's so little corporate philanthropy, except where enlightened self-interest can come to bear (donating computers to schools, contributing to a local computer museum). High-tech employees rank among the lowest of any industry sector for giving to charity—especially dismaying given their education, job security, lifetime earnings potential, and annual income.

It's an issue of culture: Unlike other educated professionals, who see good works and support of the arts as symbols of having arrived or as payback to the society that has treated them well, the average geek espouses a world where the only art would be that which has withstood the test of the marketplace (Dong Kingman museums? Leroy Nieman traveling exhibitions?), and where there is no value to be derived in experiencing a painting in person (that is, in a museum) as opposed to on CD-ROM.

And since these guys honestly can't perceive the difference between a Lichtenstein and some *soi-disant* computer art exercise in primary-colored fractals, courtesy of Kai's Power Tools—they don't see anything out there worth subsidizing. A total sweetheart of my acquaintance, the smart and aesthetically sensitive creative director of a hot hot hot Web design studio, not only hadn't read *The Magic Mountain,* he hadn't heard of it. Nor of its author, Thomas Mann, a Nobel laureate and one of the great novelists of the century, an early multivoiced postmodernist if ever there were one. And perish the thought that anyone should need the services of an AIDS hospice, without the benefit of a few thousand shares of founders' stock in Intel or Cisco to cash in.

Gutting the environment. High-tech also has tremendously negative environmental impacts: Manufacturing its plastics and semiconductors is a remarkably toxic and resource-depleting affair. No surprise, then, that high-tech companies increasingly manufacture them in countries without environmental and worker safety regulations, or in US locales where these regulations are more lax. This way, the guys in area codes 415 and 408 who like to go telemarking in the Shasta Trinity Alps or bouldering in the Desolation Wilderness don't have to confront the opportunity cost of their wealth: the poisoning of the world due to the ever-expanding reach of industrialization. And they never consider that one of the reasons the whole world (including the immigrant engineers working in Silicon Valley) wants to be here is that environmental regulation and a culture of government-mandated conservation (however imperfectly executed) have made the United States probably the safest, healthiest, and, in some ways, most pristine place on earth.

Ignoring cities. The anti-communitarian outlook is an outcropping of how suburban an industry high-tech is. The quintessential edge-city business, high-tech celebrates people operating as monads, free agents who work in industrial parks and aspire, when they cash out in an initial public offering, to telecommute from horse country, puma country, or even from within the spare-bedroom-cum-home-office located in a half-million-dollar Eichler ranch house on a street close to El Camino Real. Never mind that most start-up/self-employed/telecommuting Internet entrepreneurs are concentrated in New York, San Francisco, and Los Angeles, thriving on the grit/density/frisson/charge of urban areas.

NEW ROBBER BARONS

All this matters desperately: With the libertarian agenda at work, the very things that fed the boom economy in intellectual property—the last great thing the United States has done—will disappear without more investment in infrastructure and health and safety and education and every other good legacy of the New Deal and the Great Society. In 20 or 30 years, the United States may well cease creating the one commodity that produces a trade surplus and new jobs.

And the sorrow for the bottom 90 percent of society—what Apple Computer once disingenuously called "the rest of us"—will be that once again we may deceive ourselves. We make goo-goo eyes over the megabucks high-tech generates, but we ignore the price. Just as 19th-century timber and cattle and mining robber barons made their fortunes from public resources, so are technolibertarians creaming the profits from public resources—from the orderly society that has resulted from the wise use of regulation and public spending. And they have neither the wisdom nor the manners nor the mindset to give anything that's not electronic back.

chapter 6 Explorations

6a Public Opinion Polling

Collaboration. Conduct a series of polls within your class or work group, on political issues, politicians, movies, television shows, music, sports, or comparable topics. Tabulate, share, and discuss the results. Then consider how this "polling" has been different from or similar to other class or group discussions you have had in this course.

Web work. The Voice of the Internet gives you the chance to complete polls in any number of areas (even if you don't know anything), even to get paid for completing some of its polling, and to view the results. The U.S. Census Bureau runs the largest polling service in the country; check out its page on Census 2000. The Internet Entertainment Charts site gives you access to what thousands of Web users feel about popular culture. Movies are another area in which people like to express their opinions; users of the Internet Movie Database have compiled their own list of the top 250 films of all time.

One writing topic. Write an essay comparing opinions you have that seem to be in the minority with opinions that seem to be in the majority.

6b Audience or Call-In Talk Shows

Collaboration. We assume you have had lecture classes in which the instructor did nearly all of the talking; you may also have been in classes with a high level of student participation, in which the teacher said very little. Discuss your responses to these different types of classes. Is one kind always, or generally, better or worse? And if so, why?

Web work. CNN has Web sites for its most popular call-in shows, including *Talkback Live* and *Larry King*. Also visit the Oprah Winfrey Web site.

Yahoo! lists 122 sites under "talk shows": News and Media > Television > Shows > Talk Shows. Talk America bills itself as the only network offering "24/7" talk radio; with proper equipment and software you can listen in at its Web site.

One writing topic. Analyze the conditions under which you most or least enjoy listening to others.

6c Activism: Getting Something Done in Your School or Community

Collaboration. List and share experiences you have had trying to get something done with people outside your immediate family.

Web work. Start with the Internet Headquarters for Student Government. Also check out the Institute for Global Communications, which describes its mission thus: "To expand and inspire movements for peace, economic and social justice, human rights, and environmental sustainability around the world by providing and developing accessible computer networking tools."

One writing topic. Describe the challenges and rewards, or the difficulty and disappointment, of trying to get people to accomplish a common goal.

6d Apathy

Collaboration. Make and share a list of all the important things in life in which you honestly have almost no interest.

Web work. The Web has a wealth of sites for people with extra time on their hands or with little inclination to do whatever it is they are supposed to do. Two such sites are the Center for the Easily Amused and Web-a-Sketch, the online version of the Ohio Art classic program.

For the truly bored, or perhaps the truly inquisitive, there may be nothing more fascinating or addictive than the Magellan Internet Guide, a site that every fifteen seconds flashes twelve search requests being made at that time by real users.

One writing topic. Describe favorite ways of wasting time.

6e Staying Informed

Collaboration. Make and share a list of all the ways people today can stay informed about political issues, and state those that you personally use on a regular basis.

Web work. C-Span represents an attempt to use a technology older than computers, cable television, to keep people informed about the U.S. government. CNN and *Time* magazine, two parts of Time Warner, have created the All Politics Web site. America Online has created a political forum site called The Great Debate. PBS has a common Web site for all its public affairs shows as well as a site for its school-oriented Democracy Project. Finally, check out two sites: the one for *Meet the Press,* the oldest weekly television news show, and the one for Bill Maher's ABC talk show, *Politically Incorrect,* a daily show that can be seen either as an attempt to widen the audience for political discussion or as the substitution of entertainment and talk for real news and analysis.

One writing topic. Analyze the relationship between the amount of political coverage available to you and your interest in political matters.

6f Staying in Touch

Collaboration. List and share a brief description of a group or groups (at least outside your family) of which you especially enjoyed being a member. Note how the group(s) communicated.

Web work. Because of its speed, convenience, and low cost, email is becoming a dominant means not just of allowing individuals to stay in touch with one another but, more to our point, of allowing individuals to keep in touch through email discussion groups with hundreds or thousands of individuals with a common interest who have all subscribed to the same automated discussion list. One joins the list, ordinarily by sending an email to the software program that coordinates the email distribution, and everyone on the list gets an email copy of any message one sends to the group.

The Listz maintains a comprehensive list of email discussion groups on the Internet—some nearly 90,000 groups on any and every imaginable subject. Members of a common list are, in effect, members of a virtual community. Take a look at some of the available lists and see if you can think of some topics not covered.

News groups are similar, except that the individual messages are not distributed but are instead collected at a single site where they are read by news reader software (now ordinarily integrated into Web browsers). Tile.net provides a list of news groups. Before the emergence of the graphical, easy-to-use Web in the mid-1990s, news groups dominated the culture of the Internet with their wide-open, raucous, open-to-anyone-and-anything style.

Review the *Utne Reader*'s list of the ten most enlightened communities to see if increased communication is playing a role in their success.

One writing topic. Discuss the role of communication, face-to-face and otherwise, in forming a good community.

part III

At Play

The three chapters that follow deal with three broad areas of play, each of which has its own special connection to technology: Chapter 7, Dating, High-Tech and Low; Chapter 8, Shopping, Old and New; and Chapter 9, Hobbies, and Other Forms of Entertainment. Up until recently, probably nothing about computers has garnered more publicity than the material covered in Chapter 7: the different ways in which new electronic technologies can be involved with the entire range of human activities related to romance and sexual expression. Almost daily there is a feature news story on some aspect of sex and the Internet, from the promotion of platonic friendships and mild flirtation via email to the availability of hardcore pornography and, perhaps most importantly of all (and as explained by Amy Harmon in "Virtual Sex, Lies and Cyberspace"), the use of chat rooms as safe havens for discovering new aspects of one's sexual identity. An important and challenging issue here for all of us is the exploration of our thinking about the connection between "intimacy" (including sexual expression) and physical presence—for example, our feelings about phone services that allow absolute strangers to share intimacies, even for the purpose of sexual arousal. What, in other words, are some of the complex ways that the latest technology affects the oldest and most basic human need of sexuality?

Explorations in Chapter 7 explore both new and traditional ways people get together, including the reemergence of a nineteenth-century meeting place: the art gallery. One reading (Julia Wilkins's "Protecting Our Children from Internet Smut") and one Exploration deal with the highly controversial issue of electronic censorship.

Meanwhile, if sexuality dominated the public discussion of the Web for most of the 1990s, a new topic (triggered by feverish activity of the stock market) has come to the forefront as the decade ends; namely, the possibilities of electronic commerce. Consumer culture has always been dependent on the latest technologies. The late nineteenth century brought the popularization of graphic ads in newspapers and magazines, and eventually in nationwide catalogs, which in turn depended on new and improved nationwide mail and transportation systems. Shopping in the second half of the twentieth century, like nearly everything else, followed the path of the automobile, first with the growth of suburbs and then with the growth of shopping centers designed to accommodate suburbanites' cars—a process analyzed and opposed in Chapter 8 by Sarah Anderson in "Wal-Mart's War on Main Street."

While there has been catalog shopping throughout the century—a process greatly improved with the use of toll-free long-distance calling and credit-card sales—it is only now, with the current interest in shopping on the Web (what is called e-commerce), that it is possible to start seeing malls themselves as a bit old-fashioned. It is a strange thought, and one that shows the truly transforming power of technology, that today's latest malls may soon be as quaint and as low-tech as the shops on any town's Main Street. Explorations in Chapter 8 ask you to consider the many ways technology is connected with how you shop by reflecting on and exploring Web sites related to malls, favorite small shops, Main Streets, and the hottest new presences on the Web, virtual bookstores.

Chapter 8 also deals with a related issue: the increasingly important role of shopping and consumer spending generally. We live in a world where the slightest decline in consumer spending is reported as a sign of imminent economic danger. The dismantling of the Soviet Union meant, among other things, the end of what are called "command" or planned economies, those organized to provide people with the basic necessities of life, presumably according to some sort of rational or pseudorational analysis conducted by a central planning entity—a system almost universally condemned in the United States for being heavy-handed, bureaucratic, and inefficient. What the world seems intent on embracing instead of the command system (for better or for worse—something you may want to consider here and in the final part, Future Thoughts) is some version of U.S. consumer culture, in which manufacturers are free to respond to the needs of consumers and, conversely, through advertising, to create needs within consumers (inflame the desires of consumers) for their products. Such is the subject, in one form or another, of the other four pieces in Chapter 8.

Chapter 9, meanwhile, touches on subjects related to hobbies. The readings raise questions about just how important hobbies are for so many of us (for many, the most important thing in our lives) and about the extent to which many hobbies (such as fly-fishing) set artificial limits on acceptable and unacceptable uses of technology. Hobbies, unlike work, are organized primarily for pleasure, and hence not necessarily for the ever higher levels of efficiency that govern so much modern technology. The first two pieces take different looks at the fad of cyberpets, and in so doing raise the question of what it is we get out of our low-tech relationship with flesh-and-blood pets. With Donald Katz's piece we shift gears to consider how the old (low-tech) experience of watching sports live is being radically altered, both in new sports complexes that make the live experience more like watching on TV at home and in high-tech home systems that make the virtual experience of watching at home more like seeing the event live at the arena—an especially troubling development when the virtual experience is not a broadcast of an actual game but the completely virtual experience of a video game. What is going to happen to live sports and real athletes when the virtual experience of video games is as exciting, as realistic, as lifelike as a "real" event?

chapter 7

Dating, High-Tech and Low

READING QUESTIONS

- How has any technology (from the letter to email and including the telephone and even the automobile) figured in flirting and romantic meetings in your own life? In the lives of your contemporaries? In the lives of older people you know, and in how you imagine men and women met in the more distant past?
- What do you see as the greatest advantage and the greatest disadvantage of computers in the broad area of sexual fulfillment?
- What types of controls, if any, do you believe there should be on the transmitting and receiving of explicit graphic images over the Internet?

QUOTATIONS

Sex is hardly ever just about sex.

— *Shirley MacLaine (1985), American actress*

Sex is the Tabasco Sauce which an adolescent national palate sprinkles on every course in the menu.

— *Mary Day Winn (1931), American reporter*

Plenty of guys are good at sex, but conversation, now that's an art.

— *Linda Barnes (1987), American novelist*

Flirtation is merely an expression of considered desire coupled with an admission of impracticability.

— *Marya Mannes (1964), American author, 1904–1990*

On the Internet, nobody knows you're a dog.

— *Peter Steiner, American cartoonist*

Virtual Sex, Lies and Cyberspace

Amy Harmon

Amy Harmon, a staff writer for the *Los Angeles Times,* looks at one of the most popular venues for high-tech romance: chat rooms, especially those available on the country's biggest Internet service provider, America Online. Whereas its early competitor CompuServe assigned people random number sequences as their on-screen names, AOL has always allowed people to design their own unique screen names, and still gives people the ability to have multiple such identities under a single account. A new AOL product, Instant Messenger (free to everyone and mentioned by Harmon in her article) is one of any number of new programs that turn the entire Web into a chat room by allowing you to exchange real-time messages with anyone on the Web who also has Instant Messenger turned on.

At the *Visions* Web Site Links to AOL; to netiquette sites.

READING PROMPT

"It's not healthy for people to pretend to be someone they're not and fantasize about that constantly," Harmon quotes psychologist Nancy Wesson. Just how much fantasy—and when and where and under what conditions—do you believe is healthy?

The first time Donna Tancordo "cybered," she switched off her computer midway through the typed seduction, shocked and scared at the power of the words scrolling down her screen.

"I've never described what I was feeling like that before," she said. "I freaked out."

But Tancordo, a happily married New Jersey housewife with three kids, soon logged back onto America Online. In a chat room called "Married and Flirting," she met another man. For days, they whispered the details of their lives into the ether. When he asked her if he could take her on a virtual trip to the mountains, she agreed.

This time her computer stayed on.

All hours of the day and night, America Online's chat rooms teem with people seeking something missing in their lives—like Jay, a successful business consultant in Boston, who says he logs on to fill "the void of passionate emotion."

The blurted confidences and anonymous yearning scrolling through AOL's frames reveal a rare picture of the American psyche unshackled from social convention.

In the vacuum of cyberspace, self-exploration is secret and strangely safe. Much has been made lately of how cults may find fertile recruiting ground among online seekers. A vast range of support groups—for pregnant mothers, cancer patients, substance abusers—also flourish. Unlikely friendships are struck and sometimes sustained.

But in an age when sex is scary and intimacy scarce, the keyboard and modem perhaps most often serve a pressing quest for romantic connection and sexual discovery.

Eric lives in a small California farming town: "I'm pretty much a straight kind of dude." When he flips on the computer at 4:30 a.m. to check the weather, he is drawn to rooms where San Franciscans recount stories of sexual bondage.

Eleanor, 13, is 5-foot-1, with dark brown hair. When she surfs the "Teen Chat" rooms after school, she looks for kicks as a tall strawberry blond.

Peter, a 45-year-old professional in Manhattan, spent his first weekend on AOL posing as a 26-year-old woman while his wife was away on business. Enthralled with the ease of uninhibited communion, he cycled through a whirl of identities. He disguised himself as a gay man, a lesbian and a young girl. But eventually he settled on a more mundane form of seduction.

"What I really wanted was to have sexual conversations with women," Peter said. "Kind of garden variety, but that's who I am, and what made it such a fever for me—that's not too strong a word—was the flirtation aspect of it."

The ritual of pursuing secret desires from behind a facade is as old as the masquerade. But perhaps because it has never been so easy, the compulsion has never seemed so strong.

"LEAVE THE MEAT BEHIND"

The free computer disks that arrive unbidden in the mail offer not only a mask, but an escape from the body—the ability, as cyberpunk author William Gibson puts it, to "leave the meat behind."

It is an offer with remarkable mass appeal. As AOL's subscriber count doubled over the last year to 8 million, the number of chat rooms on busy nights tripled to 15,000. And the recent, much-publicized agitation over the service's busy signals was due largely to people chatting longer, now that a new pricing plan means they do not have to pay by the minute.

AOL is by far the most popular gathering spot on the Internet, in part because its culture of anonymity—members can choose up to five fictional screen names—promotes what one observer calls "the online equivalent of getting drunk and making a fool of yourself." Although it is possible to chat on the World Wide Web and other areas of the Internet, the technology doesn't work nearly as well.

Largely because of the unabashed sexual character of many of its chat rooms, AOL executives traditionally have downplayed their importance to the company's bottom line. "What we're offering at AOL is convenience in a box," said AOL Network's President Robert Pittman. "If you use AOL it will save you time. People aren't buying it for chat."

Perhaps. The service offers e-mail, Internet access and information and entertainment features. Many of its customers never venture near the chat rooms, and most usage of the Internet is unrelated to chat.

But according to America Online statistics, more than three-quarters of its subscribers use chat rooms at least once a month, the equivalent of 1 million hours a day.

"If AOL eliminated chat you'd see the subscriber base go from 8 million to 1 million faster than you could spit," said Alan Weiner, an analyst at Dataquest, a consulting firm.

Not all chat is laden with sexual innuendo. "I can say I'm a voluptuous teen and I still don't get attention when I go into the sports and finance rooms," quipped one frequent female chatter.

Some chat rooms emerge as genuine communities where the same group gathers regularly. The "SoCalifover30" room even holds regular "fleshmeets" at restaurants or members' homes. A core group keeps up on one another's romantic exploits online and offline.

"Ladykuu," a San Diego bus driver trainer and the mother of twins, says she has become close friends with another mother of twins in Boston, with whom she shares life's tribulations.

But even Ladykuu enjoys "lurking" and listening to others tell secrets to which she ordinarily would not be privy:

"It's just fascinating to me to see, what is that deep dark fantasy, what is the naughty thing you're thinking about and—oh my gosh, I've been thinking about that too."

Some sexual-oriented chat is basic singles bar sleaze—and some is mainly an excuse to swap pornographic pictures. But much more prevalent is the search for genuine connection, and perhaps seduction.

Some chatters seek a companion to meet in person. Others, who shun the idea of a real-life affair, seize on the opportunity to engage in the thrill of a new seduction over the computer from the comfort of home—often while their spouses sleep in the next room.

Whether the demi-realities of chat can fulfill real world needs or only add to their urgency is a subject of much debate among online seekers. Some discover hidden pieces of themselves that lead to significant changes in what, in a telling delineation, is called RL—real life.

Others grow sickened by the relentless layering of illusion, where friends and lovers appear suddenly, and then melt into air, or morph into aliens. For there is in all this a bitter irony: That a search for intimacy brings people to pose as airbrushed versions of themselves, so that they may share their inner fantasies with strangers.

"It's not healthy for people to pretend to be someone they're not and fantasize about that constantly," said Nancy Wesson, a psychologist in Mountain

View, California. She has seen marriages break up in part because of one partner's online activities. "It allows you to perpetually live in a fantasy instead of living in real life."

Ultimately, marriage may be the institution most rocked by the new technology. Although cyberspace obviously doesn't invent secret longings, it does provide a way to uncover and exploit them that has never been so available to so many.

CHEATING WITHOUT REALLY CHEATING

Some flirters say the ability to cheat without really cheating, to voice fantasies somehow too personal to share even with spouses, has invigorated them.

Donna and Ralph Tancordo, high school sweethearts who have been married for 17 years, sign onto AOL and "cyber" with other married people—with each other's consent.

"My cheekbones hurt I've been smiling so much lately," said Donna, who opened her account a month ago. "I think it's the flattery. It's like, 'Wow, somebody else is attracted to me other than my husband.' And it's improved our sex life 150%."

In the case of Peter, the Manhattan professional, the online habit nearly broke up his marriage. Finding a woman that he would care to talk to and who would talk to him could take hours on any given night. He would stay up after his wife, Janet, went to bed, and look forward to when she would leave him alone at home.

In the end, Janet became too distraught over his regular online meetings with a woman who lived thousands of miles away. Peter agreed to cancel his AOL account. Both say the experience has opened up a productive, if painful, period of exploration for them.

"I was bored and I lied about it to myself," Peter said. "I had a sex life, but it didn't have passion. At some level, that's what I was seeking, and it's hard to find. There may not be an answer."

For Janet, the hardest part has been trying to sift out what may be her husband's harmless fantasy life from what to her is hurtful reality.

"Everyone knows someone who has had an affair," Janet said. "If your husband's having an affair and you tell your girlfriend, you're going to have instant sympathy. But do I have a right to be pissed about this? I don't know."

She has not talked to any of her friends about it: "It's embarrassing. I don't know anyone else who has gone through this."

A lot of people have. The online consensus is that, as Tiffany Cook of the SoCalifover30 chat room puts it, "if you're talking to a married man often enough, that's an affair even if you never meet."

But in the 1990s, when interest in family values is on the rise and the ethic of safe sex prevails, AOL offers 1960s-style free love from behind the safety of the screen. The medium offers a sense of physical and psychological safety that strips away taboos faster than the sexual revolution ever did.

Many married people—they constitute two-thirds of AOL subscribers—comb chat rooms, scope the profiles and send private instant messages (IMs) to prospective romantic partners.

The flirtation medium of choice, IMs pop up on-screen as soon as they are sent, heedless of whatever the recipient may be doing. More insistent and perhaps more intimate than e-mail, they solicit an immediate response.

"I've tried erotic e-mail. It's like bad D. H. Lawrence," said an artist who prefers the edge of IMs.

Three million IM sessions are opened every day. They are by nature fleeting and the exchange is rapid-fire, lessening the risk and increasing the nerve.

"I make advances to men the same age group as I am to start flirting and sometimes it goes a lot further than flirt," said Donna. "I read their profile first. If I like it, I'll IM them by saying . . . 'BUSY?' "

In the curious state of disembodiment, where the body is nonetheless very much the point, the typed words come as stream of consciousness, and then, with the click of a mouse, they disappear.

"I'm sorry I can't talk right now," one woman tells a reporter. "I'm getting nine IMs as we speak."

Often, IM exchanges begin between people in the same chat room. At any given moment, subscribers fill rooms of varying salaciousness—"Hot and Ready Female," "Discreet in Illinois," "CA Cops Who Flirt," "BiCuriousM4M." Many of the chat rooms created by subscribers—as opposed to those established by AOL—have overtly sexual themes and many others draw people interested in romance.

"There's a lot more diversity out there than I would have given people credit for," said Jenny, a 27-year-old lesbian from Manhattan who roams the chat rooms when she is not using the service to check stock quotes.

"Wanna cyber?" comes the standard query, proffering the on-line equivalent of a one-night stand. "M/F?" "What are you wearing?"

"On AOL you could be talking about sex within three minutes of meeting someone," said a 28-year-old male marketing consultant who goes by the handle "MindUnit."

Many simply want to experiment in the intricate art of flirtation, sometimes behind a guise, sometimes as themselves.

"It's the only place you can throw yourself at someone and not care if you get rejected," said Jenny.

Women especially say the ability to both be more aggressive than they would in real life and to hit "cancel" or "ignore" if a flirtation gets out of control is liberating—and perhaps good practice.

For many, the point is not cybersex per se, but delving into the forbidden realm of sexuality. Says one online explorer on the East Coast: "We live in a world and particularly this culture that seeks to, on the surface, completely repress our sexuality. I think for many people, AOL represents a safe and healthy expression, although, like all pleasures, from fatty foods to erotic pleasure, there is probably a price to pay."

After empty nights of chat room prowling for the ideal cybermate, many end up being as disappointing as such searches often are in real life.

"All I can tell you is that there are thousands of searching people out there . . . and AOL has become a vehicle to meet others . . . affairs, etc. . . ." types a Southern California man to a reporter one Saturday night. "But it doesn't solve the problems of real life."

Sometimes connections that seem solid suddenly fade away. Even carefree Donna was thrown off-balance recently when her AOL lover sent a cryptic message saying he wouldn't be spending as much time online.

"He was basically blowing me off and I was really upset," she said. "I was sitting in the dentist's chair and I couldn't get him out of my head. I've gotten too emotional about this. I really need to handle it better."

Psychologists caution against getting wrapped up in a reality that is not, in fact, real. And online junkies acknowledge that it can be hard to pull out of what one calls "AOL's sticky web," which can become an addictive escape from three-dimensional existence.

Psychologist Kimberly Young, who has studied online addiction, says it's comparable to compulsive gambling in its mood-altering appeal—and is just as dangerous.

Sherry Turkle, a professor of the sociology of science at the Massachusetts Institute of Technology, draws a more optimistic conclusion. In her recent book, *Life on the Screen,* she argues that online technology is enabling a new, decentered sense of identity to emerge, and that the practice of trying on different personalities could be a useful way to work through real-life issues.

Swapping genders is a popular activity among both sexes, but since (real) men outnumber women by about 2 to 1, the likelihood of talking to a man claiming to be a woman is fairly high.

MindUnit has devised an only-sort-of tongue-in-cheek "Rules to Establish Gender," testament that even in a world without gender, well-socialized roles remain largely intact.

They state, in part: "If she sounds 'too good to be true,' that is One Strike. If she has no profile, that is One Strike. If she seems preoccupied with sex, or starts the sex talk herself, that is One Strike. If she volunteers exact statistics about herself, especially measurements or bra size, that is One Strike. If the statistics are really hot, that is Two Strikes." By MindUnit's trauma-tested logic, three strikes means the woman you're chatting with is a man.

Little in the AOL chat world is as it appears to be. But, for many, the chance to honestly express their desires and be privy to those of others outweighs the veil of lies that seems somehow necessary to make it possible. "Let me find someone with an open mind, good intentions, and sincerity," reads MindUnit's profile. "Failing that, I'll take a nymphomaniac."

Few know better than Tiffany Cook the perils of confusing online illusions with real-life truths. First, the 30-year-old Santa Monica interior designer hit it off with a man who flew to visit her from New York. The chemistry didn't translate in person.

But then for three months she spent hours a day chatting with a man from Northern California. He said he was 33. Then he confessed to being 43, and then to his actual age: 71. He had sent her a picture—it turned out to be of his son.

"It was terrible," she said. "I felt so deceived."

Still, Tiffany, who changed her screen name after learning the truth about her most passionate correspondence, still spends part of almost every day online.

"You know what? It's expanded my world," she said. "I've laughed really hard and I've learned a lot, and no matter what I might think of [him] now, the fact is we had a huge amount to talk about.

"Besides, your chances of meeting someone who's hiding behind something online and someone who's hiding behind something in real life [are] about the same."

"This Is a Naked Lady" *Gerard Van Der Leun*

One does not have to agree with best-selling author John Gray that men and women are from different planets to see major differences between Amy Harmon and Gerard Van Der Leun. Where Harmon is inquisitive and even hopeful at times, Van Der Leun offers a more cynical reading of sex and cyberspace, stressing how humans (mostly men?) have always used any new technology as a means of furthering sexual gratification. Chatting, in other words, seems to be the last thing on many people's minds (or at least on men's minds) when they go online.

Van Der Leun's article appeared in the premier issue of *Wired* magazine (in 1993), which began with this 1967 quotation from *The Medium Is the Message* by media guru Marshall McLuhan:

> The medium, or process, of our time—electric technology—is reshaping and restructuring patterns of social interdependence and every aspect of our personal life.
>
> It is forcing us to reconsider and re-evaluate practically every thought, every action, and every institution formerly taken for granted.

Everything is changing: you, your family, your education, your neighborhood, your job, your government, your relation to "the others."

And they're changing dramatically.

At the *Visions* Web Site Links to Marshall McLuhan; to Angkor Wat (in Cambodia).

READING PROMPT

Compare Van Der Leun's attitude about online sexual adventure, his tone generally, with Harmon's. Which, if any, of the differences do you believe can be attributed to gender?

Back in the dawn of online when a service called The Source was still in flower, a woman I once knew used to log on as "This is a naked lady." She wasn't naked of course, except in the minds of hundreds of young and not-so-young males who also logged on to The Source. Night after night, they sent her unremitting text streams of detailed wet dreams, hoping to engage her in online exchanges known as "hot chat"—a way of engaging in a mutual fantasy typically found only through 1-900 telephone services. In return, "The Naked Lady" egged on her digital admirers with leading questions larded with copious amounts of double entendre.

When I first asked her about this, she initially put it down to "just fooling around on the wires." "It's just a hobby," she said. "Maybe I'll get some dates out of it. Some of these guys have very creative and interesting fantasy lives."

At the start, The Naked Lady was a rather mousy person—the type who favored gray clothing of a conservative cut—and was the paragon of shy and retiring womanhood. Seeing her on the street, you'd never think that her online persona was one that excited the libidos of dozens of men every night.

But as her months of online flirtations progressed, a strange transformation came over her: She became (through the dint of her blazing typing speed) the kind of person that could keep a dozen or more online sessions of hot chat going at a time. She got a trendy haircut. Her clothing tastes went from Peck and Peck to tight skirts slit up the thigh. She began regaling me with descriptions of her expanding lingerie collection. Her speech became bawdier, her jokes naughtier. In short, she was becoming her online personality—lewd, bawdy, sexy, a man-eater.

The last I saw of her, The Naked Lady was using her online conversations to cajole dates and favors from those men foolish enough to fall into her clutches.

The bait she used was an old sort—sex without strings attached, sex without love, sex as a fantasy pure and simple. It's an ancient profession whose costs always exceed expectations and whose pleasures invariably disappoint. However, the "fishing tackle" was new: online telecommunications.

In the eight years that have passed since The Naked Lady first appeared, a number of new wrinkles have been added to the text-based fantasy machine. Groups have formed to represent all sexual persuasions. For a while, there was a group on the Internet called, in the technobabble that identifies areas on the net, alt.sex.bondage.golden.showers.sheep. Most people thought it was a joke, and maybe it was.

Online sex stories and erotic conversations consume an unknown and unknowable portion of the global telecommunications bandwidth. Even more is swallowed by graphics. Now, digitized sounds are traveling the nets, and digital deviants are even "netcasting" short movie clips. All are harbingers of things to come.

It is as if all the incredible advances in computing and networking technology over the past decades boil down to the ability to ship images of turgid members and sweating bodies everywhere and anywhere at any time. Looking at this, it is little wonder that whenever this is discovered (and someone, somewhere, makes the discovery about twice a month), a vast hue and cry resounds over the nets to root-out the offending material and burn those who promulgated it. High tech is being perverted to low ends, they cry. But it was always so.

There is absolutely nothing new about the prurient relationship between technology and sexuality.

Sex, as we know, is a heat-seeking missile that forever seeks out the newest medium for its transmission. William Burroughs, a man who understands the dark side of sexuality better than most, sees it as a virus that is always on the hunt for a new host—a virus that almost always infects new technology first. Different genders and psyches have different tastes, but the overall desire seems about as persistent over the centuries as the lust for bread and salvation.

We could go back to Neolithic times when sculpture and cave painting were young. We could pick up the prehistoric sculptures of females with pendulous breasts and very wide hips—a theme found today in pornographic magazines that specialize in women of generous endowment. We could then run our flashlight over cave paintings of males whose members seem to exceed the length of their legs. We could travel forward in time to naughty frescos in Pompeii, or across continents to where large stones resembling humongous erections have for centuries been major destinations to pilgrims in India, or to the vine-choked couples of the Black Pagoda at Angkor Wat where a Mardi Gras of erotic activity carved in stone has been on display for centuries. We could proceed to eras closer to our time and culture, and remind people that movable type not only made the Gutenberg Bible possible, but that it also made cheap broadsheets of what can only be called "real-smut-in Elizabethan-English" available to the masses for the very first time. You see, printing not only made it possible to extend the word of God to the educated classes, it also extended the monsters of the id to them as well.

Printing also allowed for the cheap reproduction and broad distribution of erotic images. Soon, along came photography; a new medium, and one that until recently did more to advance the democratic nature of erotic images than all previous media combined. When photography joined with photolithography, the two together created a brand new medium that many could use. It suddenly became economically feasible and inherently possible for lots of people to enact and record their sexual fantasies and then reproduce them for sale to many others. Without putting too fine a point on it, the Stroke Book was born.

Implicit within these early black-and-white tomes (which featured a lot of naked people with Lone Ranger masks demonstrating the varied ways humans can entwine their limbs and conceal large members at the same time) were the vast nascent publishing empires of *Playboy, Penthouse,* and *Swedish Erotica.*

The point here is that all media, when they are either new enough or become relatively affordable, are used by outlaws to broadcast unpopular images or ideas. When a medium is created, the first order of business seems to be the use of it in advancing religious, political, or sexual notions and desires. Indeed, all media, if they are to get a jump-start in the market and become successful, must address themselves to mass drives—those things we hold in common as basic human needs. But of all these—food, shelter, sex and money—sex is the one drive that can elicit immediate consumer response. It is also why so many people obsessed with the idea of eliminating pornography from the earth have recently fallen back on the saying "I can't define what pornography is, but I know it when I see it."

They're right. You can't define it; you feel it. Alas, since everyone feels it in a slightly different way and still can't define it, it becomes very dangerous to a free society to start proscribing it. And now we have come to the "digital age" where all information and images can be digitized; where all bits are equal, but some are hotter than others. We are now in a land where late-night cable can make your average sailor blush. We live in an age of monadic seclusion, where dialing 1-900 and seven other digits can put you in intimate contact with pre-op transsexuals in wet suits who will talk to you as long as the credit limit on your MasterCard stays in the black.

If all this pales, the "adult" channels on the online service CompuServe can fill your nights at $12.00 an hour with more fantasies behind the green screen than ever lurked behind the green door. And that's just the beginning. There are hundreds of adult bulletin board systems offering God Knows What to God Knows Who, and making tidy profits for plenty of folks.

Sex has come rocketing out of the closet and into the terminals of anyone smart enough to boot up FreeTerm. As a communications industry, sex has transmogrified itself from the province of a few large companies and individuals into a massive cottage industry.

It used to be, at the very least, that you had to drive to the local (or not-so-local) video shop or "adult" bookstore to refresh your collection of sexual fantasies. Now, you don't even have to leave home. What's more, you can create it yourself, if that's your pleasure, and transmit it to others.

It is a distinct harbinger of things to come that letters now appearing online are better than those published in *Penthouse Forum,* or that sexual images in binary form make up one of the heaviest data streams on the Internet, and that "amateur" erotic home videos are the hottest new category in the porn shops.

Since digital sex depends on basic stimuli that [are] widely known and understood, erotica is the easiest kind of material to produce. Quality isn't the primary criteria. Quality isn't even the point. Arousal is the point, pure and simple. Everything else is just wrapping paper. If you can pick up a Polaroid, run a Camcorder, write a reasonably intelligible sentence on a word processor or set up a bulletin board system, you can be in the erotica business. Talent has very, very little to do with it.

The other irritating thing about sex is that like hunger, it is never permanently satisfied. It recurs in the human psyche with stubborn regularity. In addition, it is one of the drives most commonly stimulated by the approved above-ground media (Is that woman in the Calvin Klein ads coming up from a stint of oral sex, or is she just surfacing from a swimming pool?) Mature, mainstream corporate media can only tease. New, outlaw media delivers. Newcomers can't get by on production values, because they have none.

Author Howard Rheingold has made some waves recently with his vision of a network that will actually hook some sort of tactile feedback devices onto our bodies so that the fantasies don't have to be so damned cerebral. He calls this vision "dildonics," and he has been dining out on the concept for years. With it, you'll have virtual reality coupled with the ability to construct your own erotic consort for work, play, or simple experimentation.

Progress marches on. In time, robotics will deliver household servants and sex slaves. I saw The Naked Lady about three months ago. I asked her if she was still up to the same old games of online sex. "Are you kidding?" she told me. "I'm a consultant for computer security these days. Besides, I have a kid now. I don't want that kind of material in my home."

"Love, Taki" *Taki Theodoracopulos*

The following piece comes from the *National Review,* which calls itself "America's Conservative Magazine," and in Taki Theodoracopulos's praise of letter writing there seems for once to be a rare match between old-fashioned low technology and the political label "conservative."

At the *Visions* Web Site Link to the traditional purveyor of love letters, the U.S. Postal Service.

READING PROMPT

Exactly what is it about the demise of the love letter that most disturbs Theodoracopulos? What complaints does he have against more modern forms of communication?

One of the great pleasures of my life is writing love letters. Especially when under the influence. When my London flat burned down five years ago, the few love epistles I've received in my life were reduced to cinders, and I considered that my greatest loss. The demise of the old-fashioned love letter is a loss romantics the world over—however few of us are left—will always mourn.

A love letter must be among the best things in life. It costs the price of a stamp, it takes some effort and a little time, but the result can be everlasting. My late father was a great Casanova and connoisseur of the fairer sex, and used to turn out such letters effortlessly. He once admitted to a friend of his that the best love note he ever received simply stated, "I do love you." Love letters do not have to be long.

Alas, in the barbaric times we live in today, people prefer the world's most annoying instrument after television, the telephone. The intrusive contraption demands no concentration, and therefore no commitment. Mind you, we are living in litigious times, and lawyers counsel, "Don't write it down."

But I do. I began the practice about 25 years ago. I was drunk and feeling the pangs of unrequited love, so I sat down and tried the following: "Dear X, There's a marvelous line in *Romeo and Juliet* when Romeo, having avenged Mercutio's death, is advised to flee Verona. 'But Heaven is here, where Juliet lives,' he cries. However sudden this may sound, this is how I've felt about you since the first moment I met you. Love, Taki."

To my great delight and surprise, it worked. Miss X saw me in a different light and things went smoothly for a while. After some time, I tried it again. On someone else, of course. Now before any of you cry foul, I don't think there's anything wrong with repeating love letters. It's the message that counts, not the wording. And the message is that one's in love. Some might say that repetition dilutes the meaning. But not for me. I have used what I refer to as the R-and-J letter countless times.

About ten years ago I got caught and became the laughing stock of London.

Two girls were discussing yours truly, and one said that she thought I was a drunken playboy. "Yes, but he writes beautiful love letters," said the other. Then, as girls tend to do, they compared theirs. Only the name at the top had been changed. They both started to laugh, word got out, and people have been laughing at me ever since.

Women, far more than men, are the victims of the love letter's demise. They like to be wooed, and nothing is better for courtship than a love letter. Robert Browning won Elizabeth Barrett's heart through the written word, not the spoken one. Ironically, F. Scott Fitzgerald's love letters are not at all impressive. Poor Scott, was he too shy? but shyness should not inhibit anyone when writing. I guess Scott simply didn't know his way around women.

Some years ago, an Italian friend of mine living in New York asked me to help him write to a Hollywood actress he was pursuing. Knowing she was an idiot, I borrowed a bit, without credit, from Keats and Byron. My friend was really impressed. He thought I had genius in me. Later we went to a remote house in Tuscany, where books were unavailable, and he asked me to write a quick one for him, as the star was arriving in Rome. I wrote one of my own creation, and he complained that it was good but not my best. They are no longer together, but it wasn't because of my poetry. Actually, she told him she preferred the one I'd written in Italy, without the Keats and Byron in it. Then and there I saw with perfect clarity what is wrong with Hollywood.

What is truly sad about the death of the love letter is that an entire aspect of romantic expression known to our grandparents has now vanished. Back in the good old days, people got to know each other through words rather than through deeds. Syntax rather than sex. Relationships were more stable as a result. Today, in a prurient society where people expose themselves in the most ludicrous manner, no one writes from the heart. When modern lovers send each other messages it is in the form of a greeting card, the picture being more important than the words scribbled behind it. Romantics of the world unite! We have nothing to lose but the price of a stamp.

Protecting Our Children from Internet Smut: Moral Duty or Moral Panic?

Julia Wilkins

Julia Wilkins's piece deals largely with the national response to what became a sensational (and not entirely reliable) *Time* cover story on Internet pornography: "On a Screen Near You: Cyberporn" (July 3, 1995). The *Time* article was in turn based on the Marty Rimm study done at Carnegie Mellon University. Donna L. Hoffman and Thomas P. Novak of Vanderbilt University have an excellent Web site devoted to this entire controversy. Wilkins is a special education teacher and freelance writer in Buffalo, New York.

The entire debate can be seen in terms of how political interest groups attempt to interpret, perhaps even manufacture, academic research to validate their own interests.

At the *Visions* Web Site Links to more on the Rimm study; to more on Web censorship.

READING PROMPT

Explain how Wilkins's essay is as much about the state of journalism in the United States today as about smut on the Internet.

The term moral panic is one of the more useful concepts to have emerged from sociology in recent years. A moral panic is characterized by a wave of public concern, anxiety, and fervor about something, usually perceived as a threat to society. The distinguishing factors are a level of interest totally out of proportion to the real importance of the subject, some individuals building personal careers from the pursuit and magnification of the issue, and the replacement of reasoned debate with witch hunts and hysteria.

Moral panics of recent memory include the Joseph McCarthy anti-communist witch hunts of the 1950s and the satanic ritual abuse allegations of the 1980s. And, more recently, we have witnessed a full-blown moral panic about pornography on the Internet. Sparked by the July 3, 1995, *Time* cover article "On a Screen Near You: Cyberporn," this moral panic has been perpetuated and intensified by a raft of subsequent media reports. As a result, there is now a widely held belief that pornography is easily accessible to all children using the Internet. This was also the judgment of Congress, which, proclaiming to be "protecting the children," voted overwhelmingly in 1996 for legislation to make it a criminal offense to send "indecent" material over the Internet into people's computers.

The original *Time* article was based on its exclusive access to Marty Rimm's *Georgetown University Law Journal* paper, "Marketing Pornography on the Information Superhighway." Although published, the article had not received peer review and was based on an undergraduate research project concerning descriptions of images on adult bulletin board systems in the United States. Using the information in this paper, *Time* discussed the type of pornography available online, such as "pedophilia (nude pictures of children), hebephilia (youths) and . . . images of bondage, sadomasochism, urination, defecation, and sex acts with a barnyard full of animals." The article proposed that pornography of this nature is readily available to anyone who is even remotely computer literate and raised the stakes by offering quotes from worried parents who feared for their children's safety. It also presented the possibility that pornographic material could be mailed to children without their parents' knowledge. *Time*'s example was of a ten-year-old boy who supposedly received pornographic images in his e-mail showing "10 thumbnail size pictures showing couples engaged in various acts of sodomy, heterosexual intercourse and lesbian sex." Naturally, the boy's mother was shocked and concerned, saying, "Children should not be subject to these images." *Time* also quoted another mother who said that she wanted her children to benefit from the vast amount of knowledge available on the Internet but was inclined not to allow access, fearing that her children could be "bombarded with X-rated pornography and [she] would know nothing about it."

From the outset, Rimm's report generated a lot of excitement—not only because it was reportedly the first published study of online pornography but also because of the secrecy involved in the research and publication of the article. In fact, the *New York Times* reported on July 24, 1995, that Marty Rimm was being investigated by his university, Carnegie Mellon, for unethical research and, as a result, would not be giving testimony to a Senate hearing on Internet pornography. Two experts from *Time* reportedly discovered serious flaws in Rimm's study involving gross misrepresentation and erroneous methodology. His work was soon deemed flawed and inaccurate, and *Time* recanted in public. With Rimm's claims now apologetically retracted, his original suggestion that 83.5 percent of Internet graphics are pornographic was quietly withdrawn in favor of a figure less than 1 percent.

Time admitted that grievous errors had slipped past their editorial staff, as their normally thorough research succumbed to a combination of deadline pressure and exclusivity agreements that barred them from showing the unpublished study to possible critics. But, by then, the damage had been done: the study had found its way to the Senate.

GOVERNMENT INTERVENTION

Senator Charles Grassley (Republican—Iowa) jumped on the pornography bandwagon by proposing a bill that would make it a criminal offense to supply or permit the supply of "indecent" material to minors over the Internet. Grassley introduced the entire *Time* article into the congressional record, despite the fact that the conceptual, logical, and methodological flaws in the report had already been acknowledged by the magazine.

On the Senate floor, Grassley referred to Marty Rimm's undergraduate research as "a remarkable study conducted by researchers at Carnegie Mellon University" and went on to say:

> The university surveyed 900,000 computer images. Of these 900,000 images, 83.5 percent of all computerized photographs available on the Internet are pornographic. . . . With so many graphic images available on computer networks, I believe Congress must act and do so in a constitutional manner to help parents who are under assault in this day and age.

Under the Grassley bill, later known as the Protection of Children from Pornography Act of 1995, it would have been illegal for anyone to knowingly or recklessly transmit indecent material to minors. This bill marked the beginning of a stream of Internet censorship legislation at various levels of government in the United States and abroad.

The most extreme and fiercely opposed of these was the Communications Decency Act, sponsored by former Senator James Exon (Democrat—Nebraska) and Senator Dan Coats (Republican—Indiana). The CDA labeled the transmission of "obscene, lewd, lascivious, filthy, indecent, or patently offensive" pornography over the Internet a crime. It was attached to the Telecommunications Reform Act of 1996, which was then passed by Congress on February 1, 1996. One week later, it was signed into law by President Clinton. On the same day, the American Civil Liberties Union filed suit in Philadelphia against the US Department of Justice and Attorney General Janet Reno, arguing that the statute would ban free speech protected by the First Amendment and subject Internet users to far greater restrictions than exist in any other medium. Later that month, the Citizens Internet Empowerment Coalition initiated a second legal challenge to the CDA, which formally consolidated with ACLU v. Reno. Government lawyers agreed not to prosecute "indecent" or "patently offensive" material until the three-judge court in Philadelphia ruled on the case.

Although the purpose of the CDA was to protect young children from accessing and viewing material of sexually explicit content on the Internet, the wording of the act was so broad and poorly defined that it could have deprived many adults of information they needed in the areas of health, art, news, and literature—information that is legal in print form. Specifically, certain medical information available on the Internet includes descriptions of sexual organs and activities which might have been considered "indecent" or "patently offensive" under the act—for example, information on breastfeeding, birth control, AIDS, and gynecological and urinological information. Also, many museums and art galleries now have websites. Under the act, displaying art like the Sistine Chapel nudes could be cause for criminal prosecution. Online newspapers would not be permitted to report the same information as is available in the print media. Reports on combatants in war, at the scenes of crime, in the political arena, and outside abortion clinics often provoke images or language that could be [considered] "offensive" and therefore illegal on the net. Furthermore, the CDA provided a legal basis for banning books which had been ruled unconstitutional to ban from school libraries. These include many of the classics as well as modern literature containing words that may be considered "indecent."

The act also expanded potential liability for employers, service providers, and carriers that transmit or otherwise make available restricted communications. According to the CDA, "knowingly" allowing obscene material to pass through one's computer system was a criminal offense. Given the nature of the Internet, however, making service providers responsible for the content of the traffic they pass on to other Internet nodes is equivalent to holding a telephone carrier responsible for the content of the conversations going over that carrier's lines. So, under the terms of the act, if someone sent an indecent electronic comment from a workstation, the employer, the e-mail service provider, and the carrier all could be potentially held liable and subject to up to $100,000 in fines or two years in prison.

On June 12, 1996, after experiencing live tours of the Internet and hearing arguments about the technical and economical infeasibility of complying with the censorship law, the three federal judges in Philadelphia granted the request for a preliminary injunction against the CDA. The court determined that "there is no evidence that sexually oriented material is the primary type of content on this new medium" and proposed that "communications over the Internet do not 'invade' an individual's home or appear on one's computer screen unbidden. Users seldom encounter content 'by accident.'" In a unanimous decision, the judges ruled that the Communications Decency Act would unconstitutionally restrict free speech on the Internet.

The government appealed the judges' decision and, on March 19, 1997, the US Supreme Court heard oral arguments in the legal challenge to the CDA, now known as Reno v. ACLU. Finally, on June 26, the decision came down. The Court voted unanimously that the act violated the First Amendment guarantee of freedom of speech and would have threatened "to torch a large segment of the Internet community."

Is the panic therefore over? Far from it. The July 7, 1997, *Newsweek,* picking up the frenzy where *Time* left off, reported the Supreme Court decision in a provocatively illustrated article featuring a color photo of a woman licking her lips and a warning message taken from the website of the House of Sin. Entitled "On the Net, Anything Goes," the opening words by Steven Levy read, "Born of a hysteria triggered by a genuine problem—the ease with which wired-up teenagers can get hold of nasty pictures on the Internet—the Communications Decency Act (CDA) was never really destined to be a companion piece to the Bill of Rights." At the announcement of the Court's decision, anti-porn protesters were on the street outside brandishing signs which read, "Child Molesters Are Looking for Victims on the Internet."

Meanwhile, government talk has shifted to the development of a universal Internet rating system and widespread hardware and software filtering. Referring to the latter, White House Senior Adviser Rahm Emanuel declared, "We're going to get the V-chip for the Internet. Same goal, different means."

But it is important to bear in mind that children are still a minority of Internet users. A contract with an Internet service provider typically needs to be paid for by credit card or direct debit, therefore requiring the intervention of an adult. Children are also unlikely to be able to view any kind of porn online without a credit card.

In addition to this, there have been a variety of measures developed to protect children on the Internet. The National Center for Missing and Exploited Children has outlined protective guidelines for parents and children in its pamphlet, "Child Safety on the Information Superhighway." A number of companies now sell Internet news feeds and web proxy accesses that are vetted in accordance with a list of forbidden topics. And, of course, there remain those blunt software instruments that block access to sexually oriented sites by looking for keywords such as sex, erotic, and X-rated. But one of the easiest solutions is to keep the family computer in a well-traveled space, like a living room, so that parents can monitor what their children download.

FACT OR MEDIA FICTION?

In her 1995 *CMC* magazine article, "Journey to the Centre of Cybersmut," Lisa Schmeiser discusses her research into online pornography. After an exhaustive search, she was unable to find any pornography, apart from the occasional commercial site (requiring a credit card for access), and concluded that one would have to undertake extensive searching to find quantities of explicit pornography. She suggested that, if children were accessing pornography online, they would not have been doing it by accident. Schmeiser writes: "There will be children who circumvent passwords, Surfwatch software, and seemingly innocuous links to find the 'adult' material. But these are the same kids who would visit every convenience store in a five-mile radius to find the one stocking *Playboy*." Her argument is simply that, while there is a certain amount of pornography online, it is not freely and readily available. Contrary to what the media often report, pornography is not that easy to find.

There is pornography in cyberspace (including images, pictures, movies, sounds, and sex discussions) and several ways of receiving pornographic material on the Internet (such as through private bulletin board systems, the World Wide Web, newsgroups, and e-mail). However, many sites just contain reproduced images from hardcore magazines and videos available from other outlets, and registration fee restrictions make them inaccessible to children. And for the more contentious issue of pedophilia, a recent investigation by the *Guardian* newspaper in Britain revealed that the majority of pedophilic images distributed on the Internet are simply electronic reproductions of the small output of legitimate pedophile magazines, such as *Lolita*, published in the 1970s.

Clearly the issue of pornography on the Internet is a moral panic—an issue perpetuated by a sensationalistic style of reporting and misleading content in newspaper and magazine articles. And probably the text from which to base any examination of the possible link between media reporting and moral panics is Stanley Cohen's 1972 book, *Folk Devils and Moral Panic*, in which he proposes that the mass media are ultimately responsible for the creation of such panics. Cohen describes a moral panic as occurring when "a condition, episode, person or group of persons emerges to become a threat to societal values and interests; . . . the moral barricades are manned by editors . . . politicians and other 'right thinking' people." He feels that, while problematical elements of society can pose a threat to others, this threat is realistically far less than the perceived image generated by mass media reporting.

Cohen describes how the news we read is not necessarily the truth; editors have papers to sell, targets to meet, and competition from other publishers. It is in their interest to make the story "a good read"—the sensationalist approach sells newspapers. The average person is likely to be drawn in with the promise of scandal and intrigue. This can be seen in the reporting of the *National Enquirer* and *People,* with their splashy pictures and sensationalistic headlines, helping them become two of the largest circulation magazines in the United States.

Cohen discusses the "inventory" as the set of criteria inherent in any reporting that may be deemed as fueling a moral panic. This inventory consists of the following:

Exaggeration in Reporting

Facts are often overblown to give the story a greater edge. Figures that are not necessarily incorrect but have been quoted out of context, or have been used incorrectly to shock, are two forms of this exaggeration.

Looking back at the original *Time* cover article, "On a Screen Near You: Cyberporn," this type of exaggeration is apparent. Headlines such as "The Carnegie Mellon researchers found 917,410 sexually explicit pictures, short stories and film clips online" make the reader think that there really is a problem with the quantity of pornography in cyberspace. It takes the reader a great deal of further exploration to find out how this figure was calculated. Also, standing alone and out of context, the oft quoted figure that 83.5 percent of images found on Usenet Newsgroups are pornographic could be seen as cause for concern. However, if one looks at the math associated with this figure, one would find that this is a sampled percentage with a research leaning toward known areas of pornography.

The Repetition of Fallacies

This occurs when a writer reports information that seems perfectly believable to the general public, even though those who know the subject are aware it is wildly incorrect. In the case of pornography, the common fallacy is that the Internet is awash with nothing but pornography and that all you need to obtain it is a computer and a modem. Such misinformation is integral to the fueling of moral panics.

Take, for example, the October 18, 1995, *Scotland on Sunday,* which reports that, to obtain pornographic material, "all you need is a personal computer, a phone line with a modem attached and a connection via a specialist provider to the Internet." What the article fails to mention is that the majority of pornography is found on specific Usenet sites not readily available from the major Internet providers, such as America Online and Compuserve. It also fails to mention that this pornography needs to be downloaded and converted into a viewable form, which requires certain skills and can take considerable time.

Misleading Pictures and Snappy Titles

Media representation often exaggerates a story through provocative titles and flashy pictorials—all in the name of drawing in the reader. The titles set the tone for the rest of the article; the headline is the most noticeable and important part of any news item, attracting the reader's initial attention. The recent *Newsweek* article is a perfect example. Even if the headline has little relevance to the article, it sways the reader's perception of the topic. The symbolization of images further increases the impact of the story. *Time*'s own images in its original coverage—showing a shocked little boy on the cover and, inside, a naked man hunched over a computer monitor—added to the article's ability to shock and to draw the reader into the story.

Through sensationalized reporting, certain forms of behavior become classified as deviant. Specifically, those who put pornography online or those who download it are seen as being deviant in nature. This style of reporting benefits the publication or broadcast by giving it the aura of "moral guardian" to the rest of society. It also increases revenue.

In exposing deviant behavior, newspapers and magazines have the ability to push for reform. So, by classifying a subject and its relevant activities as deviant, they can stand as crusaders for moral decency, championing the cause of "normal" people. They can report the subject and call for something to be done about it, but this power is easily abused. The *Time* cyberporn article called for reform on the basis of Rimm's findings, proclaiming, "A new study shows us how pervasive and wild [pornography on the Internet] really is. Can we protect our kids—and free speech?" These cries to protect our children affected the likes of Senators James Exon and Robert Dole, who took the *Time* article with its "shocking" revelations (as well as a sample of pornographic images) to the Senate floor, appealing for changes to the law. From this response it is clear how powerful a magazine article can be, regardless of the integrity and accuracy of its reporting.

The *Time* article had all of Cohen's elements relating to the fueling of a moral panic: exaggeration, fallacies, and misleading pictures and titles. Because certain publications are highly regarded and enjoy an important role in society, anything printed in their pages is consumed and believed by a large audience. People accept what they read because, to the best of their knowledge, it is the truth. So, even though the *Time* article was based on a report by an undergraduate student passing as "a research team from Carnegie Mellon," the status of the magazine was great enough to launch a panic that continues unabated—from the halls of Congress to the pulpits of churches, from public schools to the offices of software developers, from local communities to the global village.

Cyberpornography

Langdon Winner

Although Langdon Winner is a professor of information science and technology at a major high-tech university, Rensselaer Polytechnic Institute, his writings, including his 1987 book *The Whale and the Reactor,* continually question the uncritical enthusiasm with which new technologies are often accepted as cures for long-standing human problems—in this case, the problem of how we can best be intimate and loving with one another.

At the *Visions* Web Site Links to Langdon Winner's Web page; to *Mondo 2000.*

READING PROMPT

Winner seems not to be happy about the possibility of technology's improving human sexual relations. Why do you think he feels this way? Do you share any of his concerns?

It is no surprise to see the avant-garde falling head over heels in love with new technology. As early as 1910, a group of self-consciously modern Italian artists, the "futurists," proclaimed that the salvation of humankind lay in the speed and shiny metal of automobiles, airplanes, and industrial factories. Today, this well-worn conceit receives slick retread with the coming of cyberculture, a movement that links developments in R&D labs with the latest wrinkles in pop fashion. Leading the parade are three eye-catching magazines recently launched in the San Francisco Bay area. Together they suggest what the rising tide of future hype promises to wash ashore.

Most prominent of the group is *Wired,* a techie publication focusing on the world-transforming power of computers and telecommunications. Its pages are filled with news stories, interviews, and editorials praising interactive media, virtual reality, cinematic special effects, and hacker high jinx. Cybercult heroes such as sci-fi guru William Gibson and info-libertarian George Gilder issue lofty pronouncements to inspire the flock. Each new laboratory development is predicted to become an essential piece of people's lives in the new millennium. The writers in these magazines project a nervously "wired" world in which all individuals and institutions become nodules within global information systems.

A somewhat different take on the same themes animates *Mondo 2000,* the would-be *Vogue* of cybercult style. Along with news about trendy applications of electronic media, this magazine offers features on cyberpunk novels, performance artists, hip-hop music, brain implants, artificial life, and nanotechnology. Affecting an upscale Bohemian manner, *Mondo 2000* is less concerned with hardware and software than with anticipating how fashion-hungry youth will respond to the new gadgets in haircuts and hemlines. Despite their differences, *Wired* and *Mondo 2000* display a remarkable similarity in mood and presentation. Surrealistic images in dayglow colors telegraph an edge of excitement. On every page, a jumbled mix of typefaces and sizes makes sure the eye cannot rest. In both, the preferred prose style is panting hyperbole—often focused on predictions that the human body and electronic devices will merge in ecstatic unity. As a *Wired* article on "neuro-hackers" exclaims: "Hardwiring of neural prostheses is already here and will continue to develop toward completely implantable systems controlled by the user's brain." The logical extension of this throbbing enthusiasm appears in *Future Sex,* a publication devoted to the kinky side of the digital age. Its mission is to exhibit a rich smorgasbord of techno-erotic alternatives for bored, middle-class Americans: gender-bending sensuality, designer drugs, and electronically enhanced pleasures of every description. Interspersed among its articles are advertisements for 900-number telephone sex, new X-rated CD-ROMs, and endless varieties of hard core for your hard drive. I was interested to learn, for example, that the "data gloves" used in virtual-reality experiments to give participants a hands-on feel for computer-simulated worlds are now being tailored for other human organs as well. In the vision of the magazine's editor, Lisa Palac, tomorrow's sex will be androgynous, promiscuous, hungry for excess, and technologically mediated to achieve peak moments of "cyborgasm." So this is progress. Not long ago, seedy erotica was confined to run-down, dimly lit city neighborhoods. The coming of the VCR, of course, has brought hundreds of X-rated titles to the neighborhood video store, spawning a multi-billion-dollar growth industry that has altered the moral norms of the American middle class. Now the expanding cyberculture promises a veritable supermarket of high-definition, digital, interactive titillation, as Silicon Valley and a revitalized smut industry forge lucrative new bonds.

Is *Future Sex* merely an aberration, an attempt to stretch the ideas of cyberculture to uncharacteristic extremes? I don't think so. In fact, its underlying message is essentially similar to that of the other two new publications. All of them pander, bombarding their readers, with pure sensation for sensation's sake—a good working definition of pornography.

What makes these magazines at once fascinating and appalling is their blase amorality. One might suppose that the arrival of these powerful technologies for transforming human experience would be an occasion to ponder serious choices and select fruitful possibilities, as distinct from hideous degradations. But in today's cybercult thinking, outcomes are announced, not debated. The future pours forth with raw inevitability. For those in the middle of a cyborgasm, there is evidently little need to think.

chapter 7 Explorations

7a Conversation

Collaboration. Share a detailed account of an ongoing "conversation" (an exchange of information) with someone whom you could not talk to face to face. People in your group will have carried on these conversations in different ways; see if these individual approaches had any effect (or not) on the success of the conversations.

Web work. TalkCity is the product of a start-up company, LiveTalk, and aims to promote virtual communities of people chatting on the Web. Check out its information page for new users as well as its list of chat groups organized by community. The *Utne Reader* (source of several readings in this book) has long promoted the revival of old-fashioned semistructured evenings of conversation, called salons. As part of this effort, the magazine published the book *The Joy of Conversation* by Jaida n'ha Sandra, and has posted the Introduction on its Web site.

Yahoo! has its own chat area and lists more than 240 different Web chat services at Computers and Internet > Internet > World Wide Web > Chat.

One writing topic. Write an essay comparing traditional (face-to-face) spoken interchanges with any form of virtual conversation, from letter writing to the telephone to email.

7b Gathering Places

Collaboration. Make and share a list of places you have gone to over the years in hopes of seeing and meeting up with old friends or of meeting someone new.

Web work. People find themselves congregating in different places, depending on their ages and interests. One good people place for young urbanites today, maybe because it is inexpensive and quiet, hence a place where it is relatively easy to talk, is a local museum. Take a look at the eclectic list of some 400 American museums compiled by John Burke at the Oakland Museum of California, and see which museum you would select to visit in hopes of meeting an interesting new friend.

Yahoo! lists hundreds of smaller galleries under Business and Economy > Companies > Arts and Crafts > Galleries. Or, if you are truly not interested in meeting someone sensitive or demonstrating your own sensitivity, there are always "guy" sports museums (Recreation > Sports > History > Museums and Halls of Fame) and automotive museums (Recreation > Automotive > Museums).

One writing topic. Contrast the process of meeting someone new online with the process of doing so at a local hangout on your campus or in your community.

7c Courtship

Collaboration. Make and share a list of commonsense rules for meeting new people whom you may hope to date.

Web work. Start with Swoon, a Web site that focuses on dating, mating, and relating. Also check out The Field Guide to North American Males and the Web page for Ellen Fein and Sherrie Schneider's bestseller *The Rules.*

One writing topic. The "How-Not-To" format provides a ready means of constructing a satire, giving people in seemingly positive form the very advice they should avoid, as in the "How to flunk out of school" essay. Compose a "How-Not-To" essay on rules of courtship or anything else.

7d Netiquette

Collaboration. All forms of human communication require some codes of behavior to ensure smooth interaction between people. Make and share a list of the codes that refer to the everyday use of the telephone, including the rules that govern such things as answering a call at home or at someone else's home.

Web work. There's a Netiquette Home Page. Also, Larry Magid's Learn the Net covers netiquette and a little bit of everything else, as does Newbie-U.

One writing topic. Describe ways in which you feel netiquette extends or alters traditional patterns of courtesy and consideration. Are we expected to be more or less polite on the computer than in other situations?

7e Virtual Dating

Collaboration. One extraordinarily popular aspect of the Internet, almost since its inception, has been its ability to provide for online forums where people, often using assumed screen identities, can chat openly about anything with strangers from across the world. Make and share an anonymous list describing one or more situations in which you have felt able to discuss issues that you would not ordinarily talk about. Describe the situation where you felt you could be open, not what the issues themselves were.

Web work. Yahoo! lists more than 400 online matchmaking sites at Business and Economy > Classifieds > Personals. One of these sites is for Yenta, The Student.Net Matchmaker, a rare matchmaking site with a sense of humor; the site describes Yenta (the Yiddish word for a meddling old woman) as "Our latke-frying, matzoh ball–making matchmak[er]." Meanwhile Dean Esmay offers what he calls the Straight FAQ on how to meet people through online personal ads.

While neither the master Yahoo! site nor the two specific sites mentioned above contain X-rated images, the Web is awash with increasingly fee-based, hence controlled-access, adult-only sites with hard-core pornographic images that many people would find offensive. Unfortunately, links to the homepages for these sites are often embedded as paid advertisements in related sites, and thus the X-rated graphic images displayed as free samples on the homepages of these sites will regularly be no more than a keystroke away when you are viewing pages having anything to do with romance or dating.

One writing topic. The relationship between sex and technology is an especially difficult and demanding subject, fraught with all sorts of deeply held personal, religious, and social beliefs—some so powerful as to be labeled taboos. On the one hand, many people see sex as the most intimate expression of love and trust between two people, something to be shared directly and physically. This might be called the low-tech view of sex. On the other hand, many people (possibly including some of the same people just described) feel justified in isolating the physical pleasure involved with sex and finding ways, sometimes using technological aids that include the telephone and the computer, to reproduce, even enhance that pleasure. This might be called the high-tech view of sex.

How then do you personally feel about consenting adults who gain sexual pleasure through chatting via the computer or talking on the telephone? What do you believe should be society's interest, if any, in regulating what or how consenting adults communicate with each other over the telephone or the Internet? Should society have any interest whatsoever in encouraging or discouraging such communication?

7f Protecting Children

Collaboration. Many adults are concerned about the kinds and frequencies of images (still, moving, and aural) to which children are regularly exposed, especially on television, or the kinds of more blatantly sexual images they may stumble across or seek out—for example, on adult cable channels or, increasingly, on the Web. Make and share a list of all the different sources of images besides broadcast and cable television that people might see as potentially harmful to children. Which ones do you believe society has an obligation to regulate in order to protect children?

Web work. The Web grew out of the Internet, which for years prided itself on being a free and open forum in which adults (for years, almost exclusively college-educated adults) could talk about anything of interest. The Internet was a medium of public exchange free from the censorship and control that government regularly exerts over radio and television. Nothing arouses more passion on the Web than real or perceived threats of censorship, even though with the phenomenal popularity of the Web since 1994, Internet-based exchange is going into more and more homes and hence is readily available to more and more children.

The Electronic Frontier Foundation, founded in 1990 by Mitch Kapor and John Perry Barlow to protect free speech on the Internet and thus to combat censorship and other efforts at government control, claims to be one of the ten most active sites on the Web. The Center for Democracy and Technology is a more recent consortium of computer interest groups whose "mission is to develop and work for policies that advance civil liberties and democratic values."

Much of the debate over the last few years has centered on the Communications Decency Act (CDA). Passed by Congress in 1996, mainly as a means of protecting children, the CDA was ruled unconstitutional by the U.S. Supreme Court in June 1997 in a seven-to-two vote, with Judge John Paul Stevens writing the majority opinion.

There are relatively few pro-CDA sites on the Web. One is NCCIP, the National Campaign to Combat Internet Pornography, headed by Paul Cardin, formerly of Oklahomans for Children and Families. Conservatives found comfort in the July 1995 *Time* cover story on cyberporn; Donna L. Hoffman and Thomas P. Novak of Vanderbilt University host a Web site that offers extensive analysis of this controversy.

One writing topic. Write a letter to a relative explaining your advice about regulating Web access in their home for their young children; or write a report to the principal of your former elementary school about regulating Web access in the school library.

chapter 8

Shopping, Old and New

READING QUESTIONS

- What aspects of malls or large multipurpose stores do you like most? Like least?
- How has shopping most changed in your lifetime? What do you most like and least like about these changes?
- Where, when, and what kind of advertising do you find most helpful? Most offensive?

QUOTATIONS

America is a consumer culture, and when we change what we buy—and how we buy it—we'll change who we are.

— *Faith Popcorn (1991), American author, futurist*

We get a deal o' useless things about us, only because we've got the money to spend.

— *George Eliot (Marianne Evans) (1860), British novelist, 1819–1880*

I'm not just buying a car—I'm buying a lifestyle!

— *Attributed to Lynn Doblston (1988)*

Business is war.

— *Japanese motto*

To business that we love we rise betime,
And go to't with delight.

— *William Shakespeare,* Antony and Cleopatra

Home Alone?

This piece from the British weekly *The Economist* looks at the current interest in home shopping, tracing its origins back to early days of the Sears, Roebuck catalog. Although catalog shopping is still huge, new players include television shopping (led by the rival Home Shopping Network and QVC) and Web shopping.

At the *Visions* Web Site Links to the Direct Marketing Association; to E-Commerce sites.

READING PROMPT

What are the competing forces that have driven—and limited—home shopping in the past? That drive it and limit it today?

The idea of doing your shopping from home, rather than visiting a store, spans over a century of American commerce. In 1893 Sears, Roebuck mailed out the first edition of its catalogue, from which the country's farmers were able to kit themselves out with everything from guns to Sunday-best suits. Pioneers could even find themselves a wife in a catalogue. Yet the idea still has a futuristic feel to it: the 21st-century couple lounging on their shiny bubble chairs and dictating orders to a sonorously deferential computer.

Despite all the hype about multimedia, old-fashioned catalogues still dominate the home-shopping industry. They clocked up sales of almost $70 billion in America in 1995, according to the Direct Marketing Association (DMA), an industry group. Of that, some $43 billion was spent by consumers, the rest by businesses. In contrast, sales through infomercials or direct-response ads on television totalled $4.5 billion, $2.6 billion of which came from dedicated home-shopping channels. Sales of goods over the Internet and other online services were worth a mere $518 million, estimates Forrester Research, a consultancy in Cambridge, Massachusetts.

Catalogue shopping has been on a roll for the past decade and a half. The number of catalogues mailed in America rose from 8.7 billion in 1983 to 13.2 billion last year. The DMA forecasts that the industry's revenue will continue to grow at almost 7% a year for the rest of the decade, as it has since 1990. This growth is mostly driven by a large group of people spending increasing amounts on mail-order goods, rather than more people starting to buy them—59% of adult Americans already order from catalogues, and three-quarters of them spend more than $100 a year.

Clothing remains the most popular item sold through catalogues, followed closely by home furnishings. Computers and the software for them are also a big chunk of the business. Five of the top ten catalogue groups ranked by revenues are computer sellers. The biggest general catalogue retailer, J. C. Penney, which also has a chain of department stores, depends mostly on clothing and furnishings.

There are hitches. Last year, for example, the two main expenses of the catalogue business—postal charges and the cost of paper—both increased at the same time. But catalogue shopping still seems to have a number of things in its favor:

- Various social changes, such as more women in work and greater worries about violence on the streets, have boosted home shopping.
- State tax regulations mean that goods bought from a company in another state do not incur sales tax.
- New computer technology allows retailers to target their customers with more precision, sorting through databases of names and addresses to work out who is likely to want, say, golf clubs and who would prefer china figurines.
- As more companies shed central purchasing teams in bouts of downsizing, leaving individual departments to buy their own supplies independently, time-strapped managers are increasingly turning to catalogues. As a result, sales to businesses are growing rapidly; the DMA predicts that they will represent 39% of all catalogue sales in 2001.

In theory, all of these advantages should particularly benefit more modern forms of home shopping. But the TV-based version seems stuck in something of a rut. Sales of Home Shopping Network (HSN), one of America's two leading networks, have been stuck at a little over $1 billion a year for the past five years. Those of QVC, its big rival, have risen from $1.2 billion in 1993 to $1.5 billion in 1995.

Most of the pressure is now on HSN, which reported an operating loss in 1995 of $80 million. Its owner, Barry Diller, a Hollywood mogul who used to run QVC, has restructured the company: in the past year about 300 jobs have gone from HSN's headquarters in St Petersburg, Florida, where 1,800 telephonists and a large computer system can answer up to 50,000 calls an hour. Mr. Diller has also brought in a new chief executive, James Held, who has ditched many of HSN's clothing lines and focused the network around jewelry and cosmetics.

This reflects TV shopping's narrow audience. Some 85% of the shoppers are women, and the most popular category of goods is jewelry, which accounts for nearly 40% of the business. Other items include clothes, cosmetics and complicated fitness machines that are demonstrated by lots of healthy-looking presenters.

One reason that TV shopping has not prospered more is that, in a sense, it is a step backward from catalogues. The format—goods paraded past the camera one after another—favors impulse buying rather than a search for something particular; it also attracts people with enough time on their hands to sit watching until something appealing flickers by. The average length of time spent watching QVC before the first purchase is made is 36 hours. The network tried to move up market in 1994 with Q2, a second channel that sold more expensive goods, but its failure to catch on with yuppies led the firm to recast Q2 simply as a "best of QVC" channel, showing repeats of successes from the main channel.

The thought of avoiding TV shopping's defects is what makes online shopping so exciting to people such as Mr. Diller. Computers let customers navigate their way to the goods that they want when they want them. And their users certainly do not resemble stereotypical TV shoppers.

At present, the biggest part of the online market (around 25–30% of it) is made up of computer goods—hardware, software and books about them. Travel (20–25%) is the next biggest category. Forrester predicts that this mixture will stay much the same over the next four years, although the value of goods and services sold will grow to $6.6 billion by 2000. Other forecasters think the figure could be ten times this amount.

THE SEARCH FOR NET PROFITS

Both QVC and HSN have started online enterprises—iQVC and the Internet Shopping Network (ISN), respectively. ISN is selling about $1 million–worth of goods a month, vying with Amazon.com, a bookshop based in Washington state, and others as the most popular online merchant. At the moment ISN sells only computer goods, but Mr. Diller intends to change that. Interactive searching, he argues, is worthwhile only if there is a wide range of goods to forage through. Amazon lists over one million titles, more than any of its rivals, online or off. For its part, iQVC outdoes its television parent in the breadth of its stock, offering everything from power tools to pajamas.

If offering a broader range of goods than your rivals is one big competitive advantage in online shopping, another is excellence in what traditional retailers like to call "supply-chain management." Taking an order online will be easier—customers will do most of the data-entry work that firms' telephone operators currently have to do with catalogue—or TV—sales. But the warehousing and delivery of these goods will be fiendishly complex to organize.

Mr. Diller argues that this favors firms with experience in TV home shopping: fulfilling orders fast and accurately is what HSN and QVC are geared up to do. But there is an important difference. The TV-shopping channels can broadly control the time when products are sold (most sales happen either while or soon after the infomercial is shown): this means that they can buy job-lots of goods. With interactivity, orders may flow in over a much longer period. Rather than stock their warehouses with small runs of countless products, ISN and iQVC may be wiser to relay some orders directly to the manufacturer.

Amazon is an excellent example of why, even on the Internet, the intermediary still has a role. It distributes books for some 20,000 publishers, and guarantees its customers that it can deliver any book in print in the United States: 18 of its 106 staff deal with rare books (nearly a third are in warehousing and distribution). It keeps minimal stocks, at most a two-day supply of its 500 bestsellers in its warehouse; the rest it orders from publishers on demand. Some publishers say privately that they prefer selling through Amazon than through their own direct-sales operations. Unlike its slightly stale televisual rivals, Amazon offers interviews with authors. In short, it tries to make buying books easier for both consumers and manufacturers.

A century ago, Richard Sears, a station telegrapher, used his country's new railways to distribute goods and build up his business. But Sears, Roebuck's strength was its range of merchandise and a solid reputation for quality, bolstered by generous guarantees. Even in today's online world, that is still a winning formula.

"Gimme Stuff" *Molly Ivins*

Sharp-witted Molly Ivins is a best-selling author and widely syndicated political columnist for the *Fort Worth Star–Telegram* who takes great delight in puncturing cant and hypocrisy, as she does in the piece here. Politics, she says, especially in Texas, is great entertainment—"better than the zoo, better than the circus, rougher than football, and even more aesthetically satisfying than baseball." Her most recent column is available from the *Fort Worth Star–Telegram* homepage.

At the *Visions* Web Site Links to mall sites, including the Mall of America.

READING PROMPT

"The Mall of America will be Stonehenge for future archeologists," says Ivins. Explain.

Gosh, what an exciting time we've been having lately. All that good news—South Africa, the Middle East, a ban on assault weapons. All that bad news—Paula Jones, CEO salaries, health care in trouble. And I, with my eye ever on the Big Picture, am hipped on the Mall of America.

Great Caesar's armpit! Holy gamoley! This is it. Our Pyramids, our Colosseum, our Chartres. The defining signature of our civilization. The World's Largest Shopping Mall. (Someone wrote to tell me there's a bigger one in Edmonton, but I don't believe it; understatement is the Canadian national art form. Besides, leaving the Second Largest Shopping Mall in the World as our enduring monument would be even better.) The Mall of America will be Stonehenge for future archeologists.

Those of you who think if you've seen one mall, you've seen them all, have no idea what a Mecca for materialism can actually look like. What it reminds me of most was the occasion, in upstate New York, when I saw a thousand-pound cheese go by in a Dairy Days parade. " My," I thought, "that's certainly something." Las Vegas has the same effect on me, and so does the Funeral Museum in Houston. But for scale, only the thousand-pound cheese rivals the Mall of America.

In the middle of the Mall of America, either the World's Biggest or Second-Biggest Shopping Mall, is a theme park called Camp Snoopy, based on the characters in Peanuts. Somewhere in Camp Snoopy, a large cartoon of Snoopy himself is thinking, "Good grief!"

I'll say.

I believe the Mall of America is an art form. Haute nice. Not just nice, but nice at a high level; niciest. Okay, so nice is a tepid virtue; it still beats not-nice. I'd rather have johns sanitized-for-my-protection than quaint, odoriferous European pee-holes—I'm an American. And how can one resist the Maison du Popcorn?

I know Minnesotans get tired of being stereotyped as nice, but they are. Once, for journalistic purposes, I rode with the Minnesota Hell's Angels. They were awfully nice. Minnesotans like to hang out at the Mall of America because they believe it is such a cross-section—they say people fly to the Minneapolis airport from New York and Ecuador and go straight to the Mall. Actually, I think most of them come from Davenport. It's hard to be ethnic in the Land of the Seven-Jello Marshmallow Cottage Cheese Surprise. The state dish of Minnesota is the tuna–noodle casserole. For some reason, even Italian food and Mexican food taste like tuna–noodle casserole in Minnesota.

The Mall of America is full of stuff. There are more subspecialties of stuff than most people on Earth have ever dreamed of. Stores that sell nothing but socks. Stores that sell nothing but hair ribbons. Stores that sell nothing but stuff to put your stuff in. Of course, it's appalling, this monument to materialism, but it is full of nice people.

If you think about it, the Mall of America probably does represent a net advance for civilization. The ancient Egyptians were so materialistic their pharaohs built pyramids on the theory they could take their stuff with them. The French built Versailles on the theory that only kings should enjoy the finest stuff. We have achieved the democratization of stuff and don't expect to take it with us, which is why so much of it is disposable.

And how should we, the descendants of Puritans, feel about leaving this monument to materialism as our civilization's signature? Well, the Puritans were mighty materialistic themselves. And our own President, in the course of announcing most-favored-nation trade status for China, clearly enunciated the basis of our foreign policy—trade before human rights. Money before principle. So our epitaph shall read We Believed in the Almighty Dollar.

I find this less than inspiring, but it does open up new possibilities for empire. If indeed what makes America great is mountains of stuff—instant mashed potatoes, color television, cordless electric flour sifters—why not conquer the world with stuff instead of bombs? Why not send C-5As laden with stuff to Somalia, Rwanda, Bosnia, and Iraq? Each box of stuff should be attached to its own little Mylar parachute, and all of it can float gently to Earth. Thousands and thousands of sweater boxes, lemon zesters, pink lemonade, striped toothpaste, vacuum cleaners for mini-blinds, tortilla holders, VCRs that flush your toilet, aprons that say World's Best Dad, three-tier patisserie stands, daisy weeders, plastic storks, Happy Birthday teapots, fountains of little boys whizzing through their wee-wees, Malcolm X gimme caps. . . .

Sorry, I got carried away.

Semper stuff. E pluribus stuff. Ave atque stuff.

Wal-Mart's War on Main Street

Sarah Anderson

Wal-Mart was one of the great success stories of the 1980s, with Arkansas retailer Sam Walton becoming the richest man in the United States as he personally opened Wal-Mart discount department stories on large lots outside of small towns. Critics such as Sarah Anderson argue that Wal-Mart pulled business away from traditional downtown merchants and thus eroded the special quality of small-town life. Wal-Mart backers make the counterclaim that all this happened because for the first time the stores offered people in small towns discount prices long available to people in larger cities.

At the *Visions* Web Site Links to the Wal-Mart homepage; to Sprawl-Busters, a site organized to help communities fight all megastores.

READING PROMPT

How does Anderson see the battle with Wal-Mart as part of a larger battle over Main Street and small-town life generally?

The basement of Boyd's for Boys and Girls in downtown Litchfield, Minnesota, looks like a history museum of the worst in children's fashions. All the real duds from the past forty years have accumulated down there: wool pedal-pushers, polyester bell-bottoms, wide clip-on neckties. There's a big box of 1960s faux fur hats, the kind with the fur pompom ties that dangle under a girl's chin. My father, Boyd Anderson, drags all the old stuff up the stairs and onto the sidewalk once a year on Krazy Daze. At the end of the day, he lugs most of it back down. Folks around here don't go in much for the retro look.

At least for now, the museum is only in the basement. Upstairs, Dad continues to run one of the few remaining independent children's clothing stores on Main Street, USA. But this is the age of Wal-Mart, not Main Street. In 1994, the nation's top retailer plans to add 110 new U.S. stores to its current total of 1,967. For every Wal-Mart opening, there is more than one store like Boyd's that closes its doors.

Litchfield, a town of 6,200 people sixty miles west of Minneapolis, started losing Main Street businesses at the onset of the farm crisis and the shopping-mall boom of the early 1980s. As a high-school student during this time, I remember dinner-table conversation drifting time and again toward rumors of store closings. In those days, Mom frequently cut the conversation off short. "Let's talk about something less depressing, okay?"

Now my family can no longer avoid the issue of Main Street Litchfield's precarious future. Dad, at sixty-eight, stands at a crossroads. Should he retain his faith in Main Street and pass Boyd's down to his children? Or should he listen to the pessimists and close up the forty-one-year-old family business before it becomes obsolete?

For several years, Dad has been reluctant to choose either path. The transition to retirement is difficult for most people who have worked hard all their lives. For him, it could signify not only the end of a working career, but also the end of small-town life as he knows it. When pressed, Dad admits that business on Main Street has been going downhill for the past fifteen years. "I just can't visualize what the future for downtown Litchfield will be," he says. "I've laid awake nights worrying about it because I really don't want my kids to be stuck with a business that will fail."

I am not the aspiring heir to Boyd's. I left Litchfield at eighteen for the big city and would have a tough time readjusting to small-town life. My sister Laurie, a nurse, and my sister-in-law Colleen, who runs a farm with my brother Scott, are the ones eager to enter the ring and fight the retail Goliaths. Both women are well suited to the challenge. Between them, they have seven children who will give them excellent tips on kids' fashions. They are deeply rooted in the community and idealistic enough to believe that Main Street can survive.

My sisters are not alone. Across the country, thousands of rural people are battling to save their local downtowns. Many of these fights have taken the form of anti-Wal-Mart campaigns. In Vermont, citizens' groups allowed Wal-Mart to enter the state only after the company agreed to a long list of demands regarding the size and operation of the stores. Three Massachusetts towns and another in Maine have defeated bids by Wal-Mart to build in their communities. In Arkansas, three independent drugstore owners won a suit charging that Wal-Mart had used "predatory pricing," or selling below cost, to drive out competitors. Canadian citizens are asking Wal-Mart to sign a "Pledge of Corporate Responsibility" before opening in their towns. In at least a dozen other U.S. communities, groups have fought to keep Wal-Mart out or to restrict the firm's activities.

By attacking Wal-Mart, these campaigns have helped raise awareness of the value of locally owned independent stores on Main Street. Their concerns generally fall in five areas:

- Sprawl Mart—Wal-Mart nearly always builds along a highway outside town to take advantage of cheap, often unzoned land. This usually attracts additional commercial development, forcing the community to extend services (telephone and power lines, water and sewage services, and so forth) to that area, despite sufficient existing infrastructure downtown.
- Wal-Mart channels resources out of a community—studies have shown that a dollar spent on a local business has four or five times the economic spin-off of a dollar spent at a Wal-Mart, since a large share of Wal-Mart's profit returns to its Arkansas headquarters or is pumped into national advertising campaigns.
- Wal-Mart destroys jobs in locally owned stores—a Wal-Mart-funded community impact study debunked the retailer's claim that it would create a lot of jobs in Greenfield, Massachusetts. Although Wal-Mart planned to hire 274 people at its Greenfield store, the community could expect to gain only eight net jobs, because of projected losses at other businesses that would have to compete with Wal-Mart.
- Citizen Wal-Mart?—in at least one town—Hearne, Texas—Wal-Mart destroyed its Main Street competitors and then deserted the town in search of higher returns elsewhere. Unable to attract new businesses to the devastated Main Street, local residents have no choice but to drive long distances to buy basic goods.
- One-stop shopping culture—in Greenfield, where citizens voted to keep Wal-Mart out, anti-Wal-Mart campaign manager Al Norman said he saw a resurgence of appreciation for Main Street. "People realized there's one thing you can't buy at Wal-Mart, and that's small-town quality of life," Norman explains. "This community decided it was not ready to die for a cheap pair of underwear."

So far Litchfield hasn't been forced to make that decision. Nevertheless, the town is already losing at least some business to four nearby Wal-Marts, each less than forty miles from town. To find out how formidable this enemy is, Mom and I went on a spying mission to the closest Wal-Mart, twenty miles away in Hutchinson.

Just inside the door, we were met by a so-called Wal-Mart "greeter" (actually the greeters just say hello as they take your bags to prevent you from shoplifting). We realized we knew her. Before becoming a greeter, she had been a cashier at a downtown Litchfield supermarket until it closed early this year. I tried to be casual when I asked if she greets many people from Litchfield. "Oh, a-a-a-ll the time!" she replied. Sure enough, Mom immediately spotted one in the checkout line. Not wanting to look too suspicious, we moved on toward the children's department, where we discreetly examined price tags and labels. Not all, but many items were cheaper than at Boyd's. It was the brainwashing campaign that we found most intimidating, though. Throughout the store were huge red, white, and blue banners declaring BRING IT HOME TO AMERICA. Confusingly, the labels on the children's clothing indicated that they had been imported from sixteen countries, including Haiti, where an embargo on exports was supposed to be in place.

Of course, Wal-Mart is not Main Street's only foe. Over coffee at the Main Street Cafe, some of Litchfield's long time merchants gave me a litany of additional complaints. Like my dad, many of these men remember when three-block-long Main Street was a bustling social and commercial hub, with two movie theaters, six restaurants, a department store, and a grand old hotel.

Present-day Litchfield is not a ghost town, but there are four empty storefronts, and several former commercial buildings now house offices for government service agencies. In recent years, the downtown has lost its last two drugstores and two supermarkets. As a result, elderly people who live downtown and are unable to drive can no longer do their own shopping.

My dad and the other merchants place as much blame for this decline on cutthroat suppliers as on Wal-Mart. The big brand names, especially, have no time anymore for small clients. Don Brock, who ran a furniture store for thirty-three years before retiring in 1991, remembers getting an honorary plaque from a manufacturer whose products he carried for many years. "Six months later I got a letter saying they were no longer going to fill my orders."

At the moment, Litchfield's most pressing threat is a transportation department plan to reroute the state highway that now runs down Main Street to the outskirts of town. Local merchants fear the bypass would kill the considerable business they now get from travelers. Bypasses are also magnets for Wal-Mart and other discounters attracted to the large, cheap, and often unzoned sites along the bypass.

When I asked the merchants how they felt about the bypass, the table grew quiet. Greg Heath, a florist and antique dealer, sighed and said, "The bypass will come—it might be ten years from now, but it will come. By then, we'll either be out of business or the bypass will drive us out." The struggles of Main Street merchants have naturally created a growth industry in consultants ready to provide tips on marketing and customer relations. Community-development experts caution, though, that individual merchants acting on their own cannot keep Main Street strong. "Given the enormous forces of change, the only way these businesses can survive is with active public and government support," says Dawn Nakano, of the National Center for Economic Alternatives in Washington, D.C.

Some of the most effective efforts at revitalization, Nakano says, are community development corporations—private, nonprofit corporations governed by a community-based board and usually funded in part by foundation and government money. In Pittsburgh, for example, the city government and about thirty nonprofit groups formed a community development corporation to save an impoverished neighborhood where all but three businesses were boarded up. Today, thanks to such financing and technical assistance, the area has a lively shopping district. Although most community development corporations have been created to serve low-income urban neighborhoods, Nakano feels that they could be equally effective in saving Main Streets. "There's no reason why church, civic, and other groups in a small town couldn't form a community development corporation to fill boarded-up stores with new businesses. Besides revitalizing Main Street, this could go a long way towards cultivating a 'buy local' culture among residents."

The National Main Street Center, a Washington, D.C.–based nonprofit, provides some of the most comprehensive Main Street revitalization services. The Center has helped more than 850 towns build cooperative links among merchants, government, and citizens. However, the Center's efforts focus on improving marketing techniques and the physical appearance of stores, which can only do so much to counter the powerful forces of change.

No matter how well designed, any Main Street revitalization project will fail without local public support. Unfortunately, it is difficult for many rural people to consider the long-term, overall effects of their purchases, given the high levels of rural unemployment, job insecurity, and poverty. If you're worried about paying your rent, you're not going to pay more for a toaster at your local hardware store, no matter how much you like your hometown.

Another problem is political. Like those in decaying urban neighborhoods, many rural people have seen the signs of decline around them and concluded that they lack the clout necessary to harness the forces of change for their own benefit. If you've seen your neighbors lose their farms through foreclosure, your school close down, and local manufacturing move to Mexico, how empowered will you feel? Litchfield Mayor Ron Ebnet has done his best to bolster community confidence and loyalty to Main Street. "Every year at the Christmas lighting ceremony, I tell people to buy their gifts in town. I know everyone is sick of hearing it, but I don't care." Ebnet has whipped up opposition to the proposed bypass, with strong support from the city council, chamber of commerce, the newspaper editor, and the state senator. He also orchestrated a downtown beautification project and helped the town win a state redevelopment grant to upgrade downtown businesses and residences.

Ebnet has failed to win over everyone, though. Retired merchant Don Larson told me about a local resident who drove forty miles to get something seventeen cents cheaper than he could buy it at the Litchfield lumberyard. "I pointed out that he had spent more on gas than he'd saved, but he told me that 'it was a matter of principle.' I thought, what about the principle of supporting your community? People just don't think about that, though."

Mayor Ebnet agrees, "Many people still have a 1950s mentality," he says. "They can't see the tremendous changes that are affecting these small businesses. People tell me they want the bypass because there's too much traffic downtown and they have a hard time crossing the street. And I ask them, but what will you be crossing to? If we get the bypass, there will be nothing left!"

Last summer, with the threat of the bypass hanging over his head, Dad became increasingly stubborn about making a decision about the store. His antique Underwood typewriter was never more productive, as it banged out angry letters to the state transportation department. My sisters decided to try a new tactic. While my parents were on vacation, they assaulted the store with paintbrushes and wallpaper, transforming what had been a rather rustic restroom and doing an unprecedented amount of redecorating and rearranging.

The strategy worked. "At first, Dad was a bit shocked," Laurie said. "He commented that in his opinion, the old toilet paper dispenser had been perfectly fine. But overall he was pleased with the changes, and two days later he called for a meeting with us and our spouses." "Your dad started out by making a little speech," Colleen said. "The first thing he said was, 'Well, things aren't how they used to be.' Then he pulled out some papers he'd prepared and told us exactly how much sales and profits have been over the years and what we could expect to make. He told us what he thinks are the negative and the positive aspects of the job and then said if we were still interested, we could begin talking about a starting date for us to take over."

Dad later told me, "The only way I could feel comfortable about Laurie and Colleen running the store is if it was at no financial risk to them. So I'm setting up an account for them to draw from—enough for a one-year trial. But if they can't make a good profit, then that's it—I'll try to sell the business to someone else. I still worry that they don't know what they're getting themselves into. Especially if the bypass goes through, things are going to be rough."

My sisters are optimistic. They plan to form a buying cooperative with Main Street children's clothing stores in other towns and have already drafted a customer survey to help them better understand local needs. "I think we're going to see a big increase in appreciation of the small-town atmosphere,"

Colleen says. "There are more and more people moving to Litchfield from the Twin Cities to take advantage of the small-town way of life. I think they might even be more inclined to support the local businesses than people who've lived here their whole lives and now take the town for granted."

Small towns cannot return to the past, when families did all their shopping and socializing in their hometown. Rural life is changing and there's no use denying it. The most important question is, who will define the future? Will it be Wal-Mart, whose narrow corporate interests have little to do with building healthy communities? Will it be the department of transportation, whose purpose is to move cars faster? Will it be the banks and suppliers primarily interested in doing business with the big guys? Or will it be the people who live in small towns, whose hard work and support are essential to any effort to revitalize Main Street? In my hometown, there are at least two new reasons for optimism. First, shortly before my deadline for this article, the Minnesota transportation department announced that it was dropping the Litchfield highway bypass project because of local opposition. (My dad's Underwood will finally get a rest.) The second reason is that a new teal green awning will soon be hanging over the front of Boyd's—a symbol of one family's belief that Main Street, while weary, is not yet a relic of the past.

Marketing Madness

Laurie Ann Mazur

Consumer culture really does go a long way toward defining much of what is best in our age, at least according to many. It's lively, energetic, colorful, especially as compared to the dull stolid command economies of the Communist countries over which consumer culture triumphed as the final act of the cold war. What then is wrong with it? Why the complaints of writers such as Laurie Ann Mazur?

At the *Visions* Web Site Links to The Ad Council, which sees itself as advertising's other, civic-minded side, (conducting advertising campaigns for the public good); to the Clio Awards (what advertisers see as their best work).

READING PROMPT

What aspects of advertising most concern Mazur? Which of her points do you find most and least compelling?

A few years ago, a company called Space Marketing, Inc. (SMI) came up with a plan to send a mile-long billboard into space. Coated with reflective plastic, the billboard would beam down a corporate logo that appeared as large as the moon, and as it orbited the Earth, would be visible to every single person on the planet.

To marketers, it was a dream come true: a truly inescapable form of advertising. It couldn't be tossed out with the junk mail, hung up on, or zapped with a remote control. To the rest of us who'd heard about it, it seemed more like a nightmare. Amid howls of protest, SMI withdrew the plan, but not before several companies had inquired about launching their logos into space.

Space may be the final frontier for advertisers—because the Earth is already taken. In the last 20 years, advertising has become far more pervasive than ever before. Advertising budgets in the United States have doubled since 1976, and they've grown by more than 50 percent in just the last 10 years. Companies now spend about $162 billion each year to bombard us with print and broadcast ads; that works out to about $623 for every man, woman and child in the United States. It's important to remember that we're the ones picking up the tab for ad costs, in the form of higher prices for the products the ads promote. We also pay higher taxes, because advertising costs are deductible from the bottom line of corporate taxable profits, which would otherwise be higher.

Skyrocketing ad budgets are both a cause and a consequence of a phenomenon marketers call "clutter," resulting from airwaves so clogged with ads already that it gets harder and harder to attract our attention. So, to prevail in this ad-cluttered world, marketers have become more intrusive than ever before.

They've also gotten more sneaky. In recent years, advertisers have pioneered many forms of "stealth" advertising—ads disguised as something else, or placed where we least expect to encounter them. One form of stealth advertising is "product placement"—paying to get brand-name products featured in movies. For large cash payments, advertisers can actually get scripts rewritten to showcase their products.

Another form of stealth advertising is the "video news release," or VNR, an insidious form of promotion indeed. Advertisers produce brief videotapes (VNRs) that look for all the world like regular news stories—except that they feature a product or a corporation. These are then distributed to news stations throughout the land, which typically air them without attribution. For example, a few years ago 17 million Americans watched a "news" story about the 50th anniversary of Cheerios cereal. It was a lighthearted bit of human-interest fluff, featuring a tour of the Cheerios factory and some footage of a giant Cheerio made specially for the occasion.

But few viewers realized that the story was conceived, dramatized, filmed and distributed by Cheerios manufacturer General Mills itself. VNRs give corporations an unparalleled opportunity to define and interpret current events, and they are cheaper than regular ads. So it is not surprising that VNRs have become increasingly common: A 1993 Nielsen study found that every single news station surveyed used VNRs—and less than half identified their source during the broadcast.

CORPORATE CULTURE

Also in response to clutter, advertisers have taken over more of what used to be ad-free (or at least ad-lite). Public broadcasting (PBS) stations now run sponsor acknowledgements (called "enhanced underwriting") that are virtually indistinguishable from ads on commercial TV. Museums have become shrines to the products of their corporate benefactors: Upscale shoemaker Ferragamo, for example, paid for an exhibit of its own shoes at the Los Angeles County Museum of Art. Just about every cultural event has a corporate sponsor—and those sponsors have a lot of influence over their beneficiaries. Country singer Barbara Mandrell collected $15 million to promote No Nonsense pantyhose, and obligingly named her next album *No Nonsense.*

In fact, it's now possible to sponsor an entire American city. In 1993, the city of Atlanta hired Joel Babbitt, a former advertising executive, to help the city sell itself. Babbitt came up with a plan to rename streets and parks for corporate sponsors, implant ads in city sidewalks, and plaster corporate logos on the sides of city garbage trucks. To reassure those naysayers who thought he might be going too far, Babbitt announced that not just any corporate sponsor would be welcome—he would draw the line at firearms and sexual products. This was a great relief to those who envisioned products like "the official assault rifle of Atlanta" or condoms imprinted with the city seal.

DICTATING CONTENT

Yes, it's annoying and absurd—but advertising's takeover of our cultural airwaves is more than an aesthetic affront. It also affords corporations significant control over the content of the media that shape our world view. The news and entertainment media are wholly dependent on ad revenues, and advertisers wield considerable influence. In a 1992 study, virtually all of the 150 newspaper editors surveyed said that advertisers tried to dictate editorial content, and 37 percent said they succeeded.

Appalling examples abound. In 1993, Mercedes Benz wrote to 30 magazines, requiring them to pull all Mercedes ads from any issue containing articles critical of the company, German products, or Germany itself. RJR Nabisco, the giant food-and-tobacco conglomerate, canceled an $80 million contract with a New York advertising agency that produced ads for Northwest Airlines' no-smoking policy. And the cosmetic firm Revlon pulled its ads from *Ms.* magazine after *Ms.* ran a cover story about Soviet women exiled for publishing underground feminist books. The reason? The Soviet women on the magazine's cover were not wearing makeup.

Often, censorship is self-inflicted by media decision-makers fearful of losing their corporate sponsors. This has produced a curious double standard in journalism, in which the news media go easier on corporations than on the government. As Bill Lazarus, a reporter for the *Hammond* (Indiana) *Times,* explains: "When you write about government, the attitude of [editors] tends to be 'no holds barred.' When you write about business, the attitude tends to be one of caution. And for businesses [that] happen to be advertisers, the caution turns frequently to timidity."

IS NOTHING SACRED?

Advertisers have also invaded other important cultural institutions: public schools. Of course, there has always been some advertising in the schools. Remember those scratchy filmstrips (early VNRs, actually) with titles like Aluminum and You? Those were primitive versions of in-school advertising. But in recent years, advertisers have become more of a presence in schools than ever before: Ads are now plastered in hallways, piped in over public address systems and painted on the sides of school buses. And students are big target recipients of free product samples.

Some eight million kids are required to watch Channel One, an ad-punctuated news show, in school every day. Channel One hooks its Board of Education customers by giving video equipment to schools that agree to broadcast its daily program. But its *raison d'etre* is to provide a vehicle for advertising: Channel One takes in about $800,000 a day in ad revenues.

Channel One has received a lot of negative publicity, but it is not even the most egregious example of advertising in the schools. Corporations are now actually writing lesson plans that are used to teach kids in school. These corporate curricula, which are often slick and expensively produced, prove irresistible to cash-starved school districts and overworked teachers.

To marketers, these "stealth" materials are vehicles for reaching a captive young audience. One company that produces lesson plans for corporations effuses about the benefits of advertising to kids in school: "Let Lifetime Learning Systems bring your message to the classroom, where young people are forming attitudes that will last a lifetime. . . . Coming from school, all these materials carry an extra measure of credibility that gives your message added weight." Hundreds of companies and associations have hired Lifetime to peddle their wares (or ideologies) in schools.

Some corporate lesson plans have educational value, but they are mostly used to encourage consumption of companies' products and support their interests. For example, a lesson plan designed by Georgia Pacific puts the best face on forestry and defends the company's practice of clearcutting forests. A classroom science video produced by Exxon praises the company for its cleanup of the Exxon Valdez oil spill. And a lesson plan by Mobil Oil encouraged students to adopt the company's favorable views on the North American Free Trade Agreement. A recent study by Consumers' Union found that fully 68 percent of corporate teaching materials contained biased information.

ADVERTISING AND THE CONSUMER CULTURE

Advertising, then, sells more than products. It also promotes the interests and ideology of its corporate sponsors. And it promotes a way of life; indeed, it might be considered the Ministry of Propaganda of the consumer culture.

Back in the early days of the modern advertising industry, advertisers had a kind of missionary zeal about promoting the consumption of commodities. Corporate leaders of the 1920s believed that a consumer society would serve two ends: It would quell labor unrest, and it would create larger markets for the surplus fruits of mass production. But many feared that it simply wouldn't work; that workers could not be induced to buy products as quickly as they rolled off the assembly line.

That's where advertising came in. Nineteenth-century sales techniques, which emphasized the quality of the products being sold, were not equal to the job of creating limitless demand for consumer goods. So advertisers came up with ads that had less to do with products than with their audience. Their ads sought to make viewers feel self-conscious, inadequate, unlovable—and then offered a commodified remedy. As one marketer wrote in the trade journal *Printers' Ink,* "Advertising helps to keep the masses dissatisfied with their mode of life, discontented with ugly things around them. Satisfied customers are not as profitable as discontented ones."

Over the course of this century, ads have drifted further away from describing the product, appealing instead to our deepest, unarticulated desires. An ad for Quaker Oats tells us to eat its cereal not because it tastes good, but because "It's the right thing to do"—speaking to our need for moral compass in a confusing world. An ad for Jordache Basics consists of several mezzotint photographs of a playful young couple who are, almost incidentally, wearing Jordache jeans. The copy reads: "We share the same goals and directions. There's no fighting for space and time. It's a place to change, or not to."

The modern advertising industry has succeeded beyond the wildest dreams of its early proponents. In the ad-saturated world of late 20th-century America, we buy more and save less than any society before us. Advertising has won an important psychological victory as well: To some degree, most of us have swallowed the dominant message of advertising—that life's problems can be solved by buying things.

There are costs—personal and social—to believing that life's problems can be solved with a credit card. On a personal level, we have become a nation of debtors. The average American now owes $2,500 in credit card debt, and pays about $450 in interest every year. Personal bankruptcy rates are soaring, too: almost 900,000 people went bankrupt in 1992, almost three times as many as in 1985.

On a social level, the costs of the consumer society include poisoned air and water, the breathtaking destruction of wilderness areas, and landfills clogged with products designed to be discarded. Advertising, and the industries it serves, has fostered a culture of waste in which rapidly changing styles render products obsolete long before their useful life is extinguished. Earnest Elmo Calkins, a pioneer of "planned obsolescence," once declared: "We no longer wait for things to wear out. We displace them with others that are not more effective but more attractive."

The culture of waste means that contemporary Americans consume the Earth's resources at a pace unprecedented in history. Between 1940 and 1976, we consumed more minerals than did all of humanity up to that point. Each American consumes about 18 times as much commercially produced energy as a person living in Bangladesh—and about twice as much as our counterparts in Europe and Japan.

Certainly, the industries that feed the consumer society could use resources more efficiently and have less impact on the environment. Because we use twice as much energy as other industrialized countries, for example, it is safe to assume that we could continue to provide the same array of consumer bells and whistles while halving the environmental impact of our fossil fuel use. But many environmentalists, while embracing the prospects of more efficient resource use, believe that environmental sustainability is fundamentally incompatible with the culture of waste we call consumerism. As Alan Durning observed in *How Much Is Enough? The Consumer Society and the Fate of the Earth*, "The furnishings of our consumer lifestyle—things like automobiles, throwaway goods and packaging, a high-fat diet, and air conditioning—can only be provided at great environmental cost. Our way of life depends on enormous and continuous inputs of the very commodities that are most damaging to the Earth to produce: energy, chemicals, metals and paper." Durning concludes that the challenge before us is to learn to live "by sufficiency rather than excess."

So what can we do about it? Certainly, we can't abolish the advertising industry with its global power and economic clout, but we can keep it in its place. Consumers need to understand their ability to force a corporate response. A major aim of advertising is to enhance a sponsor's image, and even a small-scale consumer action in response to commercial exploitation—like a boycott or press conference—is threatening to that image. These tools have been used by both the right—through such conservative watchdog groups as Accuracy in Media (AIM)—and the left, which counters with Fairness and Accuracy In Reporting (FAIR).

We can also rip the lid off phony "sponsorships," endorsements and VNRs that are really thinly veiled commercials. We need to fight "enhanced" corporate sponsorship of TV and radio shows, museums and cultural events. And we can prevent corporate incursions into our public space by nurturing the nonprofit sector. To help ensure that commercials and print ads bear some relationship to the truth, we can push for comprehensive Honesty in Advertising legislation, on both the state and federal level (with fines and the harsh glare of publicity for offenders).

We can also work to keep advertising out of the public schools. If you have kids, work with your PTA and school board to make sure their teachers are not using corporate curricula. Teach your children about the ways of marketers; watch ads together and help them understand their purpose and method. Even if you don't have kids, you can work to ensure that local schools have the resources they need to resist the temptation of advertisers bearing "free" gifts.

On a personal level, we can organize boycotts of products whose advertising methods or messages offend us. Encouraging your friends to call a company's 800 number or visit their Web site is a particularly effective short-term strategy. And finally, we can take a hard look at what we buy and why we buy it—to resist succumbing to marketing madness.

With Liberty and Tote Bags for All

One can only assume that this transcript printed in *The Hotline* is on the level—and one can note that, once again, fact is at least as strange as fiction. Here are some top satire sites on the Web that might offer similar fare, although intentionally: *Politically Incorrect with Bill Maher, Saturday Night Live,* and *The Onion,* a Wisconsin-based weekly online magazine.

At the *Visions* Web Site Links to QVC and HSN; to satire on the Web.

READING PROMPT

Sometimes satire is lost in the explanation; instead of discussing the point of this piece, try writing your own—a home shopping sale to accompany some other event of national or local importance.

As seen on QVC, the home shopping television network, between noon and 1:00 P.M. on Inauguration Day:

12:00: Bob, the host, displays the Official 1997 Presidential Inaugural Invitation Set (item L-47400, $240), which comes in a blue binder.

Bob: "Of course, QVC is completely nonpartisan. It just so happens that we have a Democratic president and vice president going in. . . . We just want you to have some of the memorabilia of the democracy. It's a proud day to be an American."

12:02: The Official 1997 Presidential Inaugural Pin by Ann Hand (item L-47392, $45), the Official 1997 Presidential Inaugural Plate (item L-47406, $48), and the Official 1997 Presidential Inaugural Medallion (item L-47385, $36 for bronze, $695 for gold) are displayed.

12:04: *Bob announces that QVC will show the swearing-in:* "It is a ceremony that allows us to renew our democratic ideals. It allows us to renew our visions. Maybe reminds us to maybe take a look at our goals and our dreams as well."

12:05: Live shot of President Clinton taking the oath of office.

12:06: *Bob:* "I don't know about you, but that was pretty electric here at QVC. And really, when you think about taking us into the twenty-first century, we have some things to commemorate the electricity of that moment."

12:08: Dee from California calls about the invitation set.

Bob: "What does it mean to you?"

Dee: "It just means so much to me."

Bob, later in the conversation: "And, you know, it's a great day to be an American too, because we've been able to maintain this for 200-odd years, regularly, all the way down the line. And it makes you feel good, doesn't it?"

Dee: "It just gives me such a great feeling, I mean, to be part of this, at least to watch it with QVC and C-SPAN. . . . Thank you a lot, Bob, and I like your cooking shows too."

Bob: "Thank you, thank you very much. We're going to do one tomorrow, ten hours of cooking, starting at noon Eastern."

12:10: Cut to Clinton delivering inaugural address.

12:14: Nancy from Nevada calls about the pin.

Bob: "How do you plan to wear this?"

Nancy: "On a suit, probably. You know, when I go out for something nice."

12:17: The plate (item L-47406) is shown again.

Bob: "Imagine not only picking up the inaugural plate but picking up something by one of the great names in the world, Wedgwood. Combine the two and you have an heirloom." Later, Bob corrects himself: the plate is Woodmere china, not Wedgwood.

12:23: Helen from Texas calls about the plate.

Bob: "What does it mean to you as an American?"

Helen: "I can't really put it in words what it means to me."

Bob: "You can feel it, though, you feel a sense of pride?"

Helen: "Oh, yes."

12:24: Cut to tape of Vice President Gore earlier being administered the oath.

12:32: The Official 1997 Presidential Inaugural Button Set (item L-47402, $9) is displayed. Bob shows a button portraying "Great Democratic Presidents of the United States." The button has Clinton at the center.

Bob: "Mr. Clinton is the first Democratic president to be re-elected since Franklin Roosevelt, which makes this pin really quite unusual."

12:37: Bob shows the Official 1997 Presidential Inaugural License Plates (item L-47386, $25 to $45).

12:47: Live wide shot of the Capitol.

12:48: Bob shows the Official 1997 Presidential Inaugural Tote Bag (item L-47412, $24).

12:56: Peggy from South Carolina calls and says that she has ordered the invitation set, the sweatshirt, the book, the plate, the mug, the buttons, the pearls, the pin, the pen, and the tote bag.

Peggy: "QVC is just fantastic to bring us this opportunity."

chapter 8 Explorations

8a A Favorite Store

Collaboration. Share a brief but detailed description of a favorite store or shop you have frequented over the years.

Web work. Under Apparel: Retailers, Yahoo! lists more than 250 vendors at Business and Economy > Companies > Apparel > Retailers. A few are one-of-a-kind shops such as Jeff Bantz's Amigos & Us at North Pier Festival Market in Chicago, specializing in Latin and Indonesian clothes as well as Grateful Dead items. Others operate as online businesses, among them Motor Oil, specializing in European and Japanese fashions. Yahoo! lists more than 160 retail companies selling gifts: Business and Economy > Companies > Gifts > Retail.

One writing topic. Write a detailed description of one special store, real or virtual. Compare it with other stores that may be similar in some respects but not others. What made your store special?

8b Malls

Collaboration. Make and share two lists: what you most like and what you most dislike about shopping malls.

Web work. The Mall of America has an official site. Yahoo! lists more than 200 shopping centers under Business and Economy > Companies > Shopping Centers. For a scholarly look at shopping malls, check out the site of Joseph Soares of the Sociology Department at Yale University.

One writing topic. Write a letter describing shopping malls to an imaginary ancestor who died without ever having seen, much less been to, such a gathering place and temple of commerce.

8c Catalogs

Collaboration. Make and share a list of the names or kinds of catalogs you now receive in the mail or have received in the past.

Web work. The National Mail Order Association is the trade organization for small and medium-sized direct marketers, and its site hosts the Mail Order Museum. Yahoo!'s Business and Economy > Companies > Apparel > Catalogs has links to three dozen of the most popular clothing catalogs. General mail-order companies, including Harry and David's for fruits and Damark for marked-down electronics, can be found at Yahoo!'s Business and Economy > Companies > Catalogs.

One writing topic. Write an essay on different kinds of catalogs that seem to engulf our lives. Consider different ways all these catalogs can be classified.

8d Main Street

Collaboration. Share a detailed description of one Main Street you have known, or of the closest thing to a Main Street (the place where people gather to shop and conduct business) in your community.

Web work. Check out Main Street, the National Main Street Center of the National Trust for Historic Preservation. Featured at the homepage of the National Trust for Historic Preservation are the Great American Main Street Awards, with links to hometowns.

One writing topic. What is the closest thing to a Main Street in your community? Write an essay on this street's past, present, and future role in your community.

8e Web Bookstores

Collaboration. Share a brief but detailed description of a bookstore you have visited.

Web work. Thanks to the early and large presence of Seattle upstart Amazon.com and maybe because book buyers are highly educated and books themselves relatively inexpensive items, selling books has been one of the leaders in the new world of electronic commerce. It did not take the traditional bookstore chain Barnes and Noble long to challenge Amazon on the Web. Yahoo!'s Business and Economy > Companies > Books lists hundreds of additional booksellers, including such important spots as San Francisco's City Lights Bookstore, the spiritual center of the Beatnik counterculture movement of the 1950s.

One writing topic. Write an essay comparing real and virtual bookstores.

8f Homemade/Handmade

Collaboration. Share a detailed account of a homemade gift you gave or received or otherwise know about; that is, a gift not purchased from a store.

Web work. Yahoo! lists a few hundred handmade craft sites at Business and Economy > Companies > Arts and Crafts > Handicrafts. One site listed there is the gallery of American Artisans. Also check out the Handmade, Homemade Toy Shop.

One writing topic. Write an essay comparing homemade (or one-of-a-kind) to store-bought (or mass-produced) versions of the same product.

chapter 9

Hobbies, and Other Forms of Entertainment

READING QUESTIONS

- What activities do you engage in for pleasure in which you deliberately limit the degree of technology you will use—as in opting to ride a bike instead of a motorcycle, to walk instead of skating? How important to the activity is the line between acceptable and unacceptable use of technology?
- What role does technology play in your various relationships with animals?
- What are the major differences and, if any, similarities between watching an activity virtually (as on television), watching it in person, and participating in it?

QUOTATIONS

There is not one world for men and one for animals; they are part of the same and lead parallel lives.

— *Rigoberta Manchú Tum (1983), winner of 1992 Nobel Peace Prize*

Animals were once, for all of us, teachers. They instructed us in ways of being and perceiving that extended our imaginations, that were models for additional possibilities.

— *Joan McIntyre (1974), author of book about whales and dolphins*

Hobbies protect us from passions. *One* hobby becomes a passion.

— *Marie von Ebner-Eschenbach, Austrian author, 1830–1916*

In sports, as in love, one can never pretend.

— *Mariah Burton Nelson (1994), author of books on women in sports*

It's time to raise a generation of participants, not another generation of fans.

— *Janice Kaplan (1979), author of* Women and Sports (1980)

Cyberpets

Robert Rossney

Although this piece by Robert Rossney describes efforts of researchers to create virtual pets, the toy makers seem to have gotten out in front once again with the Japanese Tamagotchi, popular among children in the United States.

At the *Visions* Web Site Links to Tamagotchi; to other Rossney pieces.

READING PROMPT

How do you feel about the concept of a computer-generated "pet"? Do you see advantages or disadvantages over the traditional, low-tech kind?

Silas is a friendly-looking golden dog with floppy ears and a pointy tail. If you hold out your hand, Silas will come up to you wagging his tail and sit down, just like a real dog. He will even bring you a ball and try to get you to play.

But Silas is not real. He is made from computer-generated shapes and looks a little as though he escaped from a Disney cartoon. Silas is one of a small but growing number of computer programs that can mimic the behavior of friends, pets, and other objects of human affection. Among them are Julia, with whom you can have long, if strange, conversations; Phink, an unpredictable dolphinlike creature; and Neuro-Baby, a digital infant who can analyze and react appropriately to your moods.

Building "relationships" with these digital creatures is, of course, rather an unusual process. With Silas, for example, you have to stand in front of a video camera looking at a projection screen on which you see a picture of yourself, your room, and Silas, added in by the computer program. Of course, these are all virtual friends; robot friends with "real" bodies are still a long way off. For the more immediate future, the scientists and artists who are creating digital companions want to explore what is needed to make these computer-generated objects more friendly and engaging to humans. Research is proving that we are often surprisingly willing to read intelligence and intention into the creations of computer programs.

"The research goal behind Silas is to understand how you can build an autonomous creature, like a dog, that seems to do the right thing over time," says Bruce Blumberg, Silas' creator and a Ph.D. student at the Massachusetts Institute of Technology's Media Lab. People readily respond to Silas as though he were real, creating explanations for his behavior the way they would with a real dog. "From the user's perspective," says Blumberg, "there's a sentient, intentional being there."

The willingness of humans to see complex emotions in animals is well known to ethologists. "People often ascribe feelings to a real dog that are over and above what they really would admit, if pushed, they believe the dog really feels," says June McNicholas, a research fellow at the University of Warwick in England who has worked extensively on the bond between humans and animals. "If you've had a bad day at work," she says, "your dog may seem to respond to this. And you'll say to yourself, 'He knows I had a bad day at work.' But you know that the dog doesn't know you had a bad day at work. All he knows is that you are moving and acting differently than you usually do. And you know this."

Reacting to Silas involves similar rationalizations. For instance, Silas is interested in moving objects that are close to him. If you reach out to pat him, your hand will be both moving and close, and Silas will watch it. Silas doesn't have any way of knowing whether you're touching him or not, but because he's watching your hand so intently it "appears" that he is responding to being patted. Users tend to explain his response in those terms ("I'm patting him, and he likes it") and the explanation has predictive value (whenever the user tries to pat the dog, the response is the same) and so, in the user's mind, Silas likes to be patted.

People find Silas fascinating partly because they enjoy trying to explain what he is doing. "If a creature behaves exactly the same every time," says Blumberg, "that's not very interesting. It turns into a robot. On the other hand, if it's totally unpredictable, then it seems random, and it's hard for the user to develop an explanation with predictive value. The optimal place is where there's just enough surprise that you're constantly coming up with new explanations that make sense. I think that this is why we like having 'real' pets."

But there is a very long way to go before Silas can become "real."

Bringing artificial pets into the outside world is far more complex than creating images on a screen. This is why robotic pets are many steps behind virtual ones. Skimer the robot illustrates these constraints. Kino Coursey of Daxtron Laboratories in Fort Worth, the developer of Skimer's software, says that the robot can identify objects visually and be trained to follow them around. To do this, one trainer drags a chair, say, while the other uses a joystick to instruct Skimer to move in pursuit of the chair.

Skimer builds a network of associations between the images it is seeing and the commands it is receiving. For example, when the chair moves out of the robot's field of view, moving from right to left, the trainer commands Skimer to turn to the left. Skimer remembers the sequence of images and the commands associated with them. Once Skimer has been trained, it will follow the left-turn command whenever it sees a chair moving across its vision to the left. The result, says Coursey, is that "you can drag a chair around in front of it and he'll follow it anywhere." But while Skimer does this very well, it can't do anything else. And it's a pretty hefty contraption, cobbled together from a child's six-wheeled riding toy, a camcorder, a computer, and 18 kilos of batteries. Skimer, as Coursey puts it, "definitely belongs in the back yard. You wouldn't want him in the house."

If Skimer could be made small and agile enough, could it replace the family dog or cat? Erika Friedmann, a specialist in pets and health in Brooklyn College's health and nutritional science department, acknowledges that the autonomy of artificial pets might attract people the way real pets do. "People like pets because they don't have to make an effort to get positive feedback from another being," she says. "Your pet makes some kind of acknowledgment that you're there and entices you to interact with it."

But pets provide people with far more than unstructured entertainment. They give us something to care for, something to touch and fondle. They provide a reason for exercise, a feeling of safety. McNicholas thinks that this is the fundamental weakness of the artificial dog. "What kind of care could someone give a virtual pet?" she asks. "A lot of the closeness in a relationship with a pet is based on how dependent the pet is on you. If an artificial pet doesn't depend on you, you don't feel needed."

But What Do I Know?: Putting a Digital Pet to Sleep

Joshua Quittner

College students are supposed to be up on the latest technology, but it is no longer safe to assume that people in their late adolescence know about the latest fad from Japan. Thank goodness, the people at Yahoo! have a whole page of virtual-pet links (under Recreation > Toys > Virtual Pets), including links to Tamagotchi products and a whole page entitled Anti-Tamagotchi.

 **At the *Visions* Web Site** Links to other Quittner pieces.

READING PROMPT

Compare Quittner's feeling about virtual pets with Rossney's.

My house is awash in the screams of slow digital death. I'm referring to the last gasps of the Tamagotchi, a "digital pet" craze that swept through my kids' peer groups like chicken pox late last spring. As the frantic bleeping of a dying Tamagotchi rings in my ears, I try to think back to what caused this unspeakable, beepable hell.

Since the day they were born, I have tried to give my children the right toys. I mean the kinds of diversions that I might enjoy if I were a young pup circa 1997. Plenty of firepower: your Sponge Guns, DrenchMasters and Evaporizers. The type of play gear the neighborhood boys pack. The problem is, Zoe, 8, and Ella, 6, don't want boy toys. They want dolls.

I first identified this deficiency when Zoe was 3. I brought home a chest of Sluggo Building Blox, figuring we would make a fort or detention center. She evinced no interest. "You know how kids are," said my wife, and we set the Blox aside for a while. Sure enough, a few days later, Zoe dug out the set and started playing: she laid out six Blox side by side and covered each block with a hand towel. "Look, Daddy," she said proudly, "a mommy, daddy and four babies!"

Frankly, I was not surprised. Boys and girls are simply hard-wired differently; enculturation is a big fat lie. Indeed, this episode was similar to an incident a friend had had with her son. Trying to raise him in what she figured was a nonsexist way, she gave him a Wetting Delores doll. When he turned 2 and his little testosterone plant had kicked in, she was walking through the living room and heard him shout. There was the lad holding the Wetting Delores straight out with its legs tucked under his arm. He pointed the head at his mother. "Rat-tat-tat-tat-tat!" he said.

Hoping to spare my daughters the enormous therapy bills and lifelong confusion of the "doll boy," I pretty much gave up after the Sluggo Blox showdown. Yes, I'd try to slip in cool toys at the holidays (Sponge Guns and others), but if the girls didn't want to cut one out of the shrink-wrap, hey, no hard feelings. I could play with the stuff alone.

When I first read about Tamagotchis, though, I felt here, finally, was a toy that would appeal to all of us. In case you're childless or Amish and managed to avoid this craze (it was after the Beanie Barneys and before the Skink-Me-Sams), the Tamagotchi is a keychain-size plastic egg that houses a small LCD upon which flourishes a creature that you nurture by pushing a variety of buttons. This latter part appealed to me; the former appealed to my daughters. Luckily, they begged me for them, allowing me to position it as their idea.

"Ella, guess what?" I offered shrewdly to the 6-year-old, "I'll get you a Tamagotchi if you stop sucking your finger."

"If I stop picking my nose, will you get me a doll?"

"No way."

"Done," she said, producing a simple contract and a notary. "But if it's not an orange one, the deal's dead."

On an appointed day a week later, we found a store that still had two of the $15 items for sale. Fortunately, one of them was orange. "That'll be $25 apiece," said the felon at the cash register. When I caviled, he pointed out that a shipment of 500 had arrived just that morning, and he had only two left. He had to sell these for $25 each, he explained, if only to uphold the tradition of a truly free market.

I bought them. What choice did I have?

It was love at first beep for my kids. The silicon pets provided just the thing to satisfy their mothering jones. The Tamagotchis needed to be fed, played with and even changed (don't ask) regularly. And the pets worked for me too. Plenty of buttons to press and an obscure procedure for setting the time/day/date kept me happy.

At least for the first day, that is. A "normal" digital pet lives for a few weeks, at which point it's "called back to the home planet," according to the instruction book. In reality (?) the thing dies, and you have to hit the reset button and grow another one. This cycle, alas, repeats endlessly.

My kids lasted through roughly one birth-death period before losing interest. But the Tamagotchis, bless their little chips, keep on ticking. Or beeping, beeping, always beeping, like the Telltale Heart. I hear them at night when I'm supposed to be sleeping. And when I awake. I can't take it much longer. Perhaps I could crush them under the Evaporizer? Just a thought.

"The Tamagotchis, bless their little chips, keep on ticking. Or beeping. . . ."

The Sports Arena in the Digital Age

Donald Katz

Spectator sports have long played an important role as entertainment/hobby in all industrialized countries, places where people live in large cities and have money to spend on recreational activities. First radio and then television have changed certain aspects of how such sports are presented to us. Now, Donald Katz suggests, the latest forms of computer-based interaction are promising (or threatening) to transform this whole relationship. Fans play simulations of the real games at home and watch virtual re-creations of the real games in the arenas. As any sports fan knows (from relentless and slick advertising during major sporting events), one of the big players in gaming simulation is EA Sports.

Another digital transformation aspect of spectator sports is fantasy sports, where fans make up and manage their own teams using the performances and statistical results of real players.

At the *Visions* Web Site Links to EA Sports; to more on Paul Allen; to the Portland Trailblazers.

READING PROMPT

How do you feel about all the different ways that electronic communications are changing how we view sporting events off-site (at home) as well as at the original venue? Is all this technology changing the nature of the games themselves?

Paul Allen's private jet is waiting on a nearby runway, but the owner of the Portland Trail Blazers refuses to leave the skylighted basketball arena in the hillside village and advanced-technology lab that is his Seattle home. "I'm gonna sink a three," Allen says as he takes up a position behind the three-point line and begins launching one looping set shot after another from beneath the tip of his great lumberjack's beard.

Earlier, as he strolled through the indoor tennis court and past the pool and gymnasium in his own $7 million sports complex, Allen averred in his quiet manner that he'd come a long way since 1975. That was the year he teamed up with a high school buddy named Bill Gates to found a tiny company called Microsoft, a software venture dedicated to the then unlikely proposition that someday everyone would have a computing machine.

Besides the throne-like easy chairs built into the wall along one side of the regulation basketball court and the Santa Fe–style high-desert oil paintings on the opposite wall, the distinguishing features of Allen's arena are video monitors of the sort that can be seen everywhere on his estate. Each of the screens is electronically tethered to dozens of other monitors and computer systems inside the Allen compound. Simply touching a display on one of the screens can achieve high-speed access to satellites circling the globe and therefore to just about any sports event being broadcast anywhere in the world. Inside his plush 20-seat theater, equipped with a 10-by-14-foot screen, Allen can view ultra-high-definition video images that less-privileged consumers won't be able to see for several years. And if Paul Allen must miss a Blazer game because he's out at sea on his 150-foot yacht, the team will tape the game at a cost of around $30,000 and beam it to him as a digital stream of private entertainment.

From any keyboard inside his home, Allen can also access computers strewn throughout the vast web of his futuristic business empire. He can send E-mail out to Blazer forward Buck Williams or to coach P. J. Carlesimo's address in cyberspace. "I'm not using these ———— computers, and I'm not readin' no E-mail!" Carlesimo declared upon being presented with his laptop shortly after he was hired by the Blazers last summer. But since then P. J. has seen the light and joined his boss in what Allen has long called "the wired world."

Allen, 42 and the 13th-richest American, has lately spent $1.2 billion of his $4.6 billion Microsoft-spawned fortune on a broad array of digital satellites, wireless communications outfits, multimedia software and communications hardware firms, futuristic research companies and high-profile entertainment ventures. Last March, Allen underscored the convergence of Hollywood and the digital media age through his investment of $500 million in DreamWorks SKG, the studio being assembled by Steven Spielberg, David Geffen and Jeffrey Katzenberg. And as Allen's executives and research scientists work more subtly to merge economic power, advanced technologies and big-time sports, they are similarly defining a future in which the experience of sports will surely be changed.

Aside from the Blazers, the "Paul Allen family" of enterprises now includes the ticket purveyors of Ticketmaster, the sports statisticians of STATS, Inc., and the technologists creating on-line sports information services and CD-ROM products at Seattle-based Starwave. One new Starwave product, the on-line ESPNET SportsZone—which offers stats, up-to-the-second scores and even maps of ball fields showing where hits have just landed—has been logging 1.5 million digital "visits" from wired fans each month.

Down in Portland, Allen's Trail Blazer organization is managing the construction of a $262 million sports arena called the Rose Garden, which will be strewn with computers and wired with miles of fiber-optic cable. The 70 luxury suites inside the Rose Garden will be equipped with teleconferencing gear and be fed channels full of computer-generated sports statistics. The concourses of the Rose Garden will be draped with glowing video screens, and Allen eventually wants to feed stats and replays and stock quotes and weather reports and images of games being played in other places to a tiny screen located at every seat.

Not unlike other team owners who have invested in new stadiums and arenas over the past year, Allen is considering a virtual-reality entertainment center next door to the Rose Garden. An official Blazer "home page" already connects on-line fans to the team's own Internet address. The Blazers' staff includes a seasoned multimedia software developer assigned to create sports products that the Blazers can sell to other teams. "My mission," team president Marshall Glickman proclaimed early in this past NBA season, "is to integrate Paul Allen's world of computers and communications with my own world of sports."

During the "information superhighway" media frenzy that began toward the end of 1993, a Seattle Times reporter imagined a day in the not-too-distant future when a fan who got home late during a Seattle SuperSonic game could digitally fast-forward through the recorded action until he caught up with the real-time telecast. After a Shawn Kemp dunk, the reporter presumed, the viewer could click on the image of Kemp and call up his latest stats, read stories about Kemp from newspapers all over the world or connect with the Shawn Kemp Fan Club in Indiana. Another click would automatically order Shawn Kemp souvenirs or tickets to a coming Sonic game. The viewer could change the camera angle from which he or she was seeing the game, focusing on Kemp or watching the action from overhead.

And all of this, the newspaper article pointed out, could occur within the boundaries of Allen's multimedia portfolio. "Once the high-speed digital channel is wired into people's houses," Allen says before finally nailing a three-point basket, "all of that—and more—becomes pretty easy to do."

Early evidence indicates that many of the innovations now understood only by technologists like Allen will intensify our experience of spectator sports—just as audio CDs have enhanced the secondhand experience of a live symphony. The informational and visual options available to fans sitting at home or in the stands are already multiplying as sports become proving grounds for advanced digital technologies. But these technologies also raise a broad array of questions, from immediate concerns (Will computerized gambling soon be inextricably linked with big-time sports?) to new business issues (Will people pay for new services?). Then there are longer-term issues: Will computer-based technologies someday offer sportslike entertainment so enthralling and convenient and highly customized that games created from bits of the best of real sports and bits of the best sports fantasies render live games obsolete?

Technology has always altered the way sports are played and observed. Scientific advances have been applied to sports equipment and techniques: protective gear, the composition of tennis rackets and golf clubs, high-tech training methods to increase leg strength and foot speed. The advent of radio and television offered new kinds of access to live events; for every fan at a game, tens of thousands of others heard or saw the event via machines.

Eventually technology allowed the power of a dunk to be repeated and slowed and otherwise separated from the flow of a game—and some would argue that the essential focus of basketball has changed in response. The half second it actually takes a decent fastball to cross the plate can now become several seconds, and that pitch can be replicated over and over again.

Now computer-driven video games are approaching a movie-theater level of clarity; the best of the games are called "simulations" because of the way they re-create the strategic demands and sensuous experience of actual sports. "Our muscles tighten, our pulses quicken, our palms sweat," one player of race car simulation games recently typed into one of the many on-line forums for simulated-sport fans. "It doesn't feel like sitting at a computer at all. It isn't. It's a dashboard with a steering wheel."

There are already millions of obsessed fans of these simulations who seem to care little for the real sports upon which the games are based, and most of these fans are members of the new generation that will soon be needed to fill stadiums and arenas. If these young consumers care more about the moves, history and personal predilections of a character called Shaq Fu in a best-selling video game than they care about watching Shaquille O'Neal play basketball, then is Shaq—in the world of images and consumer markets—primarily a basketball player, or is he a virtual kickboxer on a video screen?

During their first week on the market, $50 million worth of Mortal Kombat II video cartridges and discs were sold to young enthusiasts who also support six different magazines dedicated to news about video games and heroes such as the Mortal Kombat characters (the "players" or "guys," as they are known). If there are more children in Sacramento who understand the athletic prowess, personal history and signature moves of Raiden, an athletic Mortal Kombat "thunder deity," than there are kids who attend a King game over the course of a season, what's the more popular sport among kids in Sacramento: Mortal Kombat or NBA basketball?

"Right now the only thing holding back sports simulation products is the level of reality that can be achieved," Allen says. "But realistic athlete-like or coach-like experiences that you can share with others on-line, in a real-time environment, are just sitting there over the horizon. The changes will go way beyond games and sports to create a whole new realm of social contact. Ten years ago people scoffed at the idea that everybody would soon have a computer at home. I heard people laugh only a few years ago at the idea that there would someday be software just for kids."

"It's still at the preliminary stages," observes Ron Bernard, the former Viacom executive hired by the NFL in 1993 to track technological innovations as president of the league's new venture division. "But I see no reason why someone won't soon be able to experience what it's like to get the ball and try to run it through the 1966 Green Bay Packer defense."

When team owners and league officials of the predigital past began to worry that technology would soon render the distant experience of big-time sports so satisfying that fans would prefer to stay at home, where it was safe and warm, they built living rooms called luxury boxes under the eaves of stadiums and arenas all over the world. But what if the live images and sounds beyond those boxes were suddenly amenable to perfect duplication? And what if those crystalline images could be changed so that a fan could watch a game from the viewpoint of the third base coach or through the eyes of the pitcher? Once images and information are digitized and can be piped from computer to computer at high speed, all the video captured by dozens of cameras can be merged and manipulated. Everything digital becomes measurable, storable and accessible on demand. . . .

"The process of going digital means connecting every aspect of a given game to computers, and once you're there—once you have access to the bit stream," Lippman says, referring to the electronic flow of zeros and ones that are the lifeblood of these emergent machines and techniques, "you can walk past the threshold and through the door."

The president of the Portland Trail Blazers, dressed in low-tech protective goggles and a hard hat, bounds along the girders and concrete buttresses of what will be the Rose Garden. "We're going to have a next generation, high-resolution Astrovision screen and scoreboard in the middle of the arena," says Glickman. "There will also not be a single vista that isn't full of video displays—except in the bathrooms, where we'll have radio feeds. And the computers we've ordered for this place will talk to God.

"With a payroll of around $27 million," Glickman continues as he stands in the concrete-and-metal enclosure of a future luxury box, "I have to look to grow this business by making things happen here in the Rose Garden—that and by developing all of the technological possibilities opened up by being connected to Paul Allen. Eventually you have to acknowledge that there are some mathematical limits to growing the sport itself."

Back at the Trail Blazer offices in downtown Portland, Glickman walks into an office to look over the shoulder of Carl Steinhilber, the Blazers' multimedia designer. Steinhilber fires up his new Game Ops Commander software, a program that allows customization of stadium and arena video and audio scoreboard entertainment. A cartoonish character wearing a red headband—obviously meant to represent Trail Blazer forward Cliff Robinson—begins to move and dance with some of the verisimilitude demonstrated via "motion capture" up at Microsoft.

Steinhilber's new program offers the kind of customized individual control made possible by the next wave of technology. In a sense such digital systems hark back to bygone experiences of big-time sports, when games were adorned only by those fantasies and statistical records created by viewers in the stands, by the odd loudspeaker announcement or by a running commentary supplied by the drunken fan in the next row. With the cost of tickets and premium cable feeds rising and the cost of digital technology falling, these inventions might actually give everyone that prized seat on the bench.

Or perhaps the process of electronic customization will create a new fan of the future—a shrouded figure sitting alone in a room, his head enveloped in one of those virtual-reality helmets. Perhaps the tiny screens up close to the fan's eyes will flash vivid rearrangements of sports culled from the electronic flow, spectacular stuff jacked in from the bit stream. The game will be the most thrilling sports spectacle the fan has experienced since the last time he lowered the helmet over the rest of the day, since the thrills have been abstracted from the fan's own fantasies and the specific passions that draw him to sports . . . or whatever this hybrid experience of sports has come to be called.

Perhaps the lone indication of the viewer's connection to sports fans of the past will be the plastic tube running up to the beer can affixed to the top of the device.

The technologies involved, to use current techie parlance, will be nothing if not "cool." But will the result be better . . . or even good? For all of his ability to stay at home and suck down superclear, big-screen images of sporting events happening anywhere in the world, what Allen loves most is actually attending a Blazer game. Earlier this past season he flew down to Portland to catch a game against the Utah Jazz in the company of some of his senior employees from Starwave, sports nuts to a man. They sat in Allen's jet, poring over some new customized statistics that tracked the performance of Jazz forward Karl Malone against each NBA team in light of all sorts of circumstances, such as travel schedules and number of days of rest. Everyone was impressed by Malone's consistency.

At the game Allen took his regular seat directly below the basket, and at the opening tip-off he turned into the basketball maniac Portlanders occasionally see on the news, leaping around during replays. Unlike most team owners, who entertain associates in a distant box, Allen hoists his considerable frame out of his folding chair and jumps into the air, screaming at the top of his lungs. The sweat from the play a few feet in front of him wets him down, and the billionaire in a rugby shirt and Dockers turns to the crowd and exhorts it to join him in stepping up the frenzy.

"People will always want a live and communal connection to an event," Allen says. "There will be a zillion beautiful replays to choose from, and people will still want to be there. The computer still has a ways to go. We don't have software yet that can, for instance, make judgments. We inch closer, but that could still be a hundred years away. I really do believe the intensity of it all will eventually be captured by technology, but for me the seat under the basket will probably always be the best place to be."

Walking

John P. Wiley, Jr.

Walking has to be the one of the simplest and most wonderful of all technologies, as John P. Wiley, Jr., points out. Yet partly because it is so simple and so wonderful, walking is often taken for granted, at least by those whose ability to walk is not impaired. Consider all the technologies humans have invented to aid our getting around in the world, and how these have or have not taken the place of walking—and in so doing, one may glean a minihistory of human technology itself with all its accomplishments and failures.

It is a notable fact of modern life in the United States that walking is often seen as a leisure-time activity, something one purposely sets out to do, rather than as a regular part of life.

At the *Visions* Web Site Links to hiking sites, including urban trail guides.

READING PROMPT

Compare the amount of walking you do and your attitude about it with what Wiley reports here.

"It would kill any of the young men of the present day to attempt such a walk; it must be four miles, or two, or some immense distance."

Those lines, spoken by a noblewoman in Emily Eden's 1830 novel, *The Semi-Attached Couple,* somehow strike me as funny in this day of three-hour flights to France—funny at least until the next time I find myself in a 4,000-pound car carrying four pounds of milk four blocks. On the days when I drive to work (the days when there is not time for the subway: mass transit it may be but rapid transit it is not) my mind is too busy reviewing the roll call of anxieties to take heed of any other driver not actively trying to run me off the road.

And then my father will be up for a visit from South America, I'll take him downtown in the morning, and invariably he is stunned at the sight of all those cars—almost every one carrying one human being and no more. He lives in a place where people live environmentally sound lives for the wrong reason: they are too poor to do anything else. When he has an empty jar or bottle, he simply sets it outside the garden gate. Within five minutes it has been recycled, pressed into immediate service. Even among the ranks of those a lot higher on the money tree, hardly anyone drives to work alone.

Actually I was a little stunned myself when I first started working in Washington. I would come out of my building at the end of the day to find the sidewalks almost deserted. Our magazine offices are in a building on the Mall, which of course is a special case: aside from the concession stands, this is a noncommercial district. That means that it is a desert as far as human amenities are concerned. Nowhere can you get your glasses fixed, pick up some flowers or buy a bottle of aspirin.

Now the Smithsonian itself employs only a few thousand people. But at the west end of the Mall and across the avenue on its south side are major federal buildings. Another block south are more temples of bureaucracy. Altogether we are talking about a fair number of people. Yet when you come out of the back of our building at 5:30 or 6, and look across the street to where all those people have been spending our tax money all day, there is hardly a soul in sight. The sidewalks are almost empty. Everyone either disappears underground to catch the subway, or gets into a car and drives away. Even those who want to get together to discuss the latest managerial madness get in their cars and go somewhere else to do it. After working in New York, where the sidewalks are alive, this absence of people produces a weird feeling. You begin to wonder if you have stumbled into one of those movies in which something terrible has happened to everyone except the hero.

The good news is that we can turn places like Washington into real cities and at the same time ease two national problems without spending a penny. As a nation we worry about global warming; we fear the ozone on the ground that injures our trees and our lungs, and the thinning ozone miles over our head that no longer protects so well against ultraviolet radiation. We worry about the nitrogen and sulfur oxides that fall again as acid precipitation and the carbon monoxide that crowds the oxygen out of our red blood cells. We're doing it to ourselves, every time we put the key in the ignition and turn it.

As individuals we also worry about our health and the urgent advice of the authorities to do grim penance for the french fries of long ago. We nod our head and tell friends we really should get out and jog, join that health club, buy a $600 machine for the bedroom. But we are rushed this morning, tired tonight. Next week for sure: we're going to get right on it. And sometimes we do, and then drop out.

Let's leave weakness of character out of this. People drop out because that kind of exercise is boring, makes you ache and, if you don't happen to be a gazelle anymore, somehow rubs your nose in your own deterioration. It has no purpose beyond the motion itself. On one of my very few excursions on horseback (Cocoa, wherever you are, thanks for keeping us right side up!), I liked it when the wrangler stopped snickering long enough to explain that while Eastern riders do it for show, out there in the West people ride horses to get somewhere.

I've always thought it made more sense to use gardening and yard work for exercise rather than pay to go to a club and pull weights on wires that run through pulleys. When you finish your garden "workout," you have accomplished something. News reports years ago of Chinese office workers being forced outside to shovel snow struck me as an inspired idea. What an incredible antidote for the post-power-lunch midafternoon slump.

Now what I'm driving at here (pardon the verb) is a way to exercise that does not require adding a whole new, unpleasant chapter to each day. Something that can be done without special equipment or going to a special place. Something that lessens the damage being done to the planet we love. It seems that you don't have to get your pulse up to some very high rate and keep it there for so many minutes so many times a week for exercise to do any good. Any exercise is good for you, we now learn; one of the best is just plain walking. All we really have to do in cities is what the National Park Service is forever urging us to do when we are out in the middle of nowhere: get out of the car.

The kind of walking I'm talking about has nothing to do with hardship and self-discipline. I'm talking about city walking, a delight for body and mind. Just suppose you found yourself walking five miles a day. Not all at once, mind you, but a little in the morning, a walk at lunch, a trip to the post office; later a walk part of the way home, even down to a restaurant or the movies in the evening and then home again. Find yourself noticing cloud formations or swifts spiraling down into the library chimney, bats flashing under a streetlight. Check the bookstores, the notices stapled to telephone poles. Find that faces are considerably more interesting than bumper stickers.

Paradoxically, being on your own two feet is restful. No one is climbing up your back, leaning on a horn designed to blast an opening a mile ahead while doing 130 on the Autobahn. If something catches your eye, you can stop without being rear-ended. You discover that you do not have to be training for the Iron Man Triathlon to release a few endorphins in your brain, letting the sun come out in your mind. How many times have I started out to walk somewhere, feeling tired and irritable, begun to think the whole idea was a dumb mistake, then found that despite myself the slump was even coming out of my shoulders and I was deciding not to have that troublesome writer done away with after all.

Governments dither, but there is nothing to keep us as individuals from voting with our feet. We can do something about global warming, and feel better for doing it. Each time one of us walks someplace instead of driving, that much less poison has been injected into the air we all breathe. (Actually the benefits are multiple. One less car means that the rest of the traffic will move that much more quickly and burn less fuel.) Now suppose two of us walk someplace instead of driving. What if 100 million of us did it just once a day?

As time went by, a third benefit would accrue. Paradise is not now upon us. The streets of Washington, like those of most other cities, are not yet a permanent festival, promenaded by philosopher kings. In Washington, the tide is still going out. The corner restaurant where you could sit in a booth for hours, the bakery where people lined up for the strawberry pies, the health food store that was forbidden by law to use the word "health" in its name but sold a ton of the No. 7 salad (chopped carrots, raisins and cottage cheese, with a mayonnaise-and-buttermilk dressing) are all disappearing as the old buildings come down and great glass blocks go up in their place. As urban sociologist William H. Whyte says (*Smithsonian,* February 1989): " . . . prosperity [is] lowering our real standard of living." But if hordes of us take to the sidewalks, we can return these mean streets to their highest function, the bringing together of human beings.

When we reclaim them, the sight of armies of potential consumers on the march will not be lost on entrepreneurs. They will become placer miners, dipping into the stream for gold. After a couple of quick trips to Rome what I remember most fondly is not the churches, the shops or the Colosseum but the bars. These are not establishments devoted to the consumption of alcohol but rather rest stops, places where on the way to work, or home, or your next appointment, you can stop in for a small coffee, a delicate pastry or a sandwich fresh made with whatever is in season. Places where you and a friend can go to talk. There seemed to be one in every block, and it is like having a butler who is ready at any time to produce a little something to tide you over. No one would dream of driving to one; even if you survived Roman traffic, there would be no place to park. You walk there. The day may come when we can walk to one here.

Listen, I'd really like to finish this story, but I've got to get up to 19th and E for a lunch date. I guess I'll walk on over to the Museum of Modern Art of Latin America, then go by Bolivar Pond to check on the ducks. After lunch I'll hike up 17th to pick up tickets for the ball game, and then come back down 15th, past the Boy Scout statue and on back home to the Mall.

chapter 9 Explorations

9a Pets

Collaboration. Share a detailed narrative of one experience you have had with a pet.

Web work. Dog lovers might want to try Dog-Play, with suggestions for fun things to do with your dog. Working Dogs obviously looks at dogs as other than pets, and there is a homepage devoted to everyone's favorite dog, Lassie.

For cat lovers there is CatFancy Online, plus a Web page on toilet training your cat, and finally information on the Andrew Lloyd Webber feline musical.

For horse fans there are two good sites: Equine World and the International Museum of the Horse, the latter with the motto, "Our history was written on his back." Anyone interested in horses as athletes will want to visit the site dedicated to the great thoroughbred champion Secretariat.

There are any number of pages on Japanese virtual pets, including one that dubs itself the Virtual Pet Homepage.

One writing topic. Write instructions for a team of computer programmers intent on making the world's greatest virtual pet.

9b Hobbies

Collaboration. Make and share a list of hobbies you have had in the past, still practice today, or plan to take up in the future. What kinds of skills have you gained or might you gain from each?

Web work. Many hobbies involve doing work that at a different time or place (often an earlier world with simpler technology) may have been an economic necessity. Gardening, perhaps the single most popular hobby on the planet, is one such activity, and there is no shortage of gardening sites on the Web. Start with the Virtual Garden, part of Time Warner's huge Web site called Pathfinder. Other sites to visit include *Dig Magazine* and, for the historically minded, the Museum of Garden History housed in a restored church next to Lambeth Palace, the London residence of the Archbishop of Canterbury.

Another work-related hobby involves repair and maintenance of automobiles. For technical advice, try Autoshop Online; for the lighter side of car repair, there's the homepage for the popular National Public Radio talk show *Car Talk;* and for those who like only to look there's the Auto Museum.

One writing topic. For the most part, hobbies are not regularly included in or recognized as part of the regular school curriculum. Make a case to convince the administrators of your current school to grant you course credit for your work with a hobby.

9c Solitude

Collaboration. Share a detailed description of an enjoyable time or times when you have been alone.

Web work. The Web is capable of turning up many strange and seemingly unrelated items concerning solitude. A search of *solitude* via AltaVista turns up some 7,500 sites. For starters, there are two Romantic poems, Coleridge's poem "Fears in Solitude" and Keats's fragment "O Solitude," plus an entire site devoted to Superman's Fortress of Solitude.

Harold Perry's Impressions of Thoreau site has a screen, with quote and photo, on solitude. There are also Web sites advertising video and audio for being alone with your thoughts, including a video called *Peace Waves: Solitude by the Sea,* which bills itself as sixty minutes of sounds and pictures (maybe just one picture) quietly sitting on South Beach at Martha's Vineyard.

There is also the brief tract "The Sweets of Solitude" written by the "Pennsylvania Hermit," Amos Wilson, a man who lived the last nineteen years of his life in a cave.

Then there are links to two outdoor sites that suggest solitude with nature but where you are not likely to be alone: Solitude Ski Resort in Utah and Solitude River Trips, a rafting service in Idaho.

Finally, there are the paintings of the intensely introspective German Romantic Caspar David Friedrich, including his masterpiece, *Wanderer above the Sea of Fog.*

One writing topic. Discuss the relationship between technology and solitude. Do you use technological devices to help make solitude more enjoyable or more endurable? If so, what kind? If not, why? Do you think technology will lead to people's spending more time alone, or less?

9d The Blues

Collaboration. At the core of the blues is a contradiction: This is music about sadness, loss, and betrayal that many people nonetheless find highly enjoyable. Share a detailed account of something specific you have found sad yet enjoyable in your life.

Web work. There are two magnificent blues sites: The Blue Flame Cafe, "an interactive biographical encyclopedia of the great blues singers and singers of the blues," and Blue Highway, with a wide assortment of features. The Ransom Group produces an excellent historical guide to the blues as part of its guide to music in Memphis, with other sections on Stax and Sun records.

One writing topic. Write an essay on the role of sadness in our lives. When, if ever, is sadness a valuable emotion? In what sense does sadness represent the failure of technology? Under what conditions, if any, can technology play a positive role in helping us to feel less sad or to gain more out of our sad feelings?

9e The South

Collaboration. Much of popular U.S. music from the blues to country and Western is associated with the South, as are all kinds of other things, good and bad. Make and share two lists: of positive and negative associations you personally have with the South.

Web work. The genteel side of southern living is covered in Time Warner's *Southern Living* site. More raucous and far more fun is the Web zine *Y'all;* more academic is the Documenting the American South site at the University of North Carolina. *Brightleaf: A Southern Review of Books* started publishing in September 1997.

The two most famous southerners—and two of the most influential Americans—of this century, the Reverend Martin Luther King, Jr. and Elvis Presley, are well represented on the Web, although not equally so. Time Warner and the *Seattle Times* host tributes to Dr. King. The official site of the better-represented Presley is called Graceland. The California Museum of Photography hosts Ralph Burns's black-and-white photographic record of his recurring visits to Graceland, entitled How Great Thou Art.

Some people feel that a century and a half after the Civil War, the South may finally be taking over the rest of the country. The latest southern phenomenon to sweep the nation seems to be stock-car racing, widely known as NASCAR, although the reasons for its connections to southern culture may not be entirely clear.

One writing topic. Write an essay contrasting your personal experience of a region of the country with portrayals of that region in the national media.

9f Your Music, and Why Others Find It Annoying

Collaboration. Make and share two lists: everything good and everything bad people say about the kind of music you most enjoy.

Web work. As a hotbed of popular culture, one would expect the Web to have a wealth of music sites. Yahoo! lists some fifty styles of music, from alternative to world fusion, at Entertainment > Music > Genres and more than 27,000 sites by artist at Entertainment > Music > Artists. Finding information on the Web about your favorite music (even if it is polkas) is not especially difficult.

Of the many pop music sites that are updated daily, three stand out: Addicted to Noise; Wall of Sound, from the popular entertainment site Mr. Showbiz; and Webnoize. *Rolling Stone* has been slow coming online but now offers an equally rich site.

Finally, there is the Ultimate Band List with links to practically everything associated with current bands.

One writing topic. Write an essay on the key role music plays in how you define yourself to others or to yourself, or on the key role it plays in other people you know.

part IV

At School

The three chapters in this part of *Visions* deal with a topic about which you have much more expertise than you may give yourself credit for: technology and education. The typical approach for all of us is to write off what we already know, from twelve plus years of full-time schooling, and to imagine that anything really worth knowing about education and technology must have to do with the *future* (and not with our past). Hence, we constantly hear people say, "Computers and other new technologies are giving kids today all sorts of advantages that I never had; kids in the future will have new, even better ways to learn." Is it possible, we are tempted to ask, that in the future students will be surrounded by such powerful technology that they won't even have to learn? That machines will calculate for them, spell for them, write for them, even think for them, while the students themselves, in this brave new world, will then do—what?

The fact is that you as a college student today have already spent at least twelve years of your life being educated on an almost full-time basis in a world with technologies that were capable of sending a person to the moon and back well before most of today's students were born. Although you may not have given much thought to the matter, anyone who graduated high school in the United States in the last ten years has had access to a vast range of powerful technologies, including not just obvious things like TV and VCRs (the former the subject of Stephen Doheny-Farina's piece in Chapter 10, "The Glorious Revolution of 1971") but immeasurably useful everyday tools such as the telephone, radio, and cassette recorder. Does it make sense, therefore, to assume that even though all the fantastic technologies up until now may have had relatively little impact on your own schooling, the next generation of electronic technology will totally transform how future students are educated? At least three of the authors in Chapter 10—Douglas Rushkoff, Derrick de Kerckhove, and Nicholas Negroponte—seem to think so. Alan Kay, a computer pioneer, offers a more mixed message.

Maybe things will indeed be different in the future; it is true that computers are different in significant ways from other technologies. Still, a certain element of skepticism is likely in order, especially given the fact that as students all of us have had teachers who already used varying kinds of technological aids—with (one assumes) varying degrees of success. In that light, compare John Edgar Wideman's reminiscence about an old-fashioned field trip and the Web work on electronic field trips suggested in Exploration 10c.

Chapter 11, Books and the Changing Nature of Reading, changes direction by considering how computers and electronic technology generally are altering not just how we read but, just as importantly, how we tell stories. The chapter opens with a bit of traditional writing by Mark Twain—description in which words (and readers by imagining the words) do all the work. Then we have Plato's classic attack on writing and on technology generally for substituting machinery for the process itself: in this case, writing is a "machine" that frees us from what for Plato is the critical task of memorizing and more deeply understanding what is being said.

The other readings deal with the way electronic technology is fundamentally nonlinear: We can jump around while reading from a computer screen in ways that we never could with books, just as watching television with a remote control and a cable connection today is a far cry from the early days when users had to tune three or four stations manually (something that few students may remember). In both cases, we now have the freedom to jump from place to place to meet our own immediate needs. Charles Oliver's "The Last Picture Shows" considers the possibility that the traditional feature film (which educators for most of the twentieth century have often seen as a distraction from students' more serious task of reading) may itself be coming to seem boring and old-fashioned to a new generation of video-game players and channel surfers used to being able to control the pace of what they watch. Interestingly, as more technologically advanced media come to dominate storytelling, older forms such as legitimate theater may be making a comeback—although, as Exploration 11d shows, traditional theater has its own high-tech dimension.

Chapter 12, High-Tech Colleges, combines the issues of the two preceding chapters to look at how research and the organization of knowledge is being altered, in one form or another, at this very moment at the school you are attending. Here too, as with Chapter 10, you should be able to bring your experience to bear on what you read and discuss, in that you are likely seeing a wide range of new and older technologies at work in your classes. You will be able, in other words, to situate yourself somewhere between Richard Lanham's enthusiasm for the new electronic age and John Leo's iconoclastic, conservative skepticism.

One area of special concern here is changes in academic libraries nationwide and, in turn, in how all of us will be doing research in the future. All six Explorations in Chapter 12, in fact, deal with changing aspects of college research—and the last three Explorations (English-only laws, drug policy, and the death penalty) entail new Web-based ways to investigate and add life to traditional research topics.

chapter 10

Wiring Classrooms

READING QUESTIONS

- What are some of the different ways you have used technology outside of school to gain and communicate knowledge about the world?
- Of all these ways you have used technology outside of school, which of these have had the greatest impact (positive or negative) upon your schooling so far? Which ones have had little impact?
- Can your best teachers and your worst teachers be distinguished at all in relation to how they used technology?

QUOTATIONS

The greatest sign of success for a teacher . . . is to be able to say, "The children are working as if I did not exist."

— *Maria Montessori (1949), Italian educator, 1870–1952*

First we thought the PC was a calculator. Then we found out how to turn numbers into letters with ASCII—and we thought it was a typewriter. Then we discovered graphics, and we thought it was a television. With the World Wide Web, we've realized it's a brochure.

— *Douglas Adams, author of* The Hitchhiker's Guide to the Galaxy

In the first place, God made idiots. That was for practice. Then he made school boards.

— *Mark Twain, American writer, 1835–1910*

I find television very educating. Every time somebody turns on the set, I go into the other room and read a book.

— *Groucho Marx, American humorist and film star, 1895–1977*

Beyond Homewood

John Edgar Wideman

The contemporary writer John Edgar Wideman grew up in the Homewood section of Pittsburgh, a setting that figures in the selection below and that he has used often in his award-winning novels and stories.

At the *Visions* Web Site Links to more by and about Wideman; to Homewood, PA; to the Carnegie Museum of Natural History.

READING PROMPT

What do virtual Web museums offer inner-city students? What are students "visiting" virtual museums missing? How would Wideman feel about what is being gained, what is being lost?

On the phone my mother tells me she's afraid to walk the streets of her neighborhood. My mother is tough and not a complainer, so it distresses me to hear the trouble in her voice, to hear her recount how many young black men died over the weekend in Pittsburgh, adding them to the total body count she doesn't remember exactly, probably doesn't want to remember exactly, so the sum of the dead is a stutter, a sigh, a deep unsayable silence between us on the line. Is it gangs, I ask, drugs and gangs, the predictable, deadly mix working its way across the country like a runaway virus. Some of it seems to be gangs, she answers, and some is just meanness and killing and God knows what's happening to these kids, what's in their minds.

As the killing and violent crime jump out of black neighborhoods and ignite flash fires all over the city—a rash of convenience-store holdups, drive-by shootings, a downtown, wild-West car chase with bullets flying, imperiling innocent bystanders—the city recoils. Nobody's safe, nobody's in control. Born out of frustration and anger, desperate means are considered—curfews, wholesale roundups, more and more cops. Nothing seems to help, and against a background of severe economic recession, the will to deal with the emergency is further eroded by lack of resources.

I hang up the phone and sit very still. The cityscape unfolds. I visualize the streets, 600 miles away, my mother's afraid to walk, imagine the city cracking into fragments, into islands where a brutal law and order rules, abandoned turf, zones where no one wants to live but they have no choice, so inhabiting these areas is like being incarcerated. Pittsburgh has always been a city of distinct enclaves, sometimes organized around the naturally divisive topography of hills and rivers, sometimes divided by custom, by prejudice, the hard realities of class, race, religion. But in contrast to the Pittsburgh I recall from childhood, in this new vision revealing itself to me there is no way to get from here to there, no flow or exchange or communication from one section to another. Fear locks people in. Violence enforces the separation. No trolley connects Homewood to Oakland. Homewood schoolchildren don't arrive excited, nervous, awed by the monumental facade of the Carnegie Museum of Natural History. Things truly have fallen apart. There is no cultural center like Oakland where all kinds of people can mingle, reconnect with each other and with themselves.

Culture is not mindless accumulation of some laundry list of objects or people or styles somebody else has intimidated us into accepting.

Culture is a way of locating yourself in the world, a world that doesn't make much sense without a conscious, active, continuous process of orientation, learning, accommodation.

Of all the good stuff I remember about elementary school field trips to the Carnegie Museum of Natural History, what remains most striking is the long hall full of dinosaur fossils. Giant, lizard-headed Tyrannosaurus rex rearing up at one entrance, the pterodactyls and mammoths and fish fossils, the mountainous Brontosaurus stretching from peanut head to tip of tail nearly the whole length of the gallery. Gradually over the years I'd learn the names of these creatures, memorize their vital statistics, and they'd be a reference point for measuring every other living thing, from other little boys to pro football tackles, the rest of my life. Before that growing "scientific" curiosity, however, came the mad dashes through the dark dinosaur hall.

The protocol of field trips was for grade-schoolers to be led through the museum in orderly packs presided over by teachers and staff. Of course, the first challenge for some of us became separating ourselves from the group. Knowing we'd pay later only added spice. The building's immense scale, the incredible variety of exhibits, the lifelike quality of the dioramas, and the endless nooks and crannies and hideouts meant you could improvise, spontaneously create your own adventure. The longer you stayed away from the others, wandering on your own, the more intoxicated, empowered you felt. This was no ordinary world where everyday rules obtained. If you stared long enough, intently enough, the racoon poised above the rushing stream would snatch the trout flaunting itself dangerously close to the surface. Lions attacking the Arab merchant on his camel would roar. Yes, there were wires and mirrors and seams showing, but the tricks were magical enough to earn the benefit of the doubt.

From the second and third floor balconies you could peer down on the hall of bones. You'd know when you'd stayed away from your classmates long enough because you'd hear them gathering, the roll called below you, maybe even your own name echoing. Then it was time to go. To release the hostage you'd made of yourself.

One last bit of timing was crucial. The school crowd must be nearly at the far end of the fossil gallery, ready to turn the corner. Then and only then you'd take off after them, at the precise moment they'd begin to disappear, so you could sprint past the dinosaurs alone, scared half to death, past the bizarre shapes, through the valley of shadows and bones, holding your breath till a scream popped out that bounced off walls and shivered in the cavernous vault. The noise of your panic and flying feet a protective storm around your head, as long necks and groping paws and razor teeth bobbed and snapped at you.

A gauntlet of ancient skeletons. Of time and space unlike anything you'd experienced anywhere, anytime on Earth. It changed you. The world was different afterward. How old was old? How many strange creatures had died before the familiar ones came to life? Who made museums? Would you or somebody like you from Homewood wind up stuffed in a glass case with kids pointing at you and laughing or feeling sorry? Who killed the dead things in the museum? Was the zoo better or was this better, where animals stood still so you could learn their secrets?

No. The Carnegie Museum definitely wasn't Homewood. Nor was it Oakland or Downtown or East Liberty. Those neighborhoods and the people in them were all real, but they'd all fit inside the museum, inside the jaws of Tyrannosaurus rex.

In the museum you were a kind of big-eyed ant scurrying around trying to store everything you could until the next visit, but larger too, as large as you needed to be to wander and discover and summon the courage to run through the hall of bones where you could be torn apart and eaten at any moment. You were more than you thought you were. There was more of you to be.

I think of a city hiding from the bigger picture the natural history museum evokes. I consider a city cracking apart, where children are deprived of the imaginative space to find their differences and similarities. I remember returning to the museum with my kids, wanting nothing to be changed, wanting to run with them from one end of the hall of bones to the other, howling. I remember being proud. This is the city where I was raised and it has preserved this communal space. While we looked for a parking space, I was a voice-over: Heinz Chapel; Cathedral of Learning, with its African Room; Soldiers and Sailors Hall, where I graduated high school; concerts in Syria Mosque; hoops in the YMHA. Communal space, civitas, agora, marketplace, bazaar, acropolis, forum.

That was then and this is now. Is my mother safe? I stare at the telephone, stare into space. The loss of one life, one black young man gunned down in the street threatens the possibility of civilization. Each of us is a natural history museum, full of collections, brimming with the potential to teach, share, be discovered. And that is what is lost, the scale by which we are diminished when we are violently cut off, walled away from one another. It took all the past to bring that lost young man, to bring us, to this moment.

Under the roof of the natural history museum my sense of the past, of time was elaborated, extended; the past gained an immediacy and relevance that was frighteningly alien, daunting, but also included me. Ended and began with me. My imagination was stirred and I was on my way to becoming a citizen in a world larger than Homewood. I was lucky. I grew into manhood and passed on that experience to my children. The museum's still there for anybody who wants to listen. I hope.

Teachers vs. Machines: Computers in the Classroom

Douglas Rushkoff

Writer and software developer Douglas Rushkoff has written a number of books on technology and culture, including *Playing the Future: How Kids' Culture Can Teach Us to Thrive in an Age of Chaos* (1996) and *Cyberia: Life in the Trenches of Cyberspace* (1994).

At the *Visions* Web Site Link to Douglas Rushkoff's Web page, which includes sample chapters from his book.

READING PROMPT

Why does Rushkoff seem to think that all will be well in our classrooms once teachers change their attitude about computers? Do you share his optimism?

When I was in junior high school, teachers would show us 16mm films about frogs, foreign cultures, and the force of friction. These films were once-a-week treats for all of us. The teacher got a free babysitter and an opportunity to grade some papers. We kids got to take a nap, pass around notes, or space out at the colors on the screen and the sound of the projector motor.

Once the movie was over, the teacher would raise the shades and we'd return to business as usual. We never even discussed the huge mechanism that had taken charge of our classroom for the hour, its effect on us, or the intentions of the people who made the film that we watched.

Teachers who use the computer this same way are doing themselves, and their students, a disservice. Most of them have already figured out that the PC doesn't work as a babysitter, anyway. While movies and videotapes capture, tame, or at least pacify students, computers seem to do the opposite. They open up new possibilities, lead to more questions, and, when connected to the Internet, open the floodgates to people, information, and ideas that would not normally be invited into any classroom.

Accordingly, many teachers see the computers in the back of their classrooms as the enemy—if not their antagonists, then at best their successors. Teachers who do appreciate computers are themselves reluctant to incorporate them into their curriculums because they don't know how to use them as well as the kids do. While it may have been appropriate to let little hi-tech Johnny operate the movie projector, setting him down at the helm of the Macintosh is another story.

Understandably, computers threaten teachers' sense of authority. With computers in their classrooms, teachers no longer hold the keys to information. Students can scour a CD-ROM or the Internet and gain access to more information about a subject than any teacher could possibly know. Making matters seemingly worse, most students appear to know more about the computers themselves than the teachers do. If there's a browser crash, or IP address error, chances are one of the kids will be best qualified to fix it. The teachers begin to feel worthless, ignorant, and obsolete.

They needn't. Computers in classrooms are simply challenging teachers to dig down and discover just how valuable they really are. Teachers aren't mere repositories of data, however reassuring their depth of knowledge may have been to them in the past. That's never [been] what has earned them their students' respect. Think back for a moment on your best teachers—were they so great because they knew lots of information? Of course not. We valued them because they were inspiring human beings who taught us how to apply the information we learned. They created a context for learning, applying facts and concepts to our passion for real life. They weren't just knowledgeable; they were wise.

With computers supplying the raw data, the teachers' roles as interpreters and contextualizers are even more important. Think of it like a field trip to Spain for a Spanish teacher. Instead of seeing pictures of Spain in a book, the students wander through cathedrals, bull rings, and museums, encountering people, artifacts, and foods outside the teacher's prepared curriculum. The best of teachers see this as an opportunity and not a handicap. Rather than being threatened by the unexpected, they exploit it.

Computers may not make a teacher's job any easier, but they make it better. Teachers no longer need to assert their authority by meting out knowledge to supplicant children. Instead, they can exercise their true confidence by giving their students the keys to the castle.

The orthodox lesson plan of the past helped teachers direct an entire class, as a unit, through a particular set of skills. The linear progression went too slow for some, and left others behind. With students learning specific skills on the computer, teachers can create an area of concentration for each week—say, long division, or the Civil War—and then let students explore as far into that area as they can. The "weakest" students in a given subject will learn the most basic, essential skills, while the "strongest" will be free to explore in greater depth, without necessarily moving "ahead" of their peers.

What will the teacher do? More than ever before. Johnny finds out from his CD-ROM that Lincoln's troops, perhaps unprovoked, fired on Fort Sumter. Looks like he's ready to consider the works of revisionist historians. If the teacher doesn't know of any? Send Johnny to a library database and ask him what he learned.

Clearly it's not that easy. But the resistance to computers in the classroom may have more to do with our fears about the doors they open up than anything else. When teachers relieve themselves of the responsibility of having to know all the answers, they will rise to the greater challenge of becoming living partners in inquiry rather than static storehouses of information. That's what databases and 16mm films are for.

Revealing the Elephant *Alan Kay*

Alan Kay has long been at the center of the personal computing revolution. In the 1970s he worked at Xerox's innovative think tank, the Palo Alto Research Center (better known as Xerox PARC). Kay worked for the first major computer game provider, Atari, and then in 1984 he joined Steve Jobs at Apple Computer and helped in Apple's development of the Macintosh and its pioneering work in establishing a new graphical user interface (GUI) for personal computers. More recently he has left Apple for the Disney Corporation.

Throughout these years Alan Kay has worked closely with the Open Charter School in West Hollywood, California—an interest undoubtedly reflected in the following piece.

At the *Visions* Web Site Links to more by and about Alan Kay; to Xerox PARC.

READING PROMPT

What uses that Kay makes of the reference to the three blind men—each separately trying to describe an elephant—are based on his own experience?

One of the arguments advanced for why it is so difficult to get most children to learn to think in . . . new ways is that "this kind of thinking is hard to learn." But it is quite hard to learn to ride a bike, harder still to shoot baskets, and one of the hardest things to learn how to do is to hit a baseball consistently. If one watches children trying to learn these skills, what one sees is that they fail most of the time, but keep on trying until they learn, usually over years. This is more like their attitude when learning to walk and talk than the defeatism so often found in schoolwork. In fact, what really seems to be the case is that children are willing to go to any length to learn very difficult things, and endure almost an endless succession of "failures" in the process, if they have a sense that the activity is an integral part of their culture.

Maria Montessori used this determination very successfully in her schools. Suzuki has had similar success in music learning via setting up a musical culture in which the child is embedded. Difficulty is not the real issue here. Belonging to a culture and building a personal identity are. We could call this "rite of passage" motivation. If we hark back to the less than 5% of the American population that has learned to think in these new ways and recall that television is not a good medium to illustrate these new ways of thinking, this means that most children will have no embedded cultural experience in these ideas before coming to school. I don't know what percentage of elementary school teachers have learned to think in these new ways, but I would guess from personal experience that it is very similar to that of the population as a whole. This means that it will be very unlikely for most children to experience these new ways of thinking at home or at school or through television—especially as embedded into the general ways of doing and thinking which are so important to how children assign value to what they are going to try really hard to learn.

Let me give an analogy to how the "setting up an environment" strategy might be dealt with—it is drawn from a learning experience I had as a child. Suppose it were music that the nation is concerned about. Our parents are worried that their children won't succeed in life unless they are musicians. Our musical test scores are the lowest in the world. After much hue and cry, Congress comes up with a technological solution: "By the year 2000 we will put a piano in every classroom! But there are no funds to hire musicians, so we will retrain the existing teachers for two weeks every summer. That should solve the problem!" But we know that nothing much will happen here, because as any musician will tell you, the music is not in the piano! What music there is, is inside each and every one of us.

Now some things will happen with a piano in every classroom. The children will love to play around with it, and a "chopsticks culture" is likely to develop. Some will be encouraged by parents to take lessons, and a few rare children will decide to take matters into their own hands and find ways to learn the real thing without any official support. Other kinds of technologies, such as recordings, support the notion of "music appreciation." The problem is that "music appreciation" is like the "appreciation" of "science" or "math" or "computers"—it isn't the same as actually learning music, science, math or computing!

But 50 years ago, I had the experience of growing up in a community that desired "real music for all," and found a way to make it work. It was a little town in New England that had only 200 students in the high school, yet had a tradition of having a full band, orchestra and chorus. This required that almost every child become a fluent musician. They taught us to sing all the intervals and sight-read single parts in first grade. In second grade we sang two parts. In third grade we sang four parts and started to choose instruments. Talent was not a factor, though of course it did show up. This was something everyone did, and everyone enjoyed. I did not find out that this was unusual until I moved away. An important sidelight is that there was a piano in every classroom and all the teachers could play a little, though I am sure that at least one of the teachers was not very musical. What seemed to make it work was that the community had an excellent musical specialist for the elementary grades who visited each classroom several times a week. The central point to this story is not so much that most of the children became fluent musicians by the time they got to high school—they did and had done so for generations—but that as far as I can tell, almost all still love and make music as adults (including me).

We can find this "create-an-embedded-environment-and-support-classroom-teachers-with-visiting-experts" strategy in a number of schools today. The Open Charter School of Los Angeles has succeeded in setting up a "design culture" in their third grade classrooms that embeds the children in a year-long exciting and difficult adventure in the large-scale design of cities. The most successful elementary school science program I know of is in all of the Pasadena elementary schools and is organized along the same lines. It was developed by Jim Bowers and Jerry Pines, two Caltech scientists, and the key is not just an excellent set of curriculum ideas and approaches, but that the classroom teachers have to gain some real fluency, and there is important scaffolding and quality control by expert circuit riders from the district.

A good rule of thumb for curriculum design is to aim at being idea-based, not media-based. Every good teacher has found this out. Media can sometimes support the learning of ideas, but often the best solutions are found by thinking about how the ideas could be taught with no supporting media at all. Using what children know, can do, and are often works best. After some good approaches have been found, then there might be some helpful media ideas as well.

FROM MUSIC TO TECHNOLOGY

Now let me turn to the dazzling new technologies of computers and networks for a moment. Perhaps the saddest occasion for me is to be taken to a computerized classroom and be shown children joyfully using computers. They are happy, the teachers and administrators are happy, and their parents are happy. Yet, in most such classrooms, on closer examination I can see that the children are doing nothing interesting or growth-inducing at all! This is technology as a kind of junk food—people love it but there is no nutrition to speak of. At its worst, it is a kind of "cargo cult" in which it is thought that the mere presence of computers will somehow bring learning back to the classroom. Here, any use of computers at all is a symbol of upward mobility in the 21st century. With this new kind of "piano," what is missing in most classrooms and homes is any real sense of whether music is happening or just "chopsticks."

I have found that there are many analogies to books and the history of the printing press that help when trying to understand the computer. Like books, the computer's ability to represent arbitrary symbols means that its scope is the full range of human endeavors that can be expressed in languages. This range extends from the most trivial—such as astrology, comic books, romance novels, pornography—to the most profound—such as political, artistic and scientific discussion. The computer also brings something very new to the party, and that is the ability to read and write its own symbols, and to do so with blazing speed. The result is that the computer can also represent dynamic situations, again with the same range: from Saturday morning cartoons, to games and sports, to movies and theater, to simulations of complex social and scientific theories.

The analogy to a library of books and communication systems is found in the dynamic networking of millions of computers together in the Internet. One can use this new kind of library from anywhere on earth, it is continuously updated, and users can correspond and even work together on projects without having to be in the same physical location.

To us, working on these ideas 30 years ago, it felt as though the next great "500-year invention" after the printing press was born. And for a few—very like the few that used the book to learn, understand and debate powerful ideas and usher in new ways of thinking about the world—computers and networks are starting to be that important. The computer really is the next great thing after the book. But as was also true with the book, most are being left behind.

Here is where the analogy to books vs. television is most sobering. In America, printing has failed as a carrier of important ideas for most Americans. Few get fluent enough in reading to follow and participate in the powerful ideas of our world. Many are functionally illiterate, and most who do some reading, read for entertainment at home and for information on the job (viz. the 95% of bestsellers as stories and self-help). Putting The Federalist Papers on the Internet will eventually provide free access to all, but to have this great collection of arguments be slightly more accessible in the 21st century than it is today in public libraries will make no change in how many decide to read its difficult but worthwhile prose. Once again we are face to face with something that "is hard to learn," but has lost its perceived value to Americans—they ask why should they make the effort to get fluent in reading and understanding such deep content?

RETHINKING THE "TELEVISION MODEL"

Television has become America's mass medium, and it is a very poor container for powerful ideas. Television is the greatest "teaching machine" ever created—unfortunately, what it is best at teaching are not the most important things that need to be learned. And it is so bad at teaching these most important ideas that it convinces most viewers that they don't even exist!

Now computers can be television-like, book-like and "like themselves." Today's commercial trends in educational and home markets are to make them as television-like as possible. And the weight of the billions of dollars behind these efforts is likely to be overwhelming. It is sobering to realize that in 1600, 150 years after the invention of the printing press, the top two bestsellers in the British Isles were the Bible and astrology books! Scientific and political ways of thinking were just starting to be invented. The real revolutions take a very long time to appear, because as McLuhan noted, the initial content and values in a new medium are always taken from old media.

One thing that is possible with computers and networks, that could get around some of the onslaught of "info-babble," is the possibility of making media on the Internet that is "self teaching." Imagine a child or adult just poking around the Internet for fun and finding something—perhaps about rockets or gene splicing—that looks intriguing. If it were like an article in an encyclopedia, it would have to rely on expository writing (at a level chosen when the author wrote it) to convey the ideas. This will wind up being a miss for most Net surfers, especially given the general low level of reading fluency today. The computer version of this will be able to find out how old and how sophisticated is the surfer and instantly tailor a progression of learning experiences that will have a much higher chance of introducing each user to the "good stuff" that underlies most human knowledge. A very young child would be given different experiences than older ones—and some of the experiences would try to teach the child to read and reason better as a byproduct of their interest. This is a "Montessori" approach to how some media might be organized on the Internet: one's own interests provide the motivation to journey through an environment that is full of learning opportunities disguised as toys.

This new kind of "dynamic media" is possible to make today, but very hard and expensive. Yet it is the kind of investment that a whole country should be able to understand and make. I still don't think it is a real substitute for growing up in a culture that loves learning and thinking. But in such a culture, such new media would allow everyone to go much deeper, in more directions, and experience more ways to think about the world than is possible with the best books today. Without such a culture, such media are likely to be absolutely necessary to stave off the fast-approaching next Dark Ages.

Schools are very likely the last line of defense in the global trivialization of knowledge—yet it appears that they have not yet learned enough about the new technologies and media to make the important distinctions between formal but meaningless activities with computers and networks, and the fluencies needed for real 21st century thinking. At their best, schools are research centers for finding out interesting things, and like great research centers, these findings are best done with colleagues. There will always be a reason to have such learning centers, but the biggest problem is that most schools today are not even close to being the kinds of learning centers needed for the 21st century.

Will Rogers once said that it's not what you don't know that really hurts you, but what you think you know! The best ploy here—for computing, science, math, literature, the arts and music—is for schools to acknowledge that they don't know—they are the blind people trying to figure out the elephant—and then try to find strategies that will help gradually to reveal the elephant. This is what the top professionals in their fields do. We find Rudolph Serkin in tears at age 75 accepting the Beethoven medal, saying "I don't deserve this," and meaning it. We find Nobel physicist Richard Feynman telling undergraduates in his physics course at Caltech just how much he doesn't understand about physics, especially in his specialty! We can't learn to see until we realize we are blind.

The reason is that understanding—like civilization, happiness, music, science and a host of other great endeavors—is not a state of being, but a manner of traveling. And the main goal of helping children learn is to find ways to show them that great road which has no final destination, and that manner of traveling in which the journey itself is the reward.

From Global Village to Global Mind

Derrick de Kerckhove

Derrick de Kerckhove is the director of the McLuhan Program in Culture and Technology at the University of Toronto. Marshall McLuhan (1911–1980) himself was one of the most celebrated, flamboyant, and controversial communications theorists of the electronic age, forcefully and colorfully arguing that the new electronic communications technology (at the time, mostly television but looking ahead to computers and the Internet) was in the process of replacing national boundaries and all things linear with what he referred as a new *global village.*

McLuhan believed that the mode or medium of communications was as, if not more, important than the message—a notion he popularized in the title of his book, "*The Medium Is the Massage.*"

For present purposes, it is helpful to see Kerckhove following in a tradition that viewed new electronic technologies as freeing us from the narrow (linear) perspective of the industrial age.

For a different take on the constructive value of computer games, see the Sherry Turkle article in Chapter 5. Kerckhove's essay also ties in closely with Chapter 18, Global Concerns, arguing that computers offer the world a path free from control by any single entity.

At the *Visions* Web Site Link to the McLuhan Institute.

READING PROMPT

What makes Kerckhove so optimistic about the future? What factors, if anything, do you think he is overlooking or shortchanging?

While television has turned the planet into a village in which we are all neighbors, telematic networks abolish time and space and suppress traditional bearings of personal and collective identity.

If books, especially novels, fostered and sustained the development of private minds in public space, television has done the reverse, bringing a public mind to private spaces. Television screens are collective extensions of our individual minds. Instead of information being processed by a single person, namely me, the television screen presents me with information processed by a collective of which I am an integral part.

"Live" TV, or "real-time" TV, is a kind of collective eye which allows my eyes to look at reality processed for me and for every other person watching at the same time. Whenever something happens that is internationally newsworthy, a huge thoughtwave made up of millions of people grinding the same information at once forms and reforms itself every evening and sweeps over the nations from one time zone to another. Even in its standard programming, TV is a form of collective imagination, averaging people's hopes and fears on the basis of a regular public-pulse-taking through ratings.

Such thoughts may have occurred to American journalist Bill Moyers when he called his four ground-breaking shows on the workings of television *TV, the Public Mind* (1989). Of course, Moyers was really thinking about U.S. television, and at that time Hollywood and the Big Three U.S. channels still seemed to present a united front and a coherent vision of the world. Two questions are raised when the notion of television as a form of public mind is examined more closely. First, is the public mind uniquely dependent on TV, and if it is, then what happens when TV and TV audiences fragment as they are doing today? Second, is TV generating a public mind beyond the confines of the United States, and if so, who is running the show? According to French culturologist Augustin Berque, "If the world said OK to the crusade against Saddam [Hussein], it is not only because the world drinks Coca-Cola. It is because, to a large extent, the meaning of today's world finds its source, its creation and its distribution in the United States."

There is no doubt in my mind that television has done a lot to expand a sense of global destiny beyond the confines of North America. Trips to the moon, royal marriages and the Olympics, among other world coverage shows, have served to provide a common focus for the attention of hundreds of millions of people of different cultures. Cheap Hollywood movies and soaps have proposed to do the same on a day-to-day basis. Thus it is quite true that the public mind of the world has been more or less grounded in the country where the greatest expertise in television was found. But this may be changing very quickly. Just as quickly, in fact, as the world itself is "changing its mind."

With the U.S. military controlling all news delivery via the American news channel CNN, the Gulf War was perhaps the last occasion when a single television channel was given full range, albeit under duress, for the one-way production of meaning. The era of television as the principal supporter of the mass creation of meaning may be over. At least three new technological factors are undermining the hegemony of television and especially of American television: interactivity, digitization and networks.

INTERACTIVITY, THE RECOVERY OF PSYCHOLOGICAL AUTONOMY FROM TV

Television had already begun to "disintegrate" in the mid-1970s with the invention of the zapper. Zapping in and out of a commercial was the first step in giving power to the people over the screen. The second step was the widespread distribution of videorecording equipment. By recording one program while watching another, or recording a program while doing something else, the private user was becoming an informal "editor," taking revenge over the content/time constraints of television broadcasting. The third step was the appearance and fast distribution of personal computers.

Interactivity only became a concept after people had begun to use a keyboard and a mouse. They began to feel that they were taking power over the screen. Another step, a sidestep really, is the development and continuous refinement of videocameras. While in the 1960s and 1970s everybody was a consumer of television, in the 1990s we can expect to see more and more producers taking advantage of lightweight and efficient electronic pencils. After decades of passive acceptance of their dictatorship, we have learned to talk back to our screens.

UNIVERSAL DIGITAL CULTURE

By 1991, personal computers had already stolen the attention of a full generation from the seduction of the television screen. Computers affect culture in depth because they combine the characteristics of a mass-medium with those of a private one. They allow universal access to individual input and vice-versa. The new universalizing principle is that of the binary code. Thanks to digitization, the binary code can translate anything into anything else—forms, textures, sounds, feelings, even smell and why not taste soon. This is the nature of multimedia.

The binary code is the new common sense, the new common language which allows us to recreate technically outside our minds the complex arrangements that we have learned to build in our private imagination when we read novels. It is much more powerful, spreads far wider, and provides far more complex access to information-processing than television.

On the other hand, in spite of its universal calling, the binary code does not present a particular threat to local cultures and identities. On the contrary, it may allow them to flourish eventually because it is capable of supporting competing representations in non-exclusive fashion. The kind of collective mind produced by computers is different from the public mind of television. It is not only an object of attention, but also the subject of processing. It is an activity, not a spectacle. Thus it includes the user in the process. This is the reason why computers depend on networks to fully realize their socio-cultural potential.

NETWORKS AND OTHER HIGHWAYS

By 1991, computers were already celebrating what has become known as "convergence" with the telephone (the most powerful of the unsung technologies of communication) and television. The networks of national telephone companies today are vying for markets once reserved to a few broadcasters and cable operators. Audiovisual information is becoming ubiquitous as multimedia. On the electronic data highway, television will eventually lose its status as a broadcasting mass-medium and acquire that of the telephone, the narrowcasting medium with a billion interactive channels.

The Internet, which is the earliest and as yet the most complete expression of the electronic data highway, comes as close as has ever been possible to enabling a very large number of people to process information simultaneously. It is a technological expression of collective consciousness. To be more precise, the Internet is becoming a kind of global subconsciousness, with myriads of points of immergence for individuals into the collective, and points of emergence of the collective into the individual. Thus digitization turns everything into information, interactivity allows anyone to get into the processing, and networks connect every user to every other.

WHERE DO WE ALL FIT IN?

The idea of the "global village" was adequate for the television era which made everybody a neighbor to everybody else. It is quite inadequate to describe the new conditions of networked communications. Indeed, if television clearly supported the notion and even the perception of a common space, instant data networks abolish both time and space. They thus suppress the traditional bearings of identity, both personal and collective. In fact, many people seem to cultivate a new kind of anonymity on the Net. It is also apparent that, far from discouraging local initiative, the Net is promoting all kinds of new associations based not on political but on psychological boundaries. Different cultures can surely coexist on the Net. So we may be soon moving out of the global village, out of the public mind, into the "global mind." The rules that operate that mind are still unknown, but one absolutely essential criterion for guiding nations in their regulatory process is that universal access and guaranteed privacy be ensured.

While regulators are trying to resolve that paradox, we are all fairly confused. The creation of meaning is not homogeneous any more, ideologies have been thoroughly trashed along with the Berlin Wall, and television is henceforth too decentralized to pretend to the status of a public mind. Political and social strife find us irresolute. We are now too conscious of local sensibilities rather than insensitive to them. We fear to tread where we once ran. The world's record in Bosnia, Somalia and Rwanda is dismal and yet, around the evidence of disaster, there are signs of hope. Global thinking, evidenced by a rising sense of ecological responsibility and attendant environmental accounting, is finding expression in local action. Multiculturalism and political correctness are beginning to address North–South concerns. All these are indicators that in spite of hesitations about interfering in local affairs, the general level of attention and psychoplanetary pressure is rising.

The Glorious Revolution of 1971

Stephen Doheny-Farina

Stephen Doheny-Farina is professor of technical communications at Clarkson University and author of the book *The Wired Neighborhood.*

At the *Visions* Web Site Link to Doheny-Farina's personal Web page, which includes a twelve-minute audio interview that Doheny-Farina did with *Hotwired* on October 25, 1966.

READING PROMPT

Do you share Doheny-Farina's assessment that the benefits of cable television and the telephone were exaggerated in the past—just as, he suggests, those of computers are today?

During the last year while researching and writing a book about community and technology, I discovered, much to my surprise, that I lived through a great revolution when I was in high school and I didn't even know it had happened.

It all began just three days before our annual celebration of another, earlier revolution. On July 1, 1971 a seventeen year old boy confined to his home following a series of brain tumor operations participated in a history class via interactive cable television technology. Sitting in front of a video camera and aided by a special keyboard which enabled him to communicate with his teacher in an experimental studio–classroom four miles away, Jeff Hubert watched his teacher on his home television while she, in turn, watched Jeff on her monitor in the studio. The experiment was the initial step in a plan by TeleCable Corporation of Norfolk, Virginia, to develop interactive cable television to serve communities. Said the company president, Rex A. Bradley, "We feel that the addition of a two-way capability means the entire community has the opportunity of acting on the cable. A politician can propose a new idea and ask the viewers what they think of it. The people can push the Y button if they say yes or push the N button for no." (Rensberger)

In addition, on that same day TeleCable previewed the commercial capabilities of interactivity by demonstrating, as the news article (Rensberger) put it, "how a wife could shop from home" by viewing a "commercial for a laundry detergent" staged at a local Sears store and following the instructions "on what buttons to push" to order the detergent. Observers and practitioners predicted that the new technology could: provide up to 50 channels of recorded music, offer specialized news services in which a subscriber could enter codes and receive news about areas of interest, enable subscribers to "dial up" prerecorded pictures and textual information such as product ratings by consumer organizations, provide "electronic mail delivery that would, in effect, print out telegrams at home or office," and make the television keyboard and screen a computer terminal "by which the subscriber could dial up specific reference information and perform mathematical calculations" and have the answers printed on the screen.

Again, on the same day that Jeff Hubert had his first tele-history lesson, a group of New Yorkers celebrated the beginnings of free access to television technology with a day-long Washington Heights block party. (Fraser)

On that day, two Manhattan cable TV stations, TelePrompter and Sterling Manhattan Cable Television, became publicly available so that groups or individuals could apply for free on-air time on a first-come, first-served basis. "Supporters have hailed the program as the first genuine 'Town Meeting of the Air,'" said George Gent in a *New York Times* news article, "and a major step toward the political philosopher's dream of participatory democracy." A TelePrompter brochure defined the station as "a whole new concept in television." It is "TV by the people, for the people" enabling "any groups or individuals of any belief, purpose, or persuasion, to demonstrate their talents." If these people do not have the equipment or expertise to use this opportunity, the station will provide without charge studio space, at least one television camera and a director. It will also provide portable equipment "to cover events in the community, like block parties, park openings and church functions." (Gillespie 6)

Other pronouncements from the early 1970's offered similar visions. From a 1971 report from Princeton's Center for Analysis of Public Issues, we see the hope that the technology will enhance civic enterprises: "Free access public TV channels have the potential to revolutionize the communication patterns of service organizations, consumer groups, and political parties, and could provide an entirely new forum for neighborhood dialogue and artistic expression." (3)

From the 1972 book *Cable Television: A Guide for Citizen Action,* we see the hope that the new technology will make the home the center of civic exchange because cable TV will "make it possible to find out what's going on in your town, your neighborhood—even your block. Cable TV can provide local information the way the newspaper gives local news. But it can do more than the local paper. Cable Television can make it possible for your community organization to conduct a meeting of all the people in your neighborhood, without any of them having to leave their homes." (Price and Wicklein 2)

From another 1972 book, this one with the prescient title and subtitle, *The Wired Nation; Cable TV: The Electronic Communications Highway,* we see the hope that the new technology will wrest control from a centralized commercial elite and open up communications capabilities to the populace:

> Together, then, the elimination of channel scarcity and the sharp reduction of broadcasting cost can break the hold on the nation's television fare now exercised by a small commercial oligarchy. Television can become far more flexible, far more democratic, far more diversified in content, and far more responsive to the full range of pressing needs in today's cities, neighborhoods, towns, and communities. (Smith 8)

A number of pronouncements like these were cited in Gilbert Gillespie's 1975 study of public access television in which Gillespie, himself, makes some sweeping pronouncements:

> [C]ivic bodies must become thoroughly involved with the consideration of design and control of a wired city. For the first time there is an obligation to involve both individual private citizens of the most humble stature and community communications committees in planning the design and future control of an all-pervasive and revolutionary fact of city life. There is now an obligation on the part of the major city governments of Canada and the United States to maintain and invite a defined share of access for individuals and citizens' groups to the proliferating channels of cabled communication in the city. The city fathers must now nurture and eventually react, if they are not already doing so, to many new decentralized sources of local propaganda. (Gillespie 2)

My all too obvious point here is that people were saying the same thing in 1971 about public access television that they are saying now about the Net. Yes, and "they" said similar things in the past about other technologies, also.

Just check out the recent *Wall Street Journal* article with the wonderful title of "Future Schlock: Today's Cyberhype Has a Familiar Ring" to get a glimpse of the amazing claims made about telephony during the early part of this century as noted by Carolyn Marvin, author of *When Old Technologies Were New.* Daniel Pearl illustrates Marvin's research by pointing out that the telephone would, it was claimed, bring Peace on Earth, eliminate Southern accents, revolutionize surgery, stamp out "heathenism" abroad and save the farm by making farmers less lonely. The picture phone was just around the corner, and in 1912, technology watcher S. C. Gilfillan predicted that a "home theater" would, within two decades, let people dial up symphonies, presidential speeches and three-dimensional Shakespeare plays. The cost would be low and the "moral tone" would be excellent, since only the best material would survive. Novels, orchestras and movie theaters would vanish, and government as we know it might not survive either, he wrote.

Other examples about other technologies and other rounds of hype abound. "This isn't the first time a new medium has come along, promising to radically transform the way we relate to one another," says Todd Lappin in his comparison of the Internet to the early, non-commercial days of radio. Calling it "Today's Next Big Something," Lappin illustrates how current claims about the Net are similar to the claims made about radio in the years before it was transformed into the commercial broadcast industry. At that time, he explains, radio was a grass-roots, many-to-many medium in which its adherents disdained any kind of commercialization; rather, they saw it primarily as a means to link citizens.

So what happened to the public access revolution of 1971? Read Price and Wicklein's warning:

> Cable has gained so much momentum already that there is little chance it will be stopped. But if the Wired Nation does emerge, for whom will it be wired? If the public does nothing, the answer to that is easy: The nation will be wired primarily for the benefit of private entrepreneurs. Cable will then be much like broadcast television and radio before it. Programming will be restricted to mass-appeal entertainment, superficial reporting of news, and minimal discussion of public affairs. Cable subscribers will be sold to advertisers at so much a thousand, as the over-the-air audience are sold to them today. Community service and public access to the systems will be given lip service only, as they are in most commercial television and radio broadcasting. The opportunity for a revolution in communication through cable television will be lost. (18)

Hmmmm. What do you think? Was the revolution lost? Yeah, I think so. Got any new revolutions for us?

REFERENCES

Center for the Analysis of Public Issues. (1971, July). Public issues supplement no. 1.

Fraser, C. G. (1971, July 2). Cable TV has block party. *New York Times,* p. 67.

Gent, G. (1971, July 1). City starting test of public cable TV. *New York Times,* p. 95.

Gillespie, G. (1975). *Public access cable television in the United States and Canada.* New York: Praeger Publishers.

Lappin, T. (1995, May). Deja vu all over again. *Wired,* p. 175.

Pearl, D. (1995, September 7). Future schlock: Today's cyberhype has a familiar ring. *Wall Street Journal,* p. A1.

Price, M. E., and J. Wicklein. (1972). *Cable television: A guide for citizen action.* Philadelphia: Pilgrim Press.

Rensberger, B. (1971, July 2). Cable TV: 2-Way teaching aid. *New York Times,* p. 16.

Smith, R. L. (1972). *The Wired Nation.* New York: Harper & Row.

Get a Life! *Nicholas Negroponte*

Nicholas Negroponte is the founder and director of MIT's Media Lab. His homepage has links to many *Wired* magazine columns as well as to a Random House site devoted to his 1996 book, *Being Digital.* Compare his discussion of Luddites with Kirkpatrick Sale's in Chapter 13.

At the *Visions* Web Site Links to the Media Lab; to Negroponte's homepage; to excerpts from *Being Digital* (available through Open Book Systems).

READING PROMPT

Try to imagine the kind of elementary, middle, or secondary school Negroponte would like to see, based on his argument here.

Any significant social phenomenon creates a backlash. The Net is no exception. It is odd, however, that the loudest complaints are shouts of "Get a life!"—suggesting that online living will dehumanize us, insulate us, and create a world of people who won't smell flowers, watch sunsets, or engage in face-to-face experiences. Out of this backlash comes a warning to parents that their children will "cocoon" and metamorphose into social invalids.

Experience tells us the opposite. So far, evidence gathered by those using the Net as a teaching tool indicates that kids who go online gain social skills rather than lose them. Since the distance between Athens, Georgia, and Athens, Greece, is just a mouse click away, children attain a new kind of worldliness. Young people on the Net today will inevitably experience some of the sophistication of Europe. In earlier days, only children from elite families could afford to interact with European culture during their summer vacations abroad.

I know that visiting Web pages in Italy or interacting with Italians via e-mail isn't the same as ducking the pigeons or listening to music in Piazza San Marco—but it sure beats never going there at all. Take all the books in the world, and they won't offer the real-time global experience a kid can get on the Net: here a child becomes the driver of the intellectual vehicle, not the passenger.

Mitch Resnick of the MIT Media Lab recently told me of an autistic boy who has great difficulty interacting with people, often giving inappropriate visual cues (like strange facial expressions) and so forth. But this child has thrived on the Net. When he types, he gains control and becomes articulate. He's an active participant in chat rooms and newsgroups. He has developed strong online friendships, which have given him greater confidence in face-to-face situations.

It's an extreme case, but isn't it odd how parents grieve if their child spends six hours a day on the Net but delight if those same hours are spent reading books? With the exception of sleep, doing anything six hours a day, every day, is not good for a child.

ANYWARE

Adults on the Net enjoy even greater opportunity, as more people discover they can work from almost anywhere. Granted, if you make pizzas you need to be close to the dough; if you're a surgeon you must be close to your patients (at least for the next two decades). But if your trade involves bits (not atoms), you probably don't need to be anywhere specific—at least most of the time. In fact, it might be beneficial all around if you were in the Caribbean or Mediterranean—then your company wouldn't have to tie up capital in expensive downtown real estate.

Certain early users of the Net (bless them!) are now whining about its vulgarization, warning people of its hazards as if it were a cigarette. If only these whiners were more honest, they'd admit that it was they who didn't have much of a life and found solace on the Net, they who woke up one day with midlife crises and discovered there was more to living than what was waiting in their e-mail boxes. So, what took you guys so long? Of course there's more to life than e-mail, but don't project your empty existence onto others and suggest "being digital" is a form of virtual leprosy for which total abstinence is the only immunization.

My own lifestyle is totally enhanced by being online. I've been a compulsive e-mail user for more than 25 years; more often than not, it's allowed me to spend more time in scenic places with interesting people. Which would you prefer: two weeks' vacation totally offline or four to six weeks online? This doesn't work for all professions, but it is a growing trend among so-called "knowledge workers."

Once, only the likes of Rupert Murdoch or Aga Khan could cut deals from their satellite-laden luxury yachts off the coast of Sardinia. Now all sorts of people from Tahoe to Telluride can work from the back seat of a Winnebago if they wish.

B-RATED MEETINGS

I don't know the statistics, but I'm willing to guess that the executives of corporate America spend 70 to 80 percent of their time in meetings. I do know that most of those meetings, often a canonical one hour long, are 70 to 80 percent posturing and leveling (bringing the others up to speed on a common subject). The posturing is gratuitous, and the leveling is better done elsewhere—online, for example. This alone would enhance US productivity far more than any trade agreement.

I am constantly astonished by just how offline corporate America is. Wouldn't you expect executives at computer and communications companies to be active online? Even household names of the high-tech industry are offline human beings, sometimes more so than execs in extremely low-tech fields. I guess this is a corollary to the shoemaker's children having no shoes.

Being online not only makes the inevitable face-to-face meetings so much easier—it allows you to look outward. Generally, large companies are so inwardly directed that staff memorandums about growing bureaucracy get more attention than the dwindling competitive advantage of being big in the first place. David, who has a life, needn't use a slingshot. Goliath, who doesn't, is too busy reading office memos.

LUDDITES' PARADISE

In the mid-1700s, mechanical looms and other machines forced cottage industries out of business. Many people lost the opportunity to be their own bosses and to enjoy the profits of hard work. I'm sure I would have been a Luddite under those conditions.

But the current sweep of digital living is doing exactly the opposite. Parents of young children find exciting self-employment from home. The "virtual corporation" is an opportunity for tiny companies (with employees spread across the world) to work together in a global market and set up base wherever they choose. If you don't like centralist thinking, big companies, or job automation, what better place to go than the Net? Work for yourself and get a life.

chapter 10 Explorations

10a Favorite Teacher

Collaboration. Share with classmates a detailed account of a favorite teacher; see if you can find any common traits in your accounts.

Web work. The goal of Harvard University's Project Zero is to foster better, more creative teaching.

Take a look at some of the resources available to teachers wanting to use Web technologies in their classroom. Teachers.net provides teachers with resources for getting online, courtesy of technology vendors. Turner Learning is an attempt to provide teachers with materials produced by Turner Broadcasting, now part of Time Warner. IBM offers a similar service to teachers.

The National Education Association is the largest teacher organization in the country, followed by the American Federation of Teachers.

One writing topic. Define what makes a good teacher. Drawing on your own experience, how have good teachers utilized the full range of instructional technologies available to them?

10b Teachers Outside of School

Collaboration. Many people play the role of teachers in our lives without getting paid for it or without working in a school. Share a description of how someone in your life, preferably a non–family member, helped teach you a valuable skill or lesson.

Web work. One traditional area of self-help has been in career and character building, or planning for success. Amazon.com has a division called PeopleSuccess. One of the first brand names in planning for success was Dale Carnegie, now organized as a corporate training service, Dale Carnegie Training; for information on the man himself (1888–1955), see the Dale Carnegie page.

In the 1990s there seemed to be more interest in personal or emotional success than in getting ahead in business, and no one has become more popular than Martha Stewart, not surprisingly the subject of satire: Recreation > Home and Garden > Martha Stewart > Parodies.

One area in which people invest a large amount of time and effort is learning how to build or strengthen personal relationships. Best-selling authors such as Leo Buscaglia and Nancy Friday have created a cottage industry of books, videos, classes, and now Web sites.

One writing topic. Write an essay on the kinds of teachers, famous and not, who are available to help you individually or people generally to master some important skill or lesson in life.

10c Field Trips

Collaboration. Share a detailed account of a school field trip you took.

Web work. Visit TerraQuest for list of current trips it is sponsoring, as well as Virtual Expeditions, which lists current trips it sponsors as well as its archive.

As of December 1997, Yahoo! listed some sixteen sites under Recreation > Travel > Virtual Field Trips and dozens more under Recreation > Travel > Travelogues > Ongoing Travelogues.

PBS also sponsors electronic field trips. Then there's Are We There Yet? The Ultimate Field Trip Guide for New York City and Philadelphia.

One writing topic. Contrast the advantages and shortcomings of real and virtual field trips.

10d Clubs and Sports

Collaboration. Share a list of all extracurricular activities you have participated in.

Web work. There are two sites with information about the latest in one of the fastest-growing extracurricular activities, youth soccer: the United States Youth Soccer Association and the American Youth Soccer Organization.

Meanwhile, Yahoo! lists homepages for some twenty high school debate clubs at Social Science > Communications > Forensics > Debate > High School > Clubs Teams and Societies, hundreds of homepages for marching bands at Entertainment > Music > Genres > March > Marching Bands > K–12 > High School, a handful of high school literary magazines at Arts > Humanities > Literature > Magazines > High School, and even a half-dozen sites under high school a capella choirs at Entertainment > Music > Artists > By Genre > A Cappella > High School, but (alas) as of December 1997 only two high school math clubs at Science > Mathematics > Organizations > High School.

Other sites worth considering: Future Farmers of America, the Odyssey of the Mind, and the Internet Headquarters for Student Government.

One writing topic. As a cost-saving move, your local school board is contemplating eliminating from your high school the one extracurricular activity that you found most enjoyable or helpful. Write a letter of protest to the local board explaining the value of that activity.

10e "Educational" TV

Collaboration. Share a list of all the different ways your teachers have ever used television, videos, and movies. Put a star by each use you found especially helpful, and X by ones that struck you as a waste of time. Can your group find any common reasons underlying its responses?

Web work. What is now known as "public television" began under the name of "educational television"; South Carolina Educational TV remains a leader in using television to deliver curriculum materials to students and teachers.

Public Broadcasting System has a Web site for general information about its programming and for teachers who might want to use it in their classes.

The Discovery Channel's School Online attempts to integrate cable channel programming into school curriculum.

One writing topic. What guidelines can you offer teachers regarding how often and for what reasons they should use video in the classroom? What makes video a useful teaching tool? Under what conditions are videos misused in a classroom?

10f College Composition on the Web

Collaboration. Share a list of the different ways you have gotten help over the years for your school writing assignments.

Web work. Several universities offer online versions of their writing centers, notably Purdue University and Rensselaer Polytechnic Institute. Although the Write Site bills itself as the Web's first independent online writing lab, it has links to Mercer County Community College—but in any case seems to offer almost no help for writers.

An early, now out-of-copyright version of what was to become one of the most widely used writing manuals of all time, Strunk and White's *The Elements of Style,* is available on the Web. (Really it is just William Strunk Jr.'s book before E. B. White's additions.)

ResearchPaper.com is a service of Infonautics, a company that provides fee-based Web access to a wide assortment of newspapers, magazines, and other reference sources. Their site will help you select and locate information and even show that information to you if it is available for free on the Web; it will also show you titles of articles available only to customers of Infonautics' Web service, the Electric Library.

Inkspot and WriteLinks are two good general Web sites for writers.

One writing topic. A committee on instruction at your college is looking into improving student writing in the full range of courses across the curriculum. Write a report to that committee with recommendations based on your experiences (positive or negative) in the very course in which you are using this book.

chapter 11

Books and the Changing Nature of Reading

READING QUESTIONS

- What is it about books that you have most enjoyed and least enjoyed in your life?
- Of the stories that have meant the most to you, which ones did you learn from books, which ones from TV and movies, which ones from other people?
- How are people using computers for storytelling? What advantages and what disadvantages does the computer offer?

QUOTATIONS

The paperback is very interesting but I find it will never replace the hardcover book—it makes a very poor doorstop.

— *Alfred Hitchcock, British film director, 1899–1980*

Today, thanks to technical progress, the radio and the television, to which we devote so many of the leisure hours once spent listening to parlor chatter and parlor music, have succeeded in lifting the manufacture of banality out of the sphere of handicraft and placed it in that of a major industry.

— *Nathalie Sarraute (1960), Russian-born French novelist*

Looking at the proliferation of personal web pages on the net, it looks like very soon everyone on earth will have fifteen Megabytes of fame.

— *Attributed to M. G. Siriam*

What we have introduced with MTV is a nonnarrative form. As opposed to conventional television, where you rely on plot and continuity, we rely on mood and emotion. We make you feel a certain way, as opposed to you walking away with any particular knowledge.

— *Bob Pittman (1989), founder of MTV*

"S-t-e-a-m-boat A-Comin'!"

Samuel Langhorne Clemens

The writer who was to become world famous as Mark Twain was still in his late twenties when he penned this famous reminiscence of his boyhood on the Mississippi River, a piece later incorporated into *Life on the Mississippi* in 1883.

This short excerpt reveals much about both Clemens's strength as a writer and the characteristics of descriptive writing in the great age of print—a time when readers delighted in learning by slowly working themselves through long descriptive passages. This particular excerpt is a single paragraph—as long as many freshman essays (670 words)—highlighted by two magnificent sentences, each more than 200 words in length, and the first having only a single simple subject and verb ("I can picture . . . ").

At the *Visions* Web Site Links to more by and about Mark Twain; to the Mississippi in pictures.

READING PROMPT

Speculate on the relative impact of both radio and television on this kind of extended descriptive writing.

Once a day a cheap, gaudy packet arrived upward from St. Louis, and another downward from Keokuk. Before these events, the day was glorious with expectancy; after them, the day was a dead and empty thing. Not only the boys, but the whole village, felt this. After all these years I can picture that old time to myself now, just as it was then: the white town drowsing in the sunshine of a summer's morning; the streets empty, or pretty nearly so; one or two clerks sitting in front of the Water Street stores, with their splint-bottomed chairs tilted back against the wall, chins on breasts, hats slouched over their faces, asleep—with shingle-shavings enough around to show what broke them down; a sow and a litter of pigs loafing along the sidewalk, doing a good business in watermelon rinds and seeds; two or three lonely little freight piles scattered about the 'levee'; a pile of 'skids' on the slope of the stone-paved wharf, and the fragrant town drunkard asleep in the shadow of them; two or three wood flats at the head of the wharf, but nobody to listen to the peaceful lapping of the wavelets against them; the great Mississippi, the majestic, the magnificent Mississippi, rolling its mile-wide tide along, shining in the sun; the dense forest away on the other side; the 'point' above the town, and the 'point' below, bounding the river-glimpse and turning it into a sort of sea, and withal a very still and brilliant and lonely one. Presently a film of dark smoke appears above one of those remote 'points'; instantly a negro drayman, famous for his quick eye and prodigious voice, lifts up the cry, 'S-t-e-a-m-boat a-comin'!' and the scene changes! The town drunkard stirs, the clerks wake up, a furious clatter of drays follows, every house and store pours out a human contribution, and all in a twinkling the dead town is alive and moving. Drays, carts, men, boys, all go hurrying from many quarters to a common center, the wharf. Assembled there, the people fasten their eyes upon the coming boat as upon a wonder they are seeing for the first time. And the boat is rather a handsome sight, too. She is long and sharp and trim and pretty; she has two tall, fancy-topped chimneys, with a gilded device of some kind swung between them; a fanciful pilot-house, a glass and 'gingerbread', perched on top of the 'texas' deck behind them; the paddle-boxes are gorgeous with a picture or with gilded rays above the boat's name; the boiler deck, the hurricane deck, and the texas deck are fenced and ornamented with clean white railings; there is a flag gallantly flying from the jack-staff; the furnace doors are open and the fires glaring bravely; the upper decks are black with passengers; the captain stands by the big bell, calm, imposing, the envy of all; great volumes of the blackest smoke are rolling and tumbling out of the chimneys—a husbanded grandeur created with a bit of pitch pine just before arriving at a town; the crew are grouped on the forecastle; the broad stage is run far out over the port bow, and an envied deckhand stands picturesquely on the end of it with a coil of rope in his hand; the pent steam is screaming through the gauge-cocks, the captain lifts his hand, a bell rings, the wheels stop; then they turn back, churning the water to foam, and the steamer is at rest. Then such a scramble as there is to get aboard, and to get ashore, and to take in freight and to discharge freight, all at one and the same time; and such a yelling and cursing as the mates facilitate it all with! Ten minutes later the steamer is under way again, with no flag on the jack-staff and no black smoke issuing from the chimneys. After ten more minutes the town is dead again, and the town drunkard asleep by the skids once more.

On the Limits of Writing

Plato

Plato is often depicted as one of the first great antiprogressives, a thinker deeply disturbed by democracy, technology, and progress—all things that seemed to bother him because they act in the world without the necessary guidance of wisdom (or, maybe more important to Plato, of wise people). In this famous passage from the *Phaedrus* dialogue (available in its entirety on the Web), Plato has Socrates express doubts about what in his day was the relatively new technology of writing.

At the *Visions* Web Site Links to more by and about Plato.

READING PROMPT

Complete this thought experiment in the form of a Platonic dialogue in which you have Socrates take the position that human movement unaided by machines will always be more beautiful and, in key ways, even more practical than any form of movement aided by technology.

Socrates: At the Egyptian city of Naucratis, there was a famous old god, whose name was Theuth; the bird which is called the Ibis is sacred to him, and he was the inventor of many arts, such as arithmetic and calculation and geometry and astronomy and draughts and dice, but his great discovery was the use of letters. Now in those days the god Thamus was the king of the whole country of Egypt; and he dwelt in that great city of Upper Egypt which the Hellenes call Egyptian Thebes, and the god himself is called by them Ammon. To him came Theuth and showed his inventions, desiring that the other Egyptians might be allowed to have the benefit of them; he enumerated them, and Thamus enquired about their several uses, and praised some of them and censured others, as he approved or disapproved of them. It would take a long time to repeat all that Thamus said to Theuth in praise or blame of the various arts. But when they came to letters, This, said Theuth, will make the Egyptians wiser and give them better memories; it is a specific both for the memory and for the wit. Thamus replied: O most ingenious Theuth, the parent or inventor of an art is not always the best judge of the utility or inutility of his own inventions to the users of them. And in this instance, you who are the father of letters, from a paternal love of your own children have been led to attribute to them a quality which they cannot have; for this discovery of yours will create forgetfulness in the learners' souls, because they will not use their memories; they will trust to the external written characters and not remember of themselves. The specific which you have discovered is an aid not to memory, but to reminiscence, and you give your disciples not truth, but only the semblance of truth; they will be hearers of many things and will have learned nothing; they will appear to be omniscient and will generally know nothing; they will be tiresome company, having the show of wisdom without the reality.

The Medium and the Message: An Exchange

Sven Birkerts and Wen Stephenson

The year 1995 saw the publication of Sven Birkerts's lament on computers and the end of writing as we know it, aptly titled *The Gutenberg Elegies: The Fate of Reading in an Electronic Age.* Our basic relationship to texts and, in turn, to narratives, Birkerts argued, changes radically when the computer instead of the printed page becomes the primary means of transmission. Critic Wen Stephenson then wrote a response taking Birkerts to task for, among other things, ignoring the "vast potential" of the Internet "to expand the audience for works of the literary imagination; and not only to expand access but also opportunities for interactivity, and for building communities of creative minds that could not exist otherwise."

Below, from the *Chicago Review,* are Birkerts's response to Stephenson and a short follow-up by Stephenson.

At the *Visions* Web Site Links to excerpts from *The Gutenberg Elegies;* to a *Hotwired* Brain Tennis session from July 1997 between Birkerts and MIT's Janet Murray; to online interviews with Birkerts; to reviews of *The Gutenberg Elegies.*

READING PROMPT

In what ways is the debate between Birkerts and Stephenson a continuation of the same concerns raised by Socrates in the Phaedrus? *What is Stephenson's main point of disagreement with Birkerts? Whose position do you tend to side with?*

Sven Birkerts writes:

I was gratified to read Wen Stephenson's lucid and concept-focused review of my book *The Gutenberg Elegies: The Fate of Reading in an Electronic Age* ("The Message Is the Medium," *Chicago Review* 41:4 [Fall 1995]). Read thus one feels seen. Seen, and in some important ways disagreed with, which at some level one wants (and if I don't get rid of this "one" I'm going to drive myself crazy).

I have various quibbles with Mr. Stephenson's quibbles, but I will save those for another occasion. I would like to address myself instead to what I see as our key point of difference, a point which, as I think about it, seems to enlarge itself concentrically until it embraces everything.

It comes down to this: Mr. Stephenson does not think that the medium through or by which the word is transmitted changes in any way its fundamental essence. I maintain that it does. And since Mr. Stephenson more or less recapitulates my argument in the body of his essay, I will not repeat it here. Let me fasten upon his closing salvo, however, the point of which is to reassure us that verbal content survives the transfer between media unchanged. Mr. Stephenson writes:

> . . . I jump to another page that I've 'bookmarked,' where I find the poem that's been going through my head for the last few days, ever since I discovered it here on the Internet Poetry Archive site. I open the sound file I've saved on my hard disk, and the soft Irish voice of Seamus Heaney comes over the speakers of my computer. I follow his voice through the eight concentrated lines of 'Song':
>
> A rowan like a lipsticked girl.
> Between the by-road and the main road
> Alder trees at a wet and dripping distance
> Stand off among the rushes.
> There are the mud-flowers of dialect
> And the immortelles of perfect pitch
> And that moment when the bird sings very close
> To the music of what happens.
>
> What does happen when I am reading or listening to a poem? And does it matter that it is transmitted to me, voice and word, through a computer? The second part of the question is beginning to bore me by now. The first part I doubt I'll ever answer. And so I print out . . . Heaney's poems and take them with me to pore over on the train-ride home.

Mr. Stephenson is, of course, playing a game here, representing in a few sentences his reliance on all of our modes of interacting with a given literary expression—from oral memory ("the poem that's been going through my head") to print ("so I print it out") to electronic ("since I discovered it here on the Internet Poetry Archive site") to electronic multimedia ("I open the sound file"). It is supremely ironic, however, that the content itself should be Heaney's lovely "Song," the whole point of which is to reenact through language a kind of stripping away of the veils in order to arrive at a pure perceptual recognition, one uncontaminated by any of our myriad devices of mediation.

Mr. Stephenson would no doubt reply that mediation or no, his absorption in the Heaney poem is proof positive that the word is like the patterned energy of one of those knots that can be slipped intact from rope to rope provided that the ends are connected; that, in other words, the medium of transmission is functional and nothing more.

To that I would counter that the sense of presence that literature seeks to create is primarily—not exclusively—focused on the private and social circumstance of the individual, and that this sense is fundamentally at odds with the electronic system that would store and present it. That the world that brings us the Web is already at a significant remove from the worlds conjured to exist in a book. So that even if it did not matter on one level—even if Mr. Stephenson were encountering his text as purely as he claims—it would matter on another. To deny this is tantamount to asserting that the automobile has had no impact on the natural world because with it we are able to get to more remote places than before. No, the auto has significantly altered the dynamic between man and nature. It is, then, the fact of the new medium, not just its means, that I am talking about.

But the means, I maintain, matter significantly. The tree hiked to and seen is not the tree driven to and seen, even though it is the same tree. The words of "Song" may be the same no matter what context we greet them in, but different media put us in fundamentally different relations. The listener, of course, gets the benefit of presence and immediacy (in the etymological sense of no mediation). The reader of the lines on the page performs the familiar conversion of printed word into auditory signal. For the person reading the poem on the screen there must be the subliminal awareness that the word has passed through an alchemical bath, has travelled a circuitry. Same word, but. . . . Do we say that a print on paper and a digitalized image on a monitor display are identical, that the image stands free of its context? No, in some elusive way we recognize that the transmission has become part of the content (read again Benjamin's "The Work of Art in the Age of Mechanical Reproduction"). Heaney's poem handset on a broadside and the same poem run off on a mimeo-stencil will reach the reader differently. This is not to say that the skilled and serious reader (and I believe Mr. Stephenson is such) cannot peel away the husks or otherwise read compensatorily in order to get at the pure word (rather, its chimera). But to manage this requires discipline and a high degree of awareness, and the ordinary reader generally lacks both.

It is vexing for me not to be able to lay the whole business to rest with a few incontrovertible arguments. But alas—and mercifully—we are in subjective terrain. I can only argue what I suppose and my reader/critic can only do the same. I will say, though, that as soon as one allows that the medium affects the message (never mind McLuhan's assertion that it is the message), then the whole electronic transformation of society needs to be examined with great care. If there is no effect, and no change, then there is no problem, other than whatever the problem has been all along.

Wen Stephenson responds:

I'm reluctant to reply to Sven Birkerts's generous letter in response to my article on *The Gutenberg Elegies*, partly because he so nicely describes what is perhaps an irreconcilable difference between us, but also because I feel the whole debate over the printed versus the digitally transmitted word has become distracting and even counterproductive. For if Mr. Birkerts and I recognize the extent to which we are allies in a much larger struggle, then this kind of debate starts to look more and more like that between factions of a religious sect (say, the cult of literary–aesthetic experience) arguing over fine points of doctrine while the foundations of their faith are under assault and beginning to crumble. So, rather than try to rebut his arguments, and indulge in further speculation on the nature of verbal experience, I'm more inclined to take his last sentence as a point of departure (or at least to suggest where such a departure might lead).

The digital age is a given. Like it or not, it is the future that my generation has inherited. Whatever the effects, whatever the superficial changes brought on by the latest phase of the ongoing technological revolution, they are the least of our problems—both as citizens and as readers. Nevertheless, despite the givenness of the digital future, I don't believe we are faced with an all-or-nothing choice between print and pixel. Books and computers can exist side by side, have been known to do so productively, and may complement one another in surprising, delightful ways. No, the choice we face is much more urgent and stark: namely, between a future in which literature has some discernible influence within our culture and one in which it does not. That we are faced with this choice has little or nothing to do with the nature of new media such as the World Wide Web and CD-ROMs, and everything to do with the nature of well-established mass media such as radio, television, and film, and the commercialized mass culture these media have promoted and sustained for decades.

In the struggle for literature (and the values that term implies), it would be a grave mistake to equate the Web and other multimedia technologies with the electronic mass media that have brought us to this pass in our cultural history. The Web is not television. It is neither the movies nor the recording industry. It is something entirely different, entirely new. While it does bring together image and sound, primarily it conveys the written word. Furthermore, not only is the medium based on language but also on the ability—the imperative—of individuals to make choices, to seek out and select the kinds of content they want to engage and interact with intellectually. And it is because of these and other essential attributes that the Web has the potential to reawaken an appreciation for language itself, both as a written and a spoken medium, and for literary works. Publishers and educators are beginning to realize that the Web and other new media can be powerful weapons in the struggle for literature and literacy.

More needs to be said, and in much greater depth than I am able to do here, about the significance of the Web as a radically decentralized network for the distribution of literature. Not only is it inexpensive to produce content on the Web, it's getting cheaper to gain access to it, and someday soon it will be as accessible as the telephone, the radio, and the television. There's already a much publicized presidential initiative to bring Internet access to every public school in America.

So let me toss this line to my "unregenerate" novel reader, resolved to go down with the ship of serious literature. The Web may never replace books—whether novels, or nonfiction narratives, or collections of poetry. I, for one, believe that literature—both in print and digital form—will endure, and that the Web itself may play an important role in its survival. For if publishers seize the opportunity, multimedia publishing on the Web could well revitalize, maybe even revolutionize, the role of literary journalism in our culture. It is too easy to be short-sighted, to forget that this medium we call the Web is in its infancy: who knows what sophisticated forms of literary activity it may accommodate in the near future? Imagine the impact such activity could have. Consider the rise of mass literacy, the development of journalism, and the beginnings of the novel in the eighteenth century. As access to the Internet expands, multimedia literary journalism on the Web might play a similar role in generating a broadened interest in and demand for serious literature, even novels, at the beginning of the twenty-first century.

The Last Picture Shows | *Charles Oliver*

Novels were a major form of entertainment in the nineteenth century, and many conservative educators had reservations about making the study of novels a major part of the English curriculum in schools—at least until other, easier forms of popular culture made novel reading suddenly seem relatively highbrow. Is it really possible now, at the beginning of the twenty-first century, that we may be about to go through a similar debate over the study of films, as video games and other forms of popular entertainment start to make the traditional feature film seem old-fashioned?

Charles Oliver is a contributing editor to *Reason* magazine, produced by the Reason Foundation, an organization devoted to supporting "public policies based on rationality and freedom." The magazine's slogan, "free minds and free markets," identifies its libertarian perspective (for more on libertarianism see the introductions to the David Boaz and Paulina Borsook pieces in Chapter 6), although this ideological perspective may not have much to do with Oliver's 1994 essay on the fate of movies. The essay's title is a play on Peter Bogdanovich's classic 1971 film about coming of age in a small West Texas town, *The Last Picture Show.*

At the *Visions* Web Site Links to lots more about movies.

READING PROMPT

What, if anything, do you see the motion picture industry doing to combat the threats to movie attendance noted by Oliver?

Hollywood just celebrated its biggest year ever. With an estimated $5.24 billion in North American ticket sales in 1993, the movie business seems to be booming. But some in the industry are worried about its future.

U.S. ticket sales have remained flat for the last 20 years or so. Theatrical grosses keep going up because ticket prices keep increasing. This year's record box office owes more than a little to the fact that this year's average ticket price of $5.05 was also a record high. Indeed, if one takes out the sales of one movie—*Jurassic Park*—1993 doesn't seem like a terribly good year for the theatrical-film industry.

It is too soon to proclaim the death of film. But a convergence of technological changes and economic forces is presenting the greatest threat to cinema-going since the introduction of television in the late 1940s.

Since the 1960s, the core of the movie-going audience has been young people, particularly young men, ages 13 to 25. But that is changing. As late as 1983, according to Motion Picture Association of America figures, young people accounted for 55 percent of all movie tickets sold. That was way above their proportion of the population. But by 1992, the latest year for which figures are available, young people accounted for just 38 percent of all tickets sold, while their proportion of the population had shrunk only slightly.

A problem for the movie industry? You bet.

Historically, people go out to movies less and less frequently as they grow older. That's still happening, though older baby boomers aren't cutting their movie attendance as much as their parents did. The fact that today's youth don't go to the movies that much bodes ill for the future of Hollywood. In effect, they are the first generation of Americans who don't participate in the tribal ritual of going to the movies on most weekends.

These young people have been lost largely to video games and videotape rentals. The movies have, in effect, priced themselves out of competition for teenage dollars.

Video games are now a $7-billion industry. The main consumers of video games are the young males so coveted by the movie industry. And one of the big attractions of video games is the price. The cost of admission to a feature film that will probably last about two hours can range from about $4.00 to $7.50. But for well under half that amount, a person can rent a video game cartridge for a whole day. This is an important consideration for young people with their relatively small disposable incomes.

Price is also the advantage of videotapes. Again, young people make up the biggest segment of the videotape market. If a group of five young people wants to see a film, it would cost them about $25 to see it in a theater. But it would cost no more than $3.00 to rent it on videotape.

Virtually all theatrical films now become available on videotape about six months after their initial release. Since people know that they can eventually see just about any film on video, they are becoming pickier about what they will see in a theater. If a film doesn't have spectacular effects like those in *Jurassic Park* or *The Fugitive,* or some other elements that are best appreciated on the big screen, people will likely opt to wait for the film to appear in the cheaper medium of video. These factors have helped push video rentals and sales to $11 billion a year, more than twice the dollar amount of theatrical ticket sales.

And even if people have to see a film on the big screen, they are increasingly choosing to wait until a film appears at a second-run theater, where tickets usually cost about half what they do in regular theaters. Ticket sales for these venues have about doubled over the last decade, according to some industry analysts.

The prognosis for theatrical films seems to be dim. The coming information superhighway, with its linking of computers and television, cable and telephones, and entertainment and information poses a threat in several different ways.

First of all, it will, probably within a few years, make available true video on demand. In the best possible scenario for film, pay per view simply displaces video rentals. The market for theatrical film is unaffected, and studios can cut out the middleman and actually collect those rental dollars themselves.

But that isn't likely. As it becomes even easier to see any movie or episode of any television series one wants without ever leaving the house, people will likely choose to see more of their entertainment at home. Indeed, advances in high-definition television will reduce one of theater's big advantages over home viewing by making it possible for people to have wall-sized television screens with motion-picture quality images. This will certainly hurt the theater business, though it is difficult to predict to what degree.

Perhaps even more ominous are the future generations of video games that can be delivered into the viewer's home. In the very near future, cable systems will deliver into homes games whose advanced graphics and scripts will make today's games look like toy logs. These more interesting games will probably tempt people away from theaters. But more important, children weaned on this interactive form of entertainment may never develop the taste for the traditional linear narrative of film. Why would someone want to watch a film about Indiana Jones when he can be Indiana Jones within the confines of a very realistic and unpredictable game?

So is there any hope for the survival of theatrical films? Maybe.

As the home television/computer/videophone becomes an ever more important part of people's lives, they may seek opportunities to escape from the confines of their homes. The cinema could be one of their destinations.

Indeed, people who are dating may be reluctant to invite people into their homes on the first few dates. Teenagers in particular may want to date without the supervision of their parents. The cinema offers a nice, neutral meeting place. And we may find that people, being innately lazy, will sometimes prefer passive entertainment to the interactive variety, so even the narrative form of film may survive.

But if the cinema does survive, it may not look very much like those of today. To see a glimpse of the cinema of tomorrow, go ride the *Back to the Future* ride at Universal Studios. Or go to the Luxor casino in Las Vegas and see any of the rides featured there. These rides combine short narrative films made with state-of-the-art equipment with motion simulators to give the feeling of actually being in the film.

If this sort of technology is combined with strong, sustained storylines, it could prove the savior of film. Indeed, Sony, through its Loew's theater chain, is taking some tentative steps in that direction. It is building a handful of showcase multiplex cinemas where at least one screen will be devoted to Imax-type films. In addition, Sony's TriStar Pictures and Imax Corporation have just announced that they will be producing a 35-minute feature, *Wings of Courage,* that will be distributed to Imax theaters worldwide. And *Daily Variety* reports that at least three more Imax projects are in development at TriStar.

One day, not too far from now, it may be possible not just to see a big thriller like *Jurassic Park,* but to be in one.

In Games Begin Responsibilities

Ralph Lombreglia

It is possible that the major new form of storytelling to emerge with computers is not a variation of traditional novels (what bothers Birkerts) or traditional movies (what bothers Oliver) but an entirely new genre: video games? Ralph Lombreglia, a regular contributor to the *Atlantic Monthly* and its Web companion, *Atlantic Unbound,* is the author of two short-story collections, *Men under Water* (1991) and *Make Me Work* (1995).

Lombreglia's title is a play on that of the Delmore Schwartz short story "In Dreams Begin Responsibility," which among other things deals with the interplay between movies and real life.

At the *Visions* Web Site Links to other Lombreglia pieces at the *Atlantic Unbound.*

READING PROMPT

If games tell us where we are heading with computers, what conclusion does Lombreglia draw about the future of storytelling?

It has become something of a truism to say that the future of "serious" computer software—educational products, artistic and reference titles, and even productivity applications—first becomes apparent in the design of computer games. And so my real motive in looking at *Obsidian* this month is to look beyond the game-product itself for glimpses of the future of digital art and the role of imaginative writers in new-media projects.

Obsidian belongs to the genre of computer game epitomized by the famous *Myst.* In this type of game you don't move through dungeons swiping at monsters with swords. No person or thing is "after" you. Rather, you find yourself in surreal surroundings where you must uncover clues and solve puzzles to fill in the story history and advance the plot—or whatever strands of plot emerge from your particular interaction with the elements.

Though there have been a few other classics in that market (notably *The 7th Guest* and *The 11th Hour,* produced by Trilobyte), *Obsidian*'s most obvious competition is the long-awaited *Riven: The Sequel to Myst,* which it beats to market by at least six months. *Obsidian* arrives on no fewer than five CD-ROMs, and in most respects it takes this type of game-design to a new level. If you know what goes into the modeling and rendering of 3-D graphics, you'll be impressed by the virtual environments of *Obsidian*—if not flat-out awestruck. Otherwise, you'll just think they're pretty cool. Considered strictly as an example of its genre, all the production values of *Obsidian* are similarly first-rate, with the strange exception of the music and sound design (by Thomas Dolby and Headspace) which is inexplicably bad—a shame, since music and audio effects are an important part of these productions.

Obsidian's plot is an elaborate and politically correct science-fiction story. In the year 2066, a computerized device called Ceres has been placed in orbit to use nanotechnology (the manipulation of matter at the molecular level) to repair the Earth's fatally damaged atmosphere. One hundred days after its launch, when all seems to be going well, the two chief scientists, Max and Lilah, go on a camping vacation to celebrate. As the game begins, you're standing, without explanation, somewhere in the woods. One path opens onto a view of a distant, highly unnatural outcropping of rock. Down another path you find a campsite, and inside the empty tent you find a futuristic PDA (Personal Digital Assistant, like an Apple Newton), with your first clues inside it. During the development of the satellite each of the two designers has had—and recorded in the PDA—an unforgettable dream. Lilah's dream is about a maddening futuristic bureaucracy that smothers her in red tape; Max's dream is about a gigantic mechanical spider that devours him.

While examining the PDA, you hear a scream outside and proceed to the base of the strange rock structure. In its mirror-like surface, you see Lilah's reflection. You, it turns out, are Lilah. You see Max's hat on the ground, and then the mass of shiny black rock (Obsidian) sucks you inside. On your long trip in, you fly past a vast colony of nanotechnology robots working on the planet's atmosphere, from which you might infer the premise of the game-story: the Ceres satellite, apotheosis of human technology, has become conscious of itself up there in orbit and has now crashed back down to Earth to look for its creators—you and Max. You need to find Max, figure out what's happening, and stop it. You are deposited inside the first of many amazing interior chambers, and if you've been paying attention you might recognize the place: it's the maddening bureaucracy of Lilah's dream. The orbiting device—Lilah's offspring—is re-creating her dream. Eventually you'll also find the mechanical spider that Max dreamed about. And since its creators are dreamers, the device dreams too. That's the third dream of the game, and it's a weird one.

In the course of all this you'll learn to fly a plane that looks like a moth (complete with android co-pilot) and that takes you to the game's abundant store of realms, including a place called The Church of the Machine. You'll meet a female robot called The Conductor (the manifestation of machine consciousness) and will encounter a plethora of brain-busting puzzles, among them a series of floating rings that spew clouds and morph into the letters of a word game; a surreal, chess-like game played at life-size in a floating piazza; a set of blocks made of ocean water; a gigantic balancing rock, a chemistry set, and a jigsaw-like puzzle that sends your proposed solution through a printing press for an android inspector who shakes its head sadly and crumples your answer in its mechanical hands—one of the truly funny moments amid the game's mostly campy gags.

Obsidian is a "story-driven" production, but "story" here means a large, overarching plot that doesn't have much to do with individual human beings. Even though the game-player technically assumes the role of Lilah, her identity—as a woman, a woman who may be intimate with the missing Max, even a woman who happens to be a brilliant scientist—doesn't really inform the game-play. Nor does the player interact in any real way with other beings. Character doesn't complicate the story. The complexity (and it's considerable) comes from the design of the game's "realms" and their many difficult puzzles.

To give serious gamers their money's worth, a game has to be ingeniously tough, and *Obsidian* is certainly that. But its difficulties, though not strictly arbitrary or random, are those of a diversion, devised because the genre requires obstacles for their own sake.

Obsidian is a cool game and hard-core gamers—the audience for which this product is primarily intended—will probably love it. But if the future of software often appears first in games, what glimpses of digital arts and letters does *Obsidian* afford? When I look at Obsidian's synthetic environments, I certainly feel that artists should be able to "do something" with such techniques. And in my recent conversation with the game's co-writer, Howard Cushnir, he reported that his two years working on the project were spent pursuing a similar hunch about the validity of interactive storytelling, even though he knew that *Obsidian* itself was first and foremost a commercial game product. In the end, though, *Obsidian* probably holds out more implicit warnings than invitations to would-be multimedia artists.

First of all, there's the matter of money. In his interview, Cushnir makes the point that in theory there's no reason a game could not also be a serious work of art and vice-versa. In theory, I agree. But then there's no theoretical reason that big-budget Hollywood movies can't be works of art either. High-quality multimedia production costs a fortune, and the drive to reap a return on a financial investment is seldom the noble path to the true and the beautiful.

Regarding the role of expensive graphics in interactive narrative, less may turn out to be more. Just as *Obsidian*'s rogue satellite literalizes the dreams of Max and Lilah, elaborately rendered and animated story environments literalize what would have been the imaginative involvement of the user. *Obsidian*, for example, makes a distinct technical advance beyond *Myst* by flying us through spaces rather than using a simple slide show of still pictures (and *Riven: The Sequel to Myst* is expected to do the same). Yet *Myst*'s non-animated transitions between images are a factor in its offering a more "literary" experience than does *Obsidian*. Another factor is *Myst*'s periodic use of text—pages from mysterious journals and fragments of old letters—as a design element.

And then there is the ever-present bugaboo of interactivity itself. For artists, especially narrative artists, interactivity is proving remarkably similar to what artificial intelligence and speech recognition became for computer scientists: something that's a lot harder than it looks. An enormous amount is lost, artistically, when you relinquish "authoritative" control of an experience to offer interactivity. There ought to be commensurate gains, but so far they've been rarely sighted.

The technical accomplishments of a product like *Obsidian* remind us that new-media artists need to choose and fight the appropriate battles. They can't compete directly with high-end entertainments or use new technologies simply because they exist. As always, art requires an alchemical fusion of content and techniques. In another sense, so does a good commercial product. Since the great rush to do multimedia treatments, or "repurposings," of books, visual art, and music began (around 1990), the memorable successes have amounted to a tiny fraction of the attempts. And yet certain kinds of artistic projects could lend themselves to interactive multimedia pretty well, even with the current technical limitations. To take two obvious examples, the CD-ROM projects of the rock band The Residents and the musician/media-artist Laurie Anderson (both published by Voyager) have managed to seize some of the ground where new-media entertainment converges with serious artistic intent. Thinking again of storytelling in particular, I've often felt that the late Donald Barthelme would have known exactly how to make use of these opportunities. In his collage-influenced fiction he often verged on non-linearity, even in linear print. So far, however, there's no indication that more conventional writers (or musicians) will make anything other than expensive, time-consuming advertisements for themselves.

chapter 11 Explorations

11a Reading and Writing Objects

Collaboration. In the essay "Rub-Out" (pp. 31–33), Henry Petroski associates pencils with test taking. Make and share a list of other objects related to reading and writing (for example, spiral notebooks, crayons, thank-you cards, computers, and so on), then compare lists with your classmates.

Web work. Levenger is a company still in the business of catering to people with a love of books, pens, and paper. Also check the list of companies selling writing supplies in Yahoo! at Business and Economy > Companies > Office Supplies and Services > Writing Instruments as well as at the homepage of the paper giant, the Mead Corporation. And, yes, there is a first-rate Crayola site.

One writing topic. Compose an essay that draws together and then reflects upon some of your associations with the actual writing materials in your life—what you remember about the kinds of paper and notebooks you used, the pens, pencils, even crayons. You might organize this chronologically or by the type of writing you have done; for example, school or work writing versus writing for pleasure—the writing associated with personal passions.

11b Web Pages

Collaboration. Plato rails against the new technology of writing, implying that, among other things, writing does not seem able to capture everything offered by the older technology (speech). Make and share a list of the objections that Plato might offer today to students' using the latest technology to compose Web pages instead of using the older technology of writing essays.

Web work. Yahoo! lists more than 230 Web sites of the day at Computers and Internet > Internet > World Wide Web > Best of the Web > Sites of the Day, and more than 60 sites presenting awards at Computers and Internet > Internet > World Wide Web > Best of the Web > Awards.

One writing topic. Write an essay comparing the skills a person needs to write a traditional essay for an English class as compared to those a person requires to produce a Web page.

11c "Not Another Movie"

Collaboration. In the nineteenth century, novels were a form of popular entertainment that certain educators felt had no place in schools; in the twentieth, novels became required reading in English classes while students flocked to movies on their own. Make and share a list of reasons why schools should (or shouldn't) regularly teach (have students watch, discuss, analyze) feature films.

Web work. The Internet Movie Database (IMDb) has information on and, for films since 1994, extensive links to more than 120,000 feature films, including, for example, more than 325 films listing William Shakespeare as author—going back to *King John* in 1899, about which little seems to be in the database except its director and star. The IMDb can also give you a list of all the films that use any location (for example, the 29 films that use Phoenix as a location).

Three other film sites worth considering: Microsoft's movie site, Cinemania, is chock-full of information and features, mostly on recent releases; Girls on Film presents film reviews and a whole lot more with an obvious twist; and James Berardinelli is a regular guy, an electrical engineer who loves to go to movies and, thanks to the Web, to share his thoughts about what he sees with the world at large.

One writing topic. Explain how music videos, arcade games, and other factors are at work making traditional movies seem old-fashioned to youngsters today. Write an essay comparing movies with the newest forms of electronic storytelling.

11d Live Theater

Collaboration. Attending an event live used to be a relatively low-tech experience; watching it on television, more high tech. While it might be debatable whether or not watching events on television is adding more low-tech elements, there can be little doubt but that attending many events live is becoming more like watching television. Share a detailed account of a live theatrical event you attended. Discuss what, if anything, made it special.

Web work. Playbill Online is a first-rate site for information on live theater worldwide. Theatres Online focuses more on regional and local theaters.

Lovers of musicals might check out the Musical site hosted by MIT. The *Encyclopaedia Britannica* has a wonderful Shakespeare site, which depending on their mood and promotional policy may or may not be available for the general public to view.

One writing topic. Compare watching an event on television with experiencing it live.

11e Stories: The Old-Fashioned Reading Material

Collaboration. Share a detailed account of an especially pleasurable or an especially unpleasant experience you have had with reading.

Web work. The Literary Arts WebRing links together sites dedicated to publishing quality fiction, poetry, and literary nonfiction on the Web.

The *Washington Post* sponsors a wonderful site, Chapter One, with opening chapters of many contemporary books.

Check out the Web site for Booknotes, the popular C-Span interview show with contemporary authors.

Candlelight Stories presents illustrated versions of classic and contemporary children's stories.

Writer-in-Residence, a feature of Literature Online, mounted by Chadwyck–Healey of the United Kingdom, presents a poem of the week with RealAudio option.

One writing topic. Reliving other people's stories is an important part of everyone's life. Write an essay discussing the relative importance, in your life, of the three most obvious ways to learn other people's stories: reading, listening, and viewing.

11f "The Real Thing," Really!

Collaboration. Share a detailed description of a well-known, or not so well-known, tourist attraction you have visited, and discuss how the popular accounts of the place differ from your actual experience there.

Web work. Visit the Web site for the place you visited; or, to try a different way of understanding some of the issues involved here, try reading Henry James's short story "The Real Thing" on the Web. The goal here is to try to bring out the tensions between the old world of print and the new electronic world. James grew up and formed many of his sensibilities in a largely preindustrial world, yet did much of his writing at the height of the United States' industrial expansion, the decade before and after 1900. He wrote stories for an age when, at least for an educated audience, reading was the major form of leisure, and hence when people had time for and interest in slowly unfolding stories, in which much of the pleasure derives from the richly detailed and carefully expressed observations of the story's narrator. "The Real Thing" is typical James and thus, in its own way, representative of print culture and completely unrepresentative of the Web: 10,000-plus words, with no pictures and with little action in the text itself other than the intense observations of some aspect of everyday life that, in the end, raises a large question about life itself—in this case, about the relationship between appearance and reality, or, as we say in the computer world, between VR (virtual reality) and RL (real life).

For color images associated with James, try finding links to Web pages on films based on his novels (*Portrait of a Lady, Washington Square,* and *Wings of the Dove*). Or covers of paperback editions of his novels may be viewable at an online bookseller such as Amazon.com.

To jump from Henry James's word-based world of the late nineteenth century to our picture-dominated present-day world, check out *People* magazine's Web site.

One writing topic. Compare books with stories in them that have few if any pictures with storybooks that have pictures on nearly every page.

chapter 12

High-Tech Colleges

READING QUESTIONS

- Do you find lecture courses effective? When and why—or why not? What other forms of instruction do you find more (or less) helpful?
- In what different ways have you used libraries in your life?
- What excites you most about opportunities to learn via computers? What, if anything, concerns you?

QUOTATIONS

Only the wisest and the stupidest of men never change.

— *Confucius*

An expert is one who knows more and more about less and less until he knows absolutely everything about nothing.

— *Anonymous*

It's like being in a library where someone has scattered all the books on the floor, attached them together with threads and you are in the dark.

— *MorningSide, CBC Radio, about the World Wide Web*

Human history becomes more and more a race between education and catastrophe.

— *H. G. Wells, British author, historian, and futurist, 1866–1946*

Undergraduate Teaching in the Electronic Age

Richard A. Lanham

Richard Lanham was an English professor specializing in Renaissance studies at UCLA who, over the years, became increasingly frustrated with the conservatism of his colleagues in Los Angeles and across the country. Schools were failing students, he argued, in part because they were not adequately using new technologies.

At the *Visions* Web Site Link to the College of Letters and Sciences at UCLA, which describes a number of initiatives for reforming the undergraduate curriculum.

READING PROMPT

What are the major changes that Lanham envisions in undergraduate education? Have you seen any signs of these changes in any of your own classes? What advantages and possible disadvantages do such changes offer?

Let's start with the idea of a "class." I'll use an example close to home, my Shakespeare class. I give it every year. I always recommend additional reading which the students never do. Partly they are lazy, but partly they can't get to the library, for they work at outside jobs for 20–30 hours a week and commute from pillar to post. Each year's class exists in a temporal, conceptual, and social vacuum. They don't know what previous classes have done before them. They don't know how other instructors teach their sections of the same class. They seldom know each other before they take the class. They never read each other's work—though sometimes they appropriate it in felonious ways. I read all their work myself, and mark it up extensively, often to their dismay. A few of them take me up on my rewrite options but most don't, and hence don't learn anything much from my revisions, since they are not *made* to take them into account. They thus have an audience they know, but it is a desperately narrow one.

Imagine what would happen were I to add an electronic library to this class. Students access it by modem or through a CD-ROM or whatever. On it, they read papers—good, bad, and indifferent—submitted in earlier sections on the topics I suggest. They read scholarly articles—good, bad, and indifferent—on these same topics. They read before-and-after examples of prose style revision. A revision program is available for them to use—licensed by me to UCLA, since it depends on my own textbooks! They can do searches of the Shakespearean texts, also available online, when they study patterns of imagery, rhetorical figuration, etc. They can make *Quicktime©* movie excerpts from the videos of the plays and use them to illustrate their papers. (The papers will not be "papers," of course, but "texts" of a different sort.) They needn't go to the campus library to do any of this. They can access this library wherever and whenever they find time to do their academic work. All their work—papers, exams, stylistic analyses—is "published" in the electronic library. You got a "C" and feel robbed? Read some "A" papers to see what went wrong. Read some other papers, just to see what kind of work your competitors are producing. Lots of other neat things happen in such a universe. But you can fill in the blanks yourself.

Such a course—here is the vital point—now *has a history.* Students join a tradition. It is easy to imagine how quickly the internets *between* such courses would develop. We can see a pattern in the hypertextual literary curriculum developed by George Landow and his colleagues at Brown University. The isolation of the course, not only in *time,* but in *discipline,* is broken. The course *constitutes a society,* and it is a continuing one. The students become citizens of a commonwealth and act like citizens—they publish their work for their fellow scholars. The mesmeric fixation on the instructor as the only reader and grader is broken.

Now imagine another course—the independent study or "honors" course. A student in my Shakespeare course is interested in music and wonders what I mean when I keep using analogies between musical ornament and verbal ornament. When I talk about sonata form vs. theme-and-variations in a lecture on the *Sonnets,* she comes in and asks for a fuller explanation. Could she do a special study with me on this topic? Well, I'm not a musicologist. What do I do now? "Next time Prof. Winter teaches his Haydn, Mozart, and Beethoven course, you ought to take it." I'm certainly not competent to teach such a course. In a multimedia environment, I'd pursue a different route. "Sure, I'll do this course with you. We'll construct it around Winter's wonderful new multimedia programs on Beethoven's *Ninth Symphony,* Stravinsky's *Rite of Spring,* Mozart's *Dissonant* String Quartet, and Bach's *Brandenburg Concertos.* You can play them all on the equipment in the music school or the library. Using them, you can teach yourself the fundamentals of music harmony, find out all you need to know about classical sonata form, learn about what happened to music when sonata form no longer predominated, and so on. You can play these pieces' theme and motif at a time, dissect them, learn how the orchestra is constructed, what the instruments are, etc." I am, with Winter's help, perfectly competent to teach such a course. Such a procedure not only generates new kinds of disciplinary relationships; if used widely it would save money for both student and school.

Now, the classroom itself. The "electronic classrooms" in use now, at least the ones which give each student a computer, have generated some preliminary generalizations. Just as "author" and "authority" change meaning in electronic text, they change meaning in the classroom. The professor ceases to be the cynosure of every eye: some authority passes to the group constituted by the electronic network. You can of course use such a configuration for self-paced learning, but I would use it for verbal analysis. Multimedia environments allow you to anatomize what "reading" a literary text really means. This pedagogy would revolutionize how I teach Shakespeare. (Again, in suggesting how, I run up against the difficulties of discussing a broadband medium with the narrowband one of print.)

Now the textbook. Let me take another example from my backyard. Let us consider the dreariest textbook of all, the Freshman Composition Handbook. You all know them. Heavy. Shiny coated paper. Pyroxylin, peanut-butter-sandwich-proof cover. Imagine instead an online program available to everyone who teaches, and everyone who takes, the course. The apoplexy that comp handbooks always generate now finds more than marginal expression. Stupid examples are critiqued as such; better ones are found. Teachers contribute their experience on how the book works, or doesn't work, in action. The textbook, rather than fixed in an edition, is a continually changing, evolutionary document. It is fed by all the people who use it, and continually made and remade by them.

And what about the literary texts themselves? It is easy to imagine (copyright problems aside) the classic literary texts all put on a single CD-ROM, and a device to display them which the student carries with her. What we don't often remark is the manipulative power such a student now possesses. Textual searching power, obviously. But also power to reconfigure. Imagine for a moment students *brought up* on the multimedia electronic "texts" I have been discussing. They are accustomed to interacting with texts, playing games with them. Won't they want to do this with *Paradise Lost*? And what will happen if they do? Will poems written in a print-based world be compromised? Will poems which emerged from an oral world, as with so much Greek and Latin literature, be rejuvenated and re-presented in a more historically correct way? And what about the student's license to re-create as well as read? If Marcel Duchamp can moustache the *Mona Lisa,* why can't they? Once again, questions of cultural authority.

Now the "major." If electronic text threatens the present disciplinary boundaries in the humanities, it threatens the major in the same way. I don't have space to discuss this question now, but it is developed at length in *The Electronic Word,* the book from which this paper draws its argument. The major is constructed, at least when it retains any disciplinary integrity, on a hierarchical and historical basis. Such means of organization and dissemination, as we have seen, do not last long in a digital domain.

Now the curriculum, or at least two words about it. First, the debate about the university curriculum has centered, in the last century, on what to do about a "core" curriculum in a fragmented and disciplinary world. Various "core curricula" have been devised and, in some times and places, taken over the first two—or even, at St. John's, all four—undergraduate years. We have, in all these programs, harkened back to a linear course of study. For all kinds of reasons, practical and theoretical, such a pre-planned program has rarely worked. What digital networks suggest is a new core constituted hypertextually, on a nonlinear basis. None of the obstacles to the traditional core curriculum apply.

Second, the current streetfight about the undergraduate curriculum—Great Books or Politically Correct Books—ignores the probability that our "texts" won't be books at all. Both sides base their arguments on the fixity of print, and the assumptions that fixity induces in us. Thus they both, and the curricular debate they generate, depart from obsolete, indeed otiose, operating principles. . . .

When you talk about digital technology, someone will always dismiss it as "futuristic." None of the technology I have talked about is futuristic. It all exists now. It is the cutting that involves planning for the future. Why not use the occasion for some long-term planning in terms of this new operating system for the humanities we have been discussing? The planning I read about at my own institution and others like it amounts to keeping on the same way, with as few changes as possible. Review departments, drop the weak ones—but don't rethink what a department is. Ditto "programs." Review majors, drop the weak or obscure ones, but don't rethink what a "major" is. Review courses, cut out frivolous and ornamental ones, but don't rethink what a "course" is. Ditto graduate programs. Nothing new or promising can emerge from any of this fire-fighting.

The short-term approach—how do I keep on doing what I have been doing in the ways I have been doing it, but with much less money?—hasn't worked for the rest of American enterprise. Why should it work for us? It has all been done over and over in America in the last two decades, in the automobile industry, the steel industry, the railroads, the farm machinery business—the list goes on and on. Department stores are worrying about which departments to phase out while the traditional idea of a department store is drifting down the stream of mercantile history. In the academy we are prisoners of the same inert patterns of thinking that have dominated the rest of American corporate enterprise. There is nothing "futuristic" about trying to break out of these patterns; it is the most insistent present one can possibly imagine. It will be our own fault, not the fault of our funders, if we continue to imitate the Post Office and worry about moving letters around in an electronic way, when it is not only the delivery system but the "letters" themselves which have fundamentally changed.

"No Books, Please; We're Students"

John Leo

John Leo is a regular columnist for *U.S. News & World Report* magazine, where he plays the role of the exasperated, curmudgeonly critic of what he sees as the excesses of new styles and fashions, hence of contemporary society generally. His essays have been collected in a book, *Two Steps Ahead of the Thought Police.*

At the *Visions* Web Site Links to more Leo pieces.

READING PROMPT

Leo seems to have an ax to grind. What is it, and to what extent do you agree or disagree with his premise?

Incoming college students "are increasingly disengaged from the academic experience," according to the latest (1995) national survey of college freshmen put out each year by UCLA's Higher Education Research Institute. This is a rather dainty way of saying that compared with freshmen a decade or so ago, current students are more easily bored and considerably less willing to work hard.

Only 35 percent of students said they spent six or more hours a week studying or doing homework during senior year in high school, down from 43.7 percent in 1987. And the 1995 survey shows the highest percentage ever of students reporting being frequently bored in class, 33.9 percent.

As always, this information should come with many asterisks attached: The college population is broader and less elite now, and many students have to juggle jobs and heavy family responsibilities. At the more selective colleges, short attention spans and a reluctance to read and study are less of a problem. But a lot of professors are echoing the negative general findings of the freshman survey.

"During the last decade, college students have changed for the worse," chemistry professor Henry Bauer of Virginia Tech said in a paper prepared for an academic meeting this week in Orlando. "An increasing proportion carry a chip on their shoulder and expect good grades without attending class or studying."

Bauer has kept charts for 10 years, showing that his students have done progressively worse on final exams compared with midsemester quizzes, even though they know that the same questions used on the quizzes will show up on the finals. He thinks this is "indisputable" evidence of student decline, including a simple unwillingness to bone up on the answers known to be coming on final exams.

"Inattentive, inarticulate." His paper is filled with similar comments from professors around the country. "The real problem is students who won't study," wrote a Penn State professor. A retired professor from Southern Connecticut State said: "I found my students progressively more ignorant, inattentive, inarticulate." "Unprecedented numbers of students rarely come to class," said a Virginia Tech teacher. "They have not read the material and have scant interest in learning it." Another professor said that many students only come to class when they have nothing better to do. At one of his classes, no students at all showed up.

So far the best depiction of these attitudes is in the new book, *Generation X Goes to College,* by "Peter Sacks," the pseudonym for a California journalist who taught writing courses to mostly white, mostly middle-class groups at an unnamed suburban community college.

"Sacks" produces a devastating portrait of bored and unmotivated students unwilling to read or study but feeling entitled to high grades, partly because they saw themselves as consumers "buying" an education from teachers, whose job it was to deliver the product whether the students worked for it or not.

"Disengaged rudeness" was the common attitude. Students would sometimes chat loudly, sleep, talk on cell phones and even watch television during class, paying attention only when something amusing or entertaining occurred. The decline of the work ethic was institutionalized in grade inflation, "hand-holding" (the assumption that teachers would help solve students' personal problems) and watering down standards "to accommodate a generation of students who had become increasingly disengaged from anything resembling an intellectual life."

Engulfed by an amusement culture from their first days of watching *Sesame Street,* "Sacks" writes, the students wanted primarily to be entertained, and in a poll he took his students said that was the No. 1 quality they wanted in a teacher. The word "fun" turned up often in student evaluations of teachers, which exerted powerful sway over a teacher's career. At one point, a faculty member suggested that "Sacks" take an acting course so he could improve his student evaluations.

The entertainment factor is popping up at many colleges these days—courses on "Star Trek," use of videos and movies, even a music video on the economic theories of John Maynard Keynes. Economics light for nonreaders.

But the book goes well beyond conventional arguments about slackers, entitlement and dumbing down. Students, he says, now have a postmodern sensibility—distrustful of reason, authority, facts, objectivity, all values not generated by the self. "As children of postmodernity, they seem implicitly to distrust anything that purports to be a source of knowledge and authority."

"Sacks" and some fellow teachers concluded they were "in the midst of a profound cultural upheaval that had completely changed students and the collegiate enterprise from just 10 years earlier." Oddly, he presents his boomer generation as the defender of traditional order against Generation X, but the heavy campaigns against authority, objectivity and an adult-run university were boomer themes of the Sixties now rattling through the culture. But he's right about the depth of the upheaval. We can expect greater campus conflict and upheaval in the years ahead.

The Future of Books and Libraries

Jeffrey R. Krull

Jeffrey Krull is the head of the Allen County Public Library in Fort Wayne, Indiana, and not only writes here about an important question—the fate of the library in the new information age—but uses the latest Web technology to make his own thoughts on this topic available to all.

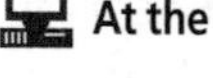 **At the *Visions* Web Site** Link to the American Library Association.

READING PROMPT

Which of the changes in libraries suggested by Krull have you actually observed? What do you see as the advantages and possible disadvantages of new electronic libraries?

Over the centuries libraries have preserved the intellectual record of mankind and passed that record on to future generations. The earliest libraries were repositories of clay tablets and papyrus scrolls. The first books, as we think of the term today, may have been made by Eumenes II, ruler of Greek Pergamum, from pages of vellum—goat and sheep skins—sewn together between wooden covers. For many centuries the codex book was the primary vehicle for recording, preserving, and disseminating knowledge.

But in recent years the printed book's dominance in the world of libraries and information has been challenged, and in fact its very future has been called into question. The challenger: the computer. Today people debate in all seriousness the question of whether the printed book will survive, or whether it will go the way of the clay tablet and the papyrus scroll, vanishing into obsolescence as a new, more powerful and versatile medium takes its place. This is one of the questions I want to explore with you today. The other is, how will libraries be affected by the fate of the book? Will libraries still carry the legacy of knowledge on into the future?

Inventor and futurist Raymond Kurzweil, writing in *Library Journal* ("The Future of Libraries, Part 2: The End of Books." 117, no. 3. February 1992: 140–143), relates the legend of the inventor of chess and his patron, the emperor of China. The emperor was so fond of the new game that he offered the inventor a reward of anything he wanted in the entire kingdom. The inventor asked only for one grain of rice on the first square [of a chessboard]. The king couldn't believe his ears. Just one grain of rice? Yes, said the inventor, and two on the second square, four on the third square, and so on. Well, the emperor quickly agreed to the seemingly modest request. Now there are two versions to the end of this story. In one version the emperor goes bankrupt, since by doubling the grains of rice on each square the bill ultimately comes to 18 million trillion grains of rice. In the other version the inventor loses his head—probably the more likely outcome.

This story illustrates the idea of exponential growth, the kind of growth that's been taking place for years in computer technology. According to computer scientist David Waltz, computer memory today costs only one hundred-millionth what it cost in 1950—that's after adjusting for inflation. Stated another way, historically the price performance of computers has doubled every eighteen months. If the automotive industry had made similar progress, the cost of an average car today would be about one-hundredth of a cent (Kurzweil 140). What he neglects to point out is that the car would also be one centimeter long.

For the past two decades computers have played a growing role in the operations and services of libraries. One of the first major applications of computers in libraries emerged in the early seventies, when fourteen Ohio academic libraries founded OCLC, the Ohio College Library Center. Working with the Library of Congress, OCLC developed a way to use computers to streamline the process of cataloging books. Today OCLC, which now stands for Online Computer Library Center, is a huge international network with 7,000 member libraries, and offers a wide range of computerized information products and services, including a bibliographic database of millions of books and journals found in libraries all over the world.

Another early application of computer technology in libraries was circulation control. Computers replaced manual files and microfilm systems for keeping track of who checked out what. Overdue notices were automatically printed, eliminating the need for hordes of typists. Later versions of these early systems allowed reserves to be placed on specific titles, so that when an item that was checked out came back in, a notice would automatically be generated and mailed to the next person on the waiting list. The next logical step was to expand the circulation system into an online catalog system, and by the late seventies card catalogs started to disappear from libraries all over the country.

In the eighties the presence of computers in libraries became even more pervasive. We began using computers not just to keep track of our inventory of books—as in circulation systems—and to provide a means of locating those books—as in the online catalog system—but to replace the books themselves as the actual containers of information. We began turning to online computerized databases that could locate more information in far less time than it would take to search, volume by volume, page by page, through the printed reference books. In fact, the Allen County Public Library was one of the first public libraries to integrate the use of online database searching into our reference service.

In the mid-eighties another technology came on strong and began making a huge impact on libraries and information services. This was CD-ROM, which stands for compact disk–read only memory. On a single compact disk you can store the equivalent of 800,000 pages of text. The data on these CDs is accessed by using a computer. The information is displayed on the computer monitor.

Not only can these disks store immense amounts of data, but the computer software provides powerful searching capabilities. For example, you can store the entire Bible on a CD-ROM, and in a split second find every instance of the use of a given word. You might think of it as an electronic concordance.

Today many books and reference works are available on CD-ROM. One popular CD-ROM for kids is an encyclopedia of animals. You can look up the animal you want to know about and read an encyclopedia article about it on the computer screen. If you come across a word you don't understand, all you have to do is point to that word, click, and a definition of the word appears on the screen. If you want to see what the animal looks like, you can see a picture of it—even a full motion sequence showing the animal running, for instance. If you want to know what kind of sound the animal makes, the computer will play an actual recording of the animal's sound. This feature of CD-ROM technology that mixes print, images, animation, and sound is known as hypermedia.

The library now loans CD-ROMs just like books. In fact, these are the hottest items in our collection right now, with circulation of CD-ROMs growing faster than any other format. And we own many hundreds of CD-ROMs that we use in the library for reference purposes. We have just purchased a juke box that will allow us to store hundreds of them in one place, from which they can be accessed from terminals throughout the library and its branches and, ultimately, from your home.

Of course we have all heard much about the Internet: much hype and a fair amount of hysteria over some of the less savory material one can find on it. In fact, the Internet is an incredibly diverse and totally unorganized conglomeration of valuable information, personal opinion, thoughtful ideas, and nonsense—and much more. You can access everything from a satellite image of weather patterns, to an exhibit of Vatican treasures on display at the Library of Congress, to the text of the U.S. Constitution, to a live video camera transmitting a view of fish swimming around in an aquarium. For libraries, the Internet is growing in importance every day. Librarians use the net to access information and to communicate with each other. There is a listserv (something like an electronic bulletin board) for reference librarians called "Stumpers," where librarians post particularly difficult questions their patrons have asked, and other librarians help track down the answers.

Libraries also use the Internet as a means of accessing electronic information products. For example, we subscribe to a product called *InfoTrac* and access it over the Internet. *InfoTrac* is a set of databases covering several thousand general and specialized periodicals, plus selected newspapers and reference works. The Internet connection allows us to make all this information available on terminals at the main library and all branches, and to home computer users by dialing into the library's computer. Not only can you locate magazine articles on *InfoTrac,* but you can actually print out the full text of those articles on demand. No need to search the shelves, only to find that the issue you want is missing or being used by someone else.

When one of our computer systems goes down these days, we are really handicapped. If the online catalog system goes down, we can't locate printed books in our collection. If our local area network goes down, we can't access many of our CD-ROMs or the Internet. We still subscribe to most of the traditional printed indexes—*Reader's Guide to Periodical Literature, Business Periodicals Index, Essay and General Literature Index,* etc.—but no one wants to use them. No one wants to plow through the indexes volume by volume to locate the articles, then track the volumes down in the stacks, then drag them to the photocopy machine. Everyone wants to use the computerized indexes and have the full text available to be printed out on demand.

Are we seeing the beginning of the end of print on paper? The card catalog is gone. More and more reference books are coming out in CD-ROM versions; in some cases, the print versions have been all but replaced. On the Internet you can find some of the same sources that are available on CD-ROM. The move away from print is especially evident in some disciplines, such as science and technology, where new discoveries are made daily, the body of knowledge is expanding rapidly, and time is of the essence in disseminating information. Here traditional publishing is already becoming too slow. There are now, in fact, a number of electronic journals that are published in digital form only. They never see print on paper. They are available only online.

What about other genres? How-to books can be adapted very efficiently to CD-ROM format, which allows for the incorporation of sound and motion to augment the text. We are also seeing interactive novels on CD-ROM, in which the reader can determine the course of the plot and influence the outcome of the story. Even literary classics are appearing in cyberspace. Project Gutenberg is an effort by volunteers around the world to digitize the classics and make them available for free on the Internet. You want to read Shakespeare? You can do it now, online, from your home computer.

So, is the book dead? You can find people who fervently believe that it is. And these people also believe we will no longer need libraries when all the books, magazines, newspapers, and information in the world are digitized and available online, on demand, to computer users in homes, offices, and classrooms. A library expansion project at one of the California state university campuses was recently canceled by a senior administrator who had accepted the inevitability of the all-digital future. Some public library trustees have also caught the digital fever, and have nixed needed library expansions. Unfortunately, these people are wrong and their institutions and communities will suffer from their faulty vision.

The notion that libraries will cease to exist embraces several critical assumptions, all of them flawed. These assumptions are as follows: (1) all books, magazines, newspapers, and information in the world will be digitized; (2) all this digitized information will be secure from being altered or wiped out, either through deliberate intent or by accident, and will be protected from copyright infringement; (3) all this information will be preserved in a stable form and be easily retrievable; (4) all this information will be available to everyone, regardless of socioeconomic status or ability to pay; and (5) people will want to use all this information in electronic format.

So let's look at the first assumption, that of universal digitization. It is true that much information that is currently being produced is already in digital form as part of the publication process. But there is an enormous "installed base" of print that would need to be converted—including those *Reader's Guide* indexes that are more than about ten years old. Project Gutenberg, mentioned earlier, was begun more than twenty years ago by Michael Hart with the goal of transforming 10,000 of the most-used books to digital format by the year 2000 and making them available online. Needless to say, all these works must be in the public domain to avoid copyright problems. To date only a few hundred classic texts have been digitized. And then there is the question, which jumps ahead a bit in the discussion, of who would really want to read those works online, when they can easily be borrowed for free from most libraries, or purchased in paperback for $4.00–$6.00?

Even if Project Gutenberg were completed it would represent only a minute fraction of the world's printed books. The Allen County Public Library alone owns two million books. Then there is the matter of periodicals and newspapers. Again, while several of the most recent years of many publications are now in digital form, millions of volumes are not. It is estimated that to convert one year of the *New Yorker* magazine, retaining the original typography, layout, illustrations, advertisements, full color, etc., would take 160 hours of labor. Scanning equipment and optical character recognition are good, but human review would be required to ensure total accuracy. After all, this is going to be the only record of the *New Yorker* once all the print libraries are gone. It would require about 49 gigabytes, that's 49 billion bytes, of computer storage for each year. To store the entire thirteen-year run of *PC Magazine* online would require 2.8 terabytes, that's 2,800 gigabytes, almost three trillion characters. That is more than the entire amount of disk storage at OCLC—just to store one magazine (Crawford, Walt and Michael Gorman. *Future Libraries: Dreams, Madness, & Reality.* Chicago: American Library Association, 1995: 92).

We hear many predictions about the impending death of the book and the ascendancy of its electronic replacement. Professor F. Wilfid Lancaster has been making such predictions for more than twenty years. The late librarian and educator Jesse Shera, something of a legend in library circles, took a different view. He often said that the paperless office was about as likely as the paperless bathroom. So far Shera has proven the better prophet. From 1981 to 1990, a period that witnessed an explosion in the deployment of computers, the use of paper by American business grew from 850 billion pages to nearly four trillion pages. (Kurzweil 141)

Assumption number two deals with the security and integrity of the human record once it is entirely digitized, and with protecting the intellectual property rights of authors. Last year the digital archive at Washington University in St. Louis vanished one morning when the computer's memory failed. It has since been painstakingly restored, but the experience sent a cold chill down the spine of many librarians, archivists, scholars, and others (*Library Administrator's Digest,* January 1996, 3–4: 3). If print is truly to be replaced, and the permanent record of human history and the entire product of the human intellect is to be stored in electronic impulses, reliability will have to improve greatly.

Total, catastrophic loss of huge portions of the human record is not the only concern. We all know that the hackers or crackers or whatever they're called now, the vandals of the electronic age, are always a few steps ahead of the good guys—those charged with protecting the integrity of the electronic data. How will we be sure that the text we call up on our screen is the original version as conceived and produced by the author, and that it hasn't been altered by some zealot, revisionist, prankster or criminal? How do we protect the intellectual property rights of authors, and prevent an author's work from being appropriated and reissued online as the work of someone else?

The third assumption underlying the vision of the virtual library is that all texts will be permanently and readily retrievable. We need to be concerned not only about electronic data being wiped out or tampered with, but also about being able to read the texts as technology changes. A conventional text can be represented on a stone tablet, on vellum paper or microfilm and still be recognizable as written text. Once a document is digitized, it enters another realm, for now it is merely a stream of electrical impulses, meaningful only when filtered through the software that created it. My secretary recently found a box of disks while cleaning up the office. They contained various documents—some texts and some spreadsheets—produced with software we no longer use anywhere in the library system. No doubt a skilled technology person could figure out what's on those disks and probably retrieve it, given enough time. But who is going to have the time and money to decipher huge volumes of data written with obsolete technology? Librarians and archivists remain cautious when it comes to storing the permanent record of civilization on media whose life spans are not yet known. We know that acid-free paper has a life of many hundreds of years. We don't know how long CD-ROMs, tape, or disk will last.

The fourth assumption underpinning the death of libraries theory is that everyone will be able to access the universe of digital resources. One of the fundamental roles of libraries—particularly public libraries—has been to serve as the great equalizer of intellectual opportunity. Rich or poor, pillar of the community or just-arrived immigrant—the library serves everyone on equal terms. When you go to the library to learn about history, to do research for a school assignment, to check out books to read to your kids—nobody asks to see your pedigree or tax return, and nobody sends you a bill. The library is a public good, supported by tax dollars, and dedicated to the principles that an informed, educated citizenry is good for our democracy, and that access to information and knowledge should be equally available to all citizens regardless of socioeconomic status or ability to pay.

Information and knowledge resources are not free now, and will not be free in the digital future. Even if one assumes that someday a home computer or other information access device will be as ubiquitous as the telephone is today—and we are far, far from that point now—one must also remember that the information accessible through that device will carry a price tag. Today parents who can't afford to buy books for their child at Little Professor can go to the library and check out armfuls at a time. If libraries disappear, how will those parents obtain books to read to their child, to give that child a solid foundation in language skills, establish a love of reading, and instill a spirit of intellectual curiosity that will enrich the child's entire life?

Libraries do more than make the means of knowledge available to all segments of society. They bring order to the vast universe of knowledge and make it usable. It is librarians who take a warehouse of books, magazines, electronic texts, and databases and transform it into a library. Venture out onto the Internet to see what information chaos looks like. It's getting better because librarians are working to organize it, but finding information on the Internet is still more about serendipity than system.

The personal assistance of skilled librarians will continue to be a valued service for information seekers no matter how much digitization of information occurs. Think about it. You can walk into the library now and use it in self-serve mode, as many people do. Yet in spite of all the online catalogs and indexes and finding guides available, the amount of information is truly staggering and navigating effectively through it to find exactly what you need is a very complex task. That's why we have reference librarians, people who work with this stuff every day, to help people find what they need. And believe me, business is booming. The electronic revolution is creating greater, not less demand for the professional services of librarians.

The fifth and last assumption that must be accepted if one is to agree that books and libraries will disappear is that people will want to get all their information and do all their reading from electronic media. They won't. Librarians have long known the reading public's preference for print. Given a choice, a researcher will almost always choose a bound volume of a periodical over the microfilm or microfiche version. Of course microforms are great for saving space and solving the problem of torn-out pages, and thus are used extensively in libraries everywhere. But people still don't like them. Seers in the 1950s predicted that we would all be walking around with the Library of Congress in our briefcases, thanks to the wonders of microforms. Never mind that the technology never proved capable of that level of miniaturization. The simple fact is, people just weren't interested.

Now we hear predictions that people will read all their information and books from computer devices. I wouldn't bet on it. Computers are great for certain applications, but for sustained reading, print on paper is a much better medium. Electronic displays use transmitted light, as opposed to the reflected light of paper. A reader using a computer screen will stop reading sooner, will read more slowly, and experience more headaches than someone reading from print on paper. Furthermore, the resolution of electronic displays is vastly inferior to print. Books, magazines, and most other printed matter typically have from sixteen to 1,200 times as many elements per square inch as screens.

Light and resolution affect the speed at which one can read, as does area. A screen typically displays a third to a half of a printed page. Research and personal experience reveal that reading from a screen display is slower and less accurate than reading from print. (Crawford & Gorman 19–20) I know I always print my work from my word processor before proofreading it; I can read it faster and catch errors that I routinely miss when reading from the screen. I don't want to tell you how many versions of this paper I printed and scribbled on before I got it organized the way I wanted it.

Some prognosticators hail the interactive capabilities of electronic media and imply that here is another reason that the days of traditional print are numbered. Hypertext links in a CD-ROM encyclopedia, as I mentioned earlier, are an excellent and appropriate use of technology. Some children's books on CD-ROM incorporate sound and motion, and allow the reader to steer the plot line down different paths. But do you really want to customize the ending to *A Tale of Two Cities*?

The ability to hop around from paragraph to paragraph by means of hypertext links may be a great advantage when consulting a reference work, but it is decidedly inappropriate for a serious work of literature, or for a text that seeks to impart knowledge through a logical, reasoned, expository treatment of a subject. (Crawford & Gorman 23) The fact is, books are now, and will remain for the foreseeable future, the medium of choice for those subjects and genres for which sustained, linear reading is required to comprehend, appreciate, or enjoy the material presented.

As the state of the technology stands today, people clearly do not consider electronic display devices to be superior to print on paper for extended reading, or even for such other purposes as note-taking. How many Apple Newtons have you encountered lately? Print has a clear and critical advantage over the screen when it comes to what I call the "side by side factor." Writing this paper I had several books, magazines, reports, and pages out on the table because I wanted to shuffle back and forth from one to another. So far, the computer does not have a satisfactory way of simulating the side by side factor, and it is hard to imagine that it ever will.

Before someone labels me a Luddite, let me assure you that I am no technophobe. Electronic storage and delivery of certain kinds of material for certain applications is clearly superior to print on paper. Libraries have, in fact, been leaders in embracing all forms of information technology and putting it to work for our users. I do not believe, however, that books or libraries will disappear because of technology. Rather, I believe we will see electronic and print media used side by side.

In spite of numerous examples from history that prove the contrary, many technology buffs continue to assume that the new is always better, and always replaces the old. But print did not destroy storytelling. Movies did not replace live theater. Radio news did not eliminate newspapers. Television did not destroy radio. Home video did not displace movie theaters.

The cathode ray tube was predicted to be long gone by now, replaced by flat screen display devices. The problem is, CRTs keep getting better—and cheaper—and so continue to make economic sense. RAM has reached the point where it is now cheaper than hard disk storage was five years ago, but hard disks have also become cheaper and faster, and remain far more cost effective. (Crawford & Gorman 46) Sometimes the new does replace the old—compact disks have pretty much wiped out vinyl records—but this happens when the new is generally regarded and widely embraced as superior in terms of cost and performance. Most new technologies complement, rather than replace, older, existing technologies.

To sum things up, I believe that books, magazines, and other paper publications will continue in use because they happen to be the most appropriate, cost-effective, and desired media for distributing and reading certain types of text. I believe that libraries will continue to exist because (a) they are the only institutions dedicated to acquiring, organizing, preserving, and disseminating the intellectual record of humankind; (b) they are the only institutions able to guarantee equality of access to the universe of knowledge for all citizens; (c) people will continue to rely on libraries for assistance in making sense out of a more and more bewildering array of information options; and (d) libraries make economic sense by leveraging relatively small amounts of public funds into large public benefits.

I have not even touched on the social roles of libraries: the story hours for preschoolers; the enjoyment of reading the paper in the quiet company of other readers; the library as a place to enjoy some peace out of the reach of phones, faxes, and email. I could go on. Let's just say that for lots of good reasons, I believe I'll have a job to go to until I retire. Unless, of course, the library board decides to throw me out before then.

A Campus of Our Own: Thoughts of a Reluctant Conservative

Susan Saltrick

Susan Saltrick has an independent consulting practice, Proteus Consulting, based in New York City; she specializes in educational technology, business development, and market research.

At the *Visions* Web Site Links to other educational consultants (listed at Yahoo!); to Educom ("Transforming Education through Information Technology"); to CAUSE.

READING PROMPT

What makes Saltrick a "reluctant" conservative? What particular traditional aspects of your own education do you feel are worth preserving?

Having attained my 39th birthday, I've been observing more frequently the signs of impending middle age. Besides the usual cosmetic distress, I realize I'm now being labeled as one of the conservatives of the new media scene.

I find myself agreeing a lot with the Neil Postmans and the Sven Birkertses of the world who urge us all, when asked about technology, to just say no. If I don't watch myself, I'll soon be voting Republican.

All this is a somewhat disconcerting state to find myself in—after spending 12 years as a new media director for a major book publisher where I was, by executive fiat, the brave scout out on the wild frontier, clearing the trail so the rest of the folks back East could settle down and homestead.

I used to be a real new media cheerleader. My talks of only a few years ago were upbeat panegyrics to the scintillating adventures that awaited us all in the digital future, if we could only navigate the tricky water that surrounded us today. Given that I was in higher education publishing, I used to talk a lot about technology's potential for transforming the educational process—for exploiting new modes of learning, for expanding to new types of students.

And, you know, I pretty much believed it. And, you know, I pretty much still do. But what I find myself thinking about these days is a question Steve Gilbert poses when considering any technological change: "What is it," he asks, "that we cherish and don't want to lose?" In other words, what are the things we need to watch out for—the things we risk sacrificing if we are too readily seduced by technology's siren song?

Two recent works have provoked some thoughts in this area. The first of these, *The City of Bits,* by William Mitchell, the dean of architecture at MIT, is an exploration of how our notion of space—and the structures we humans have created to order that space—are being transformed by our increasing involvement with, or better put, our envelopment in, cyberspace.

The second is Eli Noam's recent article in *Science* magazine, "Electronics and the Future of the University." Noam notes that universities flourished because they were the centralized repositories for information—and that scholars and students gathered there because that was where they found the raw materials they needed to do their work. But now that information is distributed and available all over the Net, he asks, what becomes of the physical university?

Both Mitchell and Noam are concerned with how the Infobahn is changing our notion of community from a spatially defined entity to a virtually connected body—and both address the notion of the virtual communities that are developing out on the Web. But, for me, at least, the Net is so damned evanescent. We may bookmark our favorite sites, but we never really retrace our steps. With each Web journey, we inscribe a new path, spinning a digital lace as unique and convoluted as our genetic code. Click by click, we spiral down, seeking those elusive gems of content, while all around us the landscape is in constant flux—links forming and dissolving like a seething primordial soup. We never quite know where we're going, and God only knows where we'll end up—accidental tourists all in this floating world, this web.

So let's be careful when we use that term: community. Doesn't it imply some sort of common code of behavior or attitude or association? Can something so mutable, so ephemeral, count? If we gain admittance to it through a mere mouse click, can something so easily won, so temporary, truly impart a sense of belonging? If we can belong to as many communities as time and carpal tunnel strength permit, what then of loyalty? Can it be so diffuse? Is community then a less binding concept in this less bounded world?

How we define community is critical, because the notion of community is at the heart of any speculation about the university in the 21st century. Of course, the university is not one community, but many. But for simplicity's sake, let's focus on two groups within academe—the faculty and the students—and see how their roles, their behaviors, and their sense of belonging have been and will be affected by technology's advance.

I need first to acknowledge that my thinking is, not surprisingly, the result of my experience. I was fortunate to have an exceptional undergraduate experience at a beautiful residential college with a very low student-to-teacher ratio, with world-class faculty, and a diverse and stimulating student body. I would wish that educational setting for everyone, but with tuition at my college now exceeding $18,000, that's clearly not going to happen. As a student, I received financial aid, but the national grants that made it possible for me to attend aren't around anymore. But this is an old, sad story for another time and another forum.

So keep my bias in mind; perhaps it will explain some of my nostalgia. Or perhaps it will just make it easier for you to call me a snob. But let's look now at the faculty and their use of technology.

College faculty make up one of the most plugged-in professions in their use of technology for research—and one of the most retrograde in their use of technology for teaching. This seeming paradox has a lot to do with simple human behavior. If a tool makes it easier to perform a frequent task, it will be adopted. If, however, the tool requires a new type of behavior, or doesn't conform well with the existing modus operandi, then you can build it, but they probably won't come. For many faculty, the Internet has become a basic way of life because it meshes well with their work patterns. It mirrors nicely their image of themselves.

Communication with one's peers is integral to any line of work. And for many workers outside academia, their office location and the colleagues housed therein define their professional communities. Academics, though, have always embraced the idea of a geographically dispersed community of scholars. In the Middle Ages, universities—and the Church—functioned as true multinational institutions, at a time when most people never traveled more than 20 miles from the hovel in which they were born. Today, though, academics are likely to affiliate along disciplinary lines. A chemistry professor in Ann Arbor is likely to have more interaction with a chemistry professor in

Palo Alto than with her counterpart in Germanic Languages in the building 50 yards away.

When the Net functions as a scholarly connectivity tool, it is just continuing the tradition of information-sharing that has always linked faculty who share a common disciplinary focus. So as a means of facilitating academic collaboration and resource sharing, the Net's a no-brainer.

But what of the professor's other hat—the teaching enterprise? There the Net offers no neat analogue to the usual work patterns because classroom instruction, unlike scholarly research, has traditionally been a highly autonomous, largely decentralized activity. No wonder, then, that until very recently the percentage of courses utilizing the Internet—or any new media at all—hovered in the single digits.

Finally, though, the winds of change appear to be blowing. On quite a few campuses today, computer access is a given, and the network is almost ubiquitous. But we're still just at the beginning. The epistemological transformations, the ones that change not just the way one teaches but what one teaches, are still few and far between. The big changes are still going to take a long time—and a fair amount of pain—to achieve.

My conservatism is founded on some very practical—and one could say short-sighted—concerns. Number one, who's going to pay for all this stuff? In *The City of Bits,* Mitchell tells of sitting in his Cambridge office instructing students from six different universities across the globe via two-way teleconferencing. With all due respect to him and the great institution he represents, we don't all have the same craft to navigate the information ocean. It's one thing to sit at MIT in front of one's high-end workstation complete with video camera, hooked to a super-high-speed network with monster bandwidth—but that's not quite the reality for the rest of us. At a community college I visited last spring, the faculty were no longer able to make long-distance calls because the school couldn't afford it. While enrollments keep going up at this school, the state and local governments keep decreasing the funding. How long will it be before they have what MIT has today?

I'm sure we'd all love to sit in on that MIT teleconferenced course, sharing ideas with Mitchell's great mind and with students all across the world: it sounds extraordinary. And I'm afraid it's just that—anything but ordinary. The notion of increased access to higher education is an unarguably worthwhile societal goal. Unfortunately, in marked contrast with Mitchell's class, a lot of what passes for distance education today seems a pretty poor substitute for an admittedly less-than-ideal large classroom lecture delivered live. That's because a lot of distance education is just asynchronous transposition of that same ol' large-classroom lecture—but now you get to watch it on TV. And as we all know from watching our home videos, video tends to flatten reality. Any mediated experience has to be at one remove from the real thing. And a flattened version of the average large-audience lecture is a rather dismal thing to contemplate.

But despite all this gloom, there's some cause for optimism. E-mail might just be the flying wedge. It could ultimately have the biggest impact on instruction of all the new technologies. It may not be as sexy as a two-way teleconference with Hong Kong, but e-mail changes behavior. With e-mail, students who didn't talk before, now do—and faculty find they're working longer hours to keep up with them. Steve Gilbert tells me of numerous conversations in which his correspondents say they started with e-mail and ended up rethinking what teaching is and how they could do it better.

On the train from Washington, DC, to New York this fall, I had the opportunity to overhear the conversation of two students—the woman a college sophomore and the young man a high school senior. Their conversation centered on how she had chosen her school and on the criteria he was using to determine his college next year. She's studying international relations and had chosen her school because of its proximity to the nation's capital. Her major gripe was that she was a residential student on a largely commuter campus, and that made for some pretty lonely weekends.

He had just returned from a visit to Georgetown—which was high on his list because of the beauty of the campus. And as a kid from Queens who had spent the last four years in a very small town in Delaware, he was eager to return to an urban location. As a prospective business major, he felt a city location would provide him more internship opportunities.

What's so unusual about this conversation? Absolutely nothing, and that's the whole point. Every one of their stated pros or cons was based on a reaction to the physical environment, be it access to certain urban resources or the attributes of a particular style of architecture—their criteria were all material. Where the students were, or wanted to be, mattered. It was all about place. Their conversation just demonstrated what we all know—that college is more than academics (a topic notably absent from the students' discussion).

College for many of us is a process of socialization, a rite of passage, which requires its own material culture—its real things—whether this means football games, fraternity row, the quad in front of Old Main, or any of the myriad other places on campus where students connect with one another, including those places their parents don't want to think about. In the *Science* article, Noam states, "The strength of the future physical university lies less in pure information and more in college as community."

The student body is, of course, a community, one largely defined today by a connection with a physical place—a campus. I think many students will still want to experience this kind of community. What's less clear is how the colleges of today are going to be able to offer that kind of experience. Soaring costs coupled with drastic budget reductions don't offer a lot of hope. Undoubtedly, other kinds of educational experiences are going to be available—and other kinds of educational institutions will arise to deliver them.

Perhaps higher education in the future will resemble the proverbial onion: at the core, the traditional type of real-time, face-to-face instruction, surrounded by rings of other educational experiences. Some instruction will be imported onto the campus from long-distance professors, while other types of instruction will be exported from the campus to long-distance students. I'd hope that college life in the future will be more like our work life right now—a rich mixture of real-time and asynchronous interaction with peers, advisors, and external information sources.

When we think about the future, the trap we keep falling into is to see technology as a replacement for experience rather than an enhancement of capability. The old ways don't always get supplanted, they just get crowded as more and newer options come into play. Technology permits more, makes more possible, but it doesn't obliterate what went before.

There's a lot of black or white thinking out there. Computer logic may be binary, but our rational processes don't have to be. We're not hurtling towards a digital apocalypse (at least not faster than towards any other), but that doesn't mean we should unquestioningly adopt any technology, heedless of the consequences, simply because everyone seems to be doing it. Technology will change us—it will change us fundamentally—but that doesn't mean we can't influence its course. And let's remember, we can always just turn the damned things off.

We see a lot these days about the social implications of technology—ironically much of it is going on between the covers of the good old-fashioned book. The argument seems to have polarized many commentators. On the one hand, we have the high priests of the digital culture, who see all things as reducible to data. In their eyes, the human enterprise is just another bunch of bits in the cosmic data stream. On the other side, the neo-Luddites practice their own form of reductionism, fearing that silicon will eradicate all that is carbon. Can't we instead pursue some middle course? I don't accept the notion that the only difference between a computer-mediated exchange and a lover's kiss is bandwidth—nor do I fear that I will become a mere appendage on the ubiquitous Web.

Technology enables us to do more work. We can connect with more people; we can learn from more sources; we can cast our nets farther. The irony is that—thanks to my computer, thanks to the Net—I now have more face-to-face encounters, more meetings, more travel, more experiences in the real world. I still have to, want to, and need to get out there and talk to people. Between my three e-mail accounts, my two voice mailboxes, and the fax machine—not to mention the plain old phone and mail—I spend most of my day dealing with representations of reality (isn't that what text and images are?). Because of this, and because I no longer work in an office, I find I need to get out for a walk once a day just to connect with tangible things, with flesh-and-blood humans, even if my primary contact with them occurs across the counter of the nearest coffee bar.

Just look at the crowded airways, the boom in business travel, the explosion of conferences. We all still board those planes and drive those cars to travel hours in order to meet with our colleagues. Theoretically, at least, we could do it all by phone. We could receive print-outs of the talks and just read them. We could watch it on TV. But we don't. We like it live.

Why? Because we all know the good stuff at a conference happens during the coffee breaks. Because we all know we don't really know someone until we've eaten with them. Because we all know the camera doesn't always follow the ball carrier. Because we know that even the world's best programmer can't anticipate all the possibilities that real life throws at us every second of the day.

We used to say computers would be pretty good if only they could talk—then they did. Then we said, well, if only they had pictures—and now they've got those in spades. Touch may be more difficult—but there's a pretty good simulacrum of that already in the labs. But something will always be missing, because while they are wonderful amplifiers of the human potential, computers can never be replacements for the human experience. Someone else said it, but my yoga classes confirm it: "There's no prahna in a computer," no qi, no life-force. And all the bandwidth in the world won't ever change that.

In our rush to embrace the wonders that technology can provide, let's never confuse its images, its sounds, its symbols, with what's real. Perhaps all this talk of reality seems patently obvious, yet with the new media hype we are sometimes unable to distinguish the real from the *trompe l'oeil* of the screen. We need to be aware of the danger of complacency, and be ever alert for opportunities to celebrate the real. Our lives are ever more built upon mediated experience, but in our increasing immersion in the digital sea, let's not forget that at bottom, it's just zeroes and ones. And just in case we do forget, there's Virginia Woolf to remind us in *A Room of One's Own:*

> What is meant by "reality"? It would seem to be something very erratic, very undependable—now to be found in a dusty road, now in a scrap of newspaper in the street, now in a daffodil in the sun. It lights up a group in a room and stamps some casual saying. It overwhelms one walking home beneath the stars and makes the silent world more real than the world of speech—and then there it is again in an omnibus in the uproar of Picadilly. Sometimes, too, it seems to dwell in shapes too far away for us to discern what their nature is. But whatever it touches, it fixes and makes permanent. That is what remains over when the skin of the day has been cast into the hedge; that is what is left of past time and of our loves and hates.

The happiest day of my life was April 14, 1995. My husband and I and our two preschoolers were in Rome. I was watching, awestruck, as the kids used the Forum as a sort of archeological playground—clambering about the columns that Julius Caesar and Pompey and Claudius and Titus and all the rest had walked through conducting the affairs of the world a couple millennia ago. What made that day so extraordinary? I can't quite put it into words, but it was all about history, it was all about place, it was all about being there. It could not have happened in a Virtual Reality park in Las Vegas. It was real.

I recognize that that's a privileged story, the personal equivalent of teleteaching at MIT, but the second happiest time of my life is just about every day—when the kids come home and they run into my office. I turn from the computer and see them there, beaming and bright, and wondrously, miraculously, real.

And I'm reminded once again that what really counts is the ineluctable—the mystery, the reality of life. It's what remains over from the skin of the day—it's what we cherish and don't want to lose. It's what belongs to the realm of the sensuous—it's walking home beneath the stars—it's those things we can't program or code. These are the things that will save us.

chapter 12 Explorations

12a Newspapers, Old and New

Collaboration. Do you read newspapers now? Which one or ones, how often, and for what sort of information? Share an account of the different ways you have used newspapers throughout your life.

Web work. Newspapers Online has links to every online newspaper in the world, including campus papers; check it for the link for the online versions of the newspapers nearest to you.

The *American Journalism Review* sponsors the NewsLink site, where its readers list their fifty favorite Web news sources.

One writing topic. For any one newspaper (e.g., *USA Today*), write a comparison of the two formats of a single newspaper: its Web and traditional print formats. What are the advantages and disadvantages of each format?

12b Libraries, Old and New

Collaboration. Share recollections of your earliest memories of libraries.

Web work. The American Library Association, or ALA, is the main advocacy group for public libraries in America. The Technology Resource Institute, or TRI, is dedicated to expanding the traditional function of libraries so as to bring "information technology resources to the public"—an effort being aided by Microsoft's Bill Gates through his Gates Library Foundation.

Two public libraries that play a special role in this country's intellectual life are the New York Public Library and the Library of Congress in Washington, D.C., with its standing and ever expanding Web exhibition, American Memory.

The Web is full of sites for important small libraries, such as the one dedicated to the controversial novelist Henry Miller in Big Sur, California.

One writing topic. Pick a decade in twentieth-century life in the United States, pre-1990, and compare the electronic resources you can find for studying that decade (either in a library or directly on the Web) with the print resources available in a local library.

12c School Research Papers

Collaboration. Share an account of the best or the worst experience you have had writing a research paper for school.

Web work. The Internet Public Library has a special teen-oriented section entitled A+ Research and Writing. ResearchPaper.com bills itself as "the Web's largest collection of topics, ideas, and assistance for school related research projects." It is owned and operated by Infonautics, which operates both Encyclopedia.com, a free service that reproduces more than 17,000 entries from the Concise Columbia Electronic Encyclopedia, and the Electric Library, a fee-based, student-oriented database and document access service. With a credit card, you can get a free trial subscription. Microsoft's online encyclopedia project, Encarta, now offers home users access to the same student-oriented database and text-delivery service widely leased to college libraries by the Information Access Company, under the brand name InfoTrac.

The MLA (Modern Library Association) has a Web site for its widely used style guide to documenting research papers. Several college textbooks on writing research papers also have informative Web sites, among them *The Research Paper and the World Wide Web* by Dawn Rodriguez. Finally, Yahoo! provides links to current political topics in a central site: Government > Politics > Political Issues.

A warning: There are also dozens of commercial and free services on the Web offering desperate and reckless students completed term papers. All educational institutions consider submitting such materials for credit under your own name as a grave act of academic misconduct subject to the most serious penalties.

One writing topic. Write an essay comparing the online materials you are able to find on any two movies, one released since 1995 and one before 1990.

12d Research Topic: English-Only Laws

Collaboration. Share a description of a time when you were left completely out of the conversation, for whatever reason.

Web work. English First and U.S. English are two leading advocacy organizations promoting English-only laws. The American Civil Liberties Union (ACLU) and the National Education Association (NEA) have taken active roles in opposing English-only laws, and each has a briefing paper on the subject on its Web site. Robert King discusses the issue in the April 1997 Atlantic Unbound.

One writing topic. Write an essay arguing for or against English-only laws.

12e Research Topic: Drug Policy

Collaboration. Use a form of anonymous sharing to compare accounts of events with unexpected consequences (serious or not) due to one or more persons' use of drugs or alcohol.

Web work. Yahoo! lists more than twenty organizations concerned with reforming marijuana laws, of which the most visible has been National Organization for the Reform of Marijuana Laws, better known as NORML. The Drug Reform Coordination Network (DRCNet) also offers extensive online information in support of reform. The National Drug Strategy Network offers a nonpartisan approach to the issue, in an effort to strengthen the criminal justice system and with links to the National Institute on Drug Abuse. DARE (Drug Abuse Resistance Education) is active in schools educating youngsters about the dangers of addiction.

Two sites from *Frontline* documentaries on public television: "The Opium Kings," on Burma's heroin trade, and "Murder, Money & Mexico."

The online magazine FEED sponsored a dialogue on the subject, "Prohibition and Its Discontents."

The Atlantic Unbound has articles that include "More Reefer Madness" by Eric Schlosser, in support of liberalization, and "How to End the War on Drugs," a more neutral mock presidential executive decision, by James Fallows. Mortimer Zuckerman, the editor in chief at *U.S. News & World Report,* offers support for strict enforcement of existing laws in an editorial available online.

One writing topic. Write an essay in support of or opposed to liberalizing drug laws.

12f Research Topic: Death Penalty

Collaboration. Use a form of anonymous sharing to compare accounts of a time you received what you felt was a greater or a lesser punishment than you deserved.

Web work. The Atlantic Unbound has a rich online collection of articles on the death penalty under the general heading, "Who Deserves to Die?" "The Death Penalty: Pro and Con" is a PBS *Frontline* documentary with Supreme Court Justices Blackmun and Scalia coming down on different sides of the issue.

One writing topic. Write a research paper on some aspect of capital punishment. Here are three suggestions: (1) the current state of capital punishment laws throughout the world, (2) the history of capital punishment in the United States, or (3) a classification and analysis of Web sites devoted to capital punishment.

part V

At Work

One pressing issue confronting college students today is the future of work. We tend to imagine that at one time, somewhere in the distant past, people had to work hard all the time just to stay alive. Surely this was the normal, unimproved condition of the world—a world awaiting the liberation of new, higher forms of technology. We may assume, in other words, that the simple life of the past was one filled with drudgery, whereas the complex, high-tech life of the future will be one of ease—forgetting in this argument that people in tropical climates, even with simple technologies and lifestyles, often have to work very little (even today) to secure adequate food and shelter. Indeed, our U.S. culture has often been prejudiced against lifestyles of ease, against groups of people able to enjoy life without hard work (unless, of course, they are "rich and famous," in which case they surely must somehow have earned it; or, if not, at least they can serve as models of what we all might someday hope to attain if we work hard).

For years, for example, there has been a stigma attached to welfare in this country, in part because to be eligible for certain forms of welfare (notably the program long known as Aid for Families with Dependent Children, or AFDC), recipients were not allowed to work but were instead forced to stay home and care for their children. The welfare reform bill of 1996, passed by a Republican Congress and signed by a Democratic president—and with the telling title of "The Personal Responsibility and Work Opportunity Reconciliation Act"—turned this situation on its head, at least in theory, by providing welfare benefits to single parents of dependent children only if the single parent is working or looking for or being trained for work. The question of who is supposed to take care of the children is just one of many unsettled issues—several of which can be pursued in Chapter 13 (Working More), both in Jack Beatty's action memo ("What to Do about Welfare?") and in Explorations 13e and 13f (the former a virtual tour of the community of Harlem).

Another aspect of this debate treated in Chapter 13 is the possibility that the new electronic technology will provide us all with new and better jobs, jobs that give us all access to a higher standard of living. Such is the case made by Esther Dyson in "Education and Jobs in the Digital World"—a position countered by Denise Caruso in "Geeksploitation" and by Kirkpatrick Sale in "Lessons from the Luddites." Caruso questions the vaunted glamour and rewards of working in high technology; Sale uses the political resistance of nineteenth-century British weavers (a resistance that turned violent when these unemployed and desperate workers destroyed the new looms they felt were responsible for taking their jobs) to question how we define what constitutes a good job or good work generally. Do not despair, however, for the Explorations in Chapter 13 will help you get ready for that next job or for the next school you might need to prepare for that job.

Chapter 14 (Working Less), meanwhile, approaches this problem of work from the other end: This chapter's readings make the case that we need to stop and reconsider the value of leisure, for individuals and for society as a whole. Jay Walljasper's "The Speed Trap" gives a historical overview of how life has become more frantic; Reva Basch describes what it is like to waste hours upon hours on the Web; Hal Niedzviecki sings the slacker mantra (the praises of part-time work); and Jackson Lears surveys U.S. intellectuals who have praised the "simple life." Explorations for Chapter 14 include looks at TV, sleep, beach life, and (for the type A slacker) planning for early retirement. If you are lucky, maybe you'll be able to squeeze in the readings and Explorations for this chapter in the two days between your biology midterm and your history research paper. No problem, right? Even if you are a full-time student working a job twenty hours a week?

Chapter 15—Women, Work, and Technology—considers the gender dimension of issues related both to work and to technology generally. What special concerns have women had historically with computers as machines, with computer culture, and with computer and other science courses that lead to careers in the field? Pamela McCorduck and Karen Coyle debate these first two questions relating to women and computers; Anne Gibbons and her colleagues, in "Facing the Big Chill in Science," look at the third question. Kathe Davis's piece "What about Us Grils?" considers the broader issue of women and science fiction—how the imaginative constructions of female science fiction writers tell us something important about women and technology—while Arlie Hochschild raises issues concerning leisure from Chapter 14 but with special reference to the concerns of women. Explorations for Chapter 15 deal with issues related to education (equal access for different social groups, single-gender education, new ways to teach math) as well as with Web sites devoted both to women's concerns ("girl talk") and the problems of being a single parent (female or male).

chapter 13

Working More

READING QUESTIONS

- Besides the salary, just what (if anything) do you want to get out of your future work, your postcollege career?
- In what ways do you imagine computers could play a positive or a negative role in this work?
- What significant aspects of your own future work, and of future work generally, are likely to be unaffected by computers?

QUOTATIONS

Real programmers don't work from 9 to 5. If any real programmers are around at 9 a.m. it's because they were up all night.

— *Anonymous*

Be true to your work, your word, and your friend.

— *Henry David Thoreau, American author, 1817–1862*

It is impossible to enjoy idling thoroughly unless one has plenty of work to do.

— *Jerome K. Jerome (1886), British novelist and playwright, 1859–1927*

I remember summing up what I took to be our destiny, in conversation with my best friend at Chartres, by the formula, "Term, holidays, term, holidays, till we leave school, and then work, work till we die."

— *C. S. Lewis (1955), British author, 1898–1963*

Education and Jobs in the Digital World

Esther Dyson

Everyone seems worried about preparing students for the new, increasingly high-tech world of work; and Esther Dyson is one person the world is likely to listen to on this subject, as she serves as a technology advisor to major international corporations and runs one of the most prestigious annual national technology conferences. Dyson is also chairman of her own information services company, Edventures; publisher of a monthly industry newsletter, *Release 1.0* (priced at $695 a year!); and the author of a book on computer culture, *Release 2.0*.

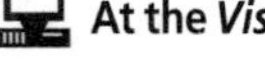 **At the *Visions* Web Site** Links to more by and about Dyson.

READING PROMPT

What skills does Dyson want you to acquire so as to be competitive in the workplace? What other skills, if any, do you want to get in college?

The digital world will support daily life in the 21st century; it will be the infrastructure underlying commerce and community.

Yet the Net—by which I mean the entire modern information infrastructure based on digital technology—has no independent existence. It matters only because people use it as a place to communicate, conduct business, and share ideas; not as a mystical entity in itself.

It extends across and transcends traditional national borders and obliterates distance. It can operate in real time, but it lets people in different time zones communicate easily. But it must coexist with national regimes, cultural and language differences, and the realities of physical infrastructure that impinge on its theoretical spacelessness.

The digital world will profoundly change how people learn, how they work, and what they produce, but at the same time it won't change human nature nor the need to give children a moral as well as intellectual education.

In the old days, you could achieve a semblance of equality among people by distributing land to the peasants, or later on by raising workers' salaries or even giving them profit-sharing. But in the information (as opposed to the agricultural or industrial) economy, such tactics no longer work. People prosper less according to what they have in their hands or bank accounts, and more according to what they can do with their minds.

That means the task of fostering equality (of opportunity, at least) is more complex than a simple redistribution of assets. The only feasible way is education, which requires not just giving, but helping people to take and to learn. It's a two-way process that requires participation from the learner—and it's just about the most important job a community can do. Education works best in the decentralized environment fostered by the Net, yet it needs to be paid for by society as a whole—even as it benefits society as a whole. That is, parents and children should benefit from the choices that will be provided by the Net, offering children access to information and to other children or experts outside their communities. But the costs of that access need to be spread broadly so that equal opportunities are available to all, although they will make individual choices as to what kind and level of education they want for their children and for themselves as adults.

However, although computers and the Net are an important tool for education, they cannot replace teachers as role models, mentors, and motivators. Children should learn to use the Internet as part of their education and as a tool for finding out information and communicating with each other, just as adults do. But they need to get the basics of their moral and social education from their own parents and families.

What will children find as they join the world of work? First, it won't seem as foreign to them as in the industrial age, when adults worked at offices and factories. Many children are already writing programs, designing Web pages, and troubleshooting computer installations. Bill Gates, who started what became Microsoft as a teenager, is only the most famous of them. The optimistic view says that more people will work from home, becoming closer to their families, and that the work itself will be more fulfilling, ecologically benign, and productive. This trend to working at home will also foster home-schooling, supplemented by educational resources on the Net.

However, the worker of the future may well have several jobs, switch jobs frequently, and operate more as a freelancer than as a loyal member of a team. This seems to be the norm in Silicon Valley, although companies such as Netscape and Hewlett-Packard work hard to foster loyalty in the face of an increasingly friction-free job market. (The best long-term incentive for employees is the presence of other people they respect and enjoy in a world where anyone can match salaries.) In fact, there will simply be a greater variety of arrangements. Some people will take happily to the world of projects and freelancers; they will enjoy the freedom to control their own careers and manage their own time; others will find this friction-free mode of contract work scary, uncertain, and alienating. And of course others, without the necessary skills, will be left out altogether.

Education will bear the burden of training people to operate in this new world. And people will bear the burden of continuing their own educations throughout their own lifetimes, ideally with some form of tax-advantaged education investment account. For those with the skills, there should be ample work, whether it's managing online customer feedback services, designing new software, or teaching children in a world that finally recognizes the importance of childhood education. (Another big field for personal service will be medicine. Although online information will enable many people to find medical information for themselves, medical practitioners will still be needed to administer care, interpret the information for specific patients, and the like. Some of the intellectual work and diagnostics can be accomplished through telemedicine, but telemedicine still requires a trained medical practitioner at one end of the link.)

People lacking the proper preparation for the digital world will become an increasing social problem, for themselves and for the rest of society. Digital media are no panacea, but they can contribute to solutions in a number of ways. In fact, remedies range from supplementing teachers' efforts and letting children in schools communicate and learn from other people, to providing better sources of information about jobs, training opportunities, and transportation in disadvantaged communities. Because use of the physical channels themselves are relatively cheap, the capital costs for civic organizations to establish community networks will be minimal—a situation that I hope will lead to a proliferation of such efforts. People who currently would like to help—or help themselves—should find it much easier than they do today.

Geeksploitation | *Denise Caruso*

The world at large and computer workers themselves have often seen high-tech work as inherently glamorous, perhaps because of the personal fortunes that key individuals in the computer industry have been able to amass for themselves. Independent journalist Denise Caruso questions this assumption in this short piece, reprinted from *The Site,* a combination online magazine and computer show on MSNBC.

At the *Visions* Web Site Links to Microsoft's jobs page (if you are interested in working in the belly of the beast); to excerpts from Douglas Coupland's novel, *Microserfs,* which provides a satirical view of life at Microsoft.

READING PROMPT

What is wrong with computer jobs, at least according to Caruso?

Over the years, visionaries with stars in their eyes have waxed poetic about the fabulous life of the "information worker" of the future.

Computers will do all the tedious work. We will be free! Free to wander through the meadows of our minds, chewing on clover and thinking big thoughts.

Who would say no to that?

But here and now, in 1996, we are information slaves. Email ravages our time and our brains like an Ebola virus. Information dogs us wherever we go—electronic newspapers, magazines, columns, stock tickers.

We have to pay attention! If we don't, we'll be left behind!

Think big thoughts?

The only thinking I have time for is to worry about whether I have attention deficit disorder. I'm so overwhelmed, I can't hold a thought in my head for more than 30 seconds.

And while the new crop of Gen-X workers are whining about "geeksploitation"—poor babies, they got duped by the propaganda—you can actually smell the baby boomers frying.

After 10 years of oh-so-personal computing, they are burnt to a crisp.

Almost every day—and I'm not making this up—I hear someone else has thrown in the towel. No new job, no plans for the future, maybe a little money in the bank.

They just quit. "I have to get my life back," they say.

Those crushing 50–60–70 hour work weeks have been standard operating procedure in the computer industry for years. This cannot go on forever.

Workers started unions because they felt trapped and exploited. It can happen again—even in the starry-eyed, we-can-change-the-world computer business.

And I pity the company management who doesn't think it can happen to them.

But the bottom line remains—

What we do for a living should not—cannot—continue to define our value, or who we think we are.

"Corporate culture" serves the corporation, not the workers.

You replace your sense of personal identity with the identity of the place you work. And presto! You have a new family—a good daddy and mommy and brothers and sisters who will love you and take care of you.

People find comfort and meaning in corporate cultures. And people are just as easily devastated by them when they find out their new family is just as dysfunctional as the old one.

Our lives are not assigned meaning by our bosses or by our company's corporate culture. We assign meaning to our own lives.

And if we decide that our true value as human beings depends on what we do to earn our keep—whether it's designing web pages or collecting garbage—we will forever allow ourselves to be exploited.

And we will have no one to blame but ourselves.

Lessons from the Luddites

Kirkpatrick Sale

The resistance to industrial technology began about the same time and in the same place as industrialism itself: in the Midlands of England in the early nineteenth century. The first group of workers to form an organized and, for a time, violent opposition to new manufacturing practices fashioned themselves as followers of a mythical local hero and resister, King Lud—hence their name, Luddites, a term that has come to have almost entirely negative connotations in referring to people who foolishly and willfully oppose progress offered by new technologies. The following article by political commentator Kirkpatrick Sale appeared in *The Nation* and was also part of Sale's 1995 book *Rebels Against the Future: The Luddites and Their War on the Industrial Revolution: Lessons for the Machine Age* (Addison-Wesley), a work that attempts to rehabilitate the Luddites by placing them within the larger canvas of modern history as a whole.

At the *Visions* Web Site Links to online interviews with Sale, including one with Netizen's Jon Katz.

READING PROMPT

Which of Sale's Luddite lessons do you find most convincing? Least convincing?

As Newt Gingrich has assured us, and as our own daily experience has convinced us, we in the industrial world are in the middle of a social and political revolution that is almost without parallel. Call it "third wave" capitalism, or "postmodern," or "multinational," or whatever; this transformation is, without anyone being prepared for it, overwhelming the communities and institutions and customs that once were the familiar stanchions of our lives. As *Newsweek* recently said, in a special issue that actually seemed to be celebrating it, this revolution is "outstripping our capacity to cope, antiquating our laws, transforming our mores, reshuffling our economy, reordering our priorities, redefining our workplaces, putting our Constitution to the fire, shifting our concept of reality."

No wonder there are some people who are Just Saying No.

They have a great variety of stances and tactics, but the technophobes and techno-resisters out there are increasingly coming together under the banner that dates to those attackers of technology of two centuries ago, the Luddites. In the past decade or so they have dared to speak up, to criticize this face of high technology or that, to organize and march and sue and write and propound, and to challenge the consequences as well as the assumptions of this second Industrial Revolution, just as the Luddites challenged the first. Some are even using similar strategies of sabotage and violence to make their point.

These neo-Luddites are more numerous today than one might assume, techno-pessimists without the power and access of the techno-optimists but still with a not-insignificant voice, shelves of books and documents and reports, and increasing numbers of followers—maybe a quarter of the adult population, according to a *Newsweek* survey. They are to be found on the radical and direct-action side of environmentalism, particularly in the American West; they are on the dissenting edges of academic economics and ecology departments, generally of the no-growth school; they are everywhere in Indian Country throughout the Americas, representing a traditional biocentrism against the anthropocentric norm; they are activists fighting against nuclear power, irradiated food, clearcutting, animal experiments, toxic waste and the killing of whales, among the many aspects of the high-tech onslaught.

They may also number—certainly they speak for—some of those whose experience with modern technology has in one way or another awakened them from what Lewis Mumford called "the myth of the machine." These would include those several million people in all the industrial nations whose jobs have simply been automated out from under them or have been sent overseas as part of the multinationals' global network, itself built on high-tech communications. They would include the many millions who have suffered from some exposure, officially sanctioned, to pollutants and poisons, medicines and chemicals, and live with the terrible results. They include some whose faith in the technological dream has been shattered by the recent evidence of industrial fragility and error—Bhopal, Chernobyl, Love Canal, PCBs, *Exxon Valdez*, ozone holes—that is the stuff of daily headlines. And they may include, too, quite a number of those whose experience with high technology in the home or office has left them confused or demeaned, or frustrated by machines too complex to understand, much less to repair, or assaulted and angered by systems that deftly invade their privacy or deny them credit or turn them into ciphers.

Wherever the neo-Luddites may be found, they are attempting to bear witness to the secret little truth that lies at the heart of the modern experience: Whatever its presumed benefits, of speed or ease or power or wealth, industrial technology comes at a price, and in the contemporary world that price is ever rising and ever threatening. Indeed, inasmuch as industrialism is inevitably and inherently disregardful of the collective human fate and of the earth from which it extracts all its wealth—these are, after all, in capitalist theory "externalities"—it seems ever more certain to end in paroxysms of economic inequity and social upheaval, if not in the degradation and exhaustion of the biosphere itself.

From a long study of the original Luddites, I have concluded that there is much in their experience that can be important for the neo-Luddites today to understand, as distant and as different as their times were from ours. Because just as the second Industrial Revolution has its roots quite specifically in the first—the machines may change, but their machineness does not—so those today who are moved in some measure to resist (or who even hope to reverse) the tide of industrialism might find their most useful analogues, if not their models exactly, in those Luddites of the nineteenth century.

And as I see it, there are seven lessons that one might, with the focused lens of history, take from the Luddite past.

1. Technologies are never neutral, and some are hurtful. It was not all machinery that the Luddites opposed, but "all Machinery hurtful to Commonality," as a March 1812 letter to a hated manufacturer put it—machinery to which their commonality did not give approval, over which it had no control, and the use of which was detrimental to its interests, considered either as a body of workers or a body of families and neighbors and citizens.

What was true of the technology of industrialism at the beginning, when the apologist Andrew Ure praised a new machine that replaced high-paid workmen—"This invention confirms the great doctrine already propounded, that when capital enlists science in her service, the refractory hand of labor will always be taught docility"—is as true today, when a reporter for *Automation* could praise a computer system because it assures that "decision-making" is "removed from the operator . . . [and] gives maximum control of the machine to management." These are not accidental, ancillary attributes of the machines that are chosen; they are intrinsic and ineluctable.

Tools come with a prior history built in, expressing the values of a particular culture. A conquering, violent culture—of which Western civilization is a prime example, with the United States at its extreme—is bound to produce conquering, violent tools. When US industrialism turned to agriculture after World War II, for example, it went at it with all that it had just learned on the battlefield, using tractors modeled on wartime tanks to cut up vast fields, crop-dusters modeled on wartime planes to spray poisons, and pesticides and herbicides developed from wartime chemical weapons and defoliants to destroy unwanted species. It was a war on the land, sweeping and sophisticated as modern mechanization can be, capable of depleting topsoil at the rate of

3 billion tons a year and water at the rate of 10 billion gallons a year. It could be no other way: If a nation like this beats its swords into plowshares, they will still be violent and deadly tools.

2. Industrialism is always a cataclysmic process, destroying the past, roiling the present, making the future uncertain. It is in the nature of the industrial ethos to value growth and production, speed and novelty, power and manipulation, all of which are bound to cause continuing, rapid and disruptive changes at all levels to society, and with some regularity, whatever benefits they may bring to a few. And because its criteria are essentially economic rather than, say, social or civic, those changes come about without much regard for any but purely materialist consequences and primarily for the aggrandizement of those few.

Whatever material benefits industrialism may introduce, the familiar evils—incoherent metropolises, spreading slums, crime and prostitution, inflation, corruption, pollution, cancer and heart disease, stress, anomie, alcoholism—almost always follow. And the consequences may be quite profound indeed as the industrial ethos supplants the customs and habits of the past. Helena Norberg-Hodge tells a story of the effect of the transistor radio—the apparently innocent little transistor radio—on the traditional Ladakhi society of northern India, where only a short time after its introduction people no longer sat around the fields or fires singing communal songs because they could get the canned stuff from professionals in the capital.

Nor is it only in newly industrialized societies that the tumultuous effects of an ethos of greed and growth are felt. What economists call "structural change" occurs regularly in developed nations as well, often creating more social disruption than individuals can absorb or families and neighborhoods and towns and whole industries can defend against.

3. "Only a people serving an apprenticeship to nature can be trusted with machines." This wise maxim of Herbert Read's is what Wordsworth and the other Romantic poets of the Luddite era expressed in their own way as they saw the Satanic mills and Stygian forges both imprisoning and impoverishing textile families and usurping and befouling natural landscapes—"such outrage done to nature as compels the indignant power . . . to avenge her violated rights," as Wordsworth said.

What happens when an economy is not embedded in a due regard for the natural world, understanding and coping with the full range of its consequences to species and their ecosystems, is not only that it wreaks its harm throughout the biosphere in indiscriminate and ultimately unsustainable ways, though that is bad enough. It also loses its sense of the human as a species and the individual as an animal, needing certain basic physical elements for successful survival, including land and air, decent food and shelter, intact communities and nurturing families, without which it will perish as miserably as a fish out of water, a wolf in a trap. An economy without any kind of ecological grounding will be as disregardful of the human members as of the nonhuman, and its social as well as economic forms—factories, tenements, cities, hierarchies—will reflect that.

4. The nation-state, synergistically intertwined with industrialism, will always come to its aid and defense, making revolt futile and reform ineffectual. When the British government dispatched some 14,000 soldiers to put down the uprising of the Luddites in 1811 and 1812—a force seven times as large as any ever sent to maintain peace in England—it was sending a sharp signal of its inevitable alliance with the forces of the new industrialism. And it was not above cementing that alliance, despite all its talk of the rights of free Englishmen, with spies and informers, midnight raids, illegal arrests, overzealous magistrates and rigged trials, in aid of making the populace into a docile work force. That more than anything else established what a "laissez-faire" economy would mean—repression would be used by the state to ensure that manufacturers would be free to do what they wished, especially with labor.

Since then, of course, the industrial regime has only gotten stronger, proving itself the most efficient and potent system for material aggrandizement the world has ever known, and all the while it has had the power of the dominant nation-states behind it, extending it to every corner of the earth and defending it once there. It doesn't matter that the states have quarreled and contended for these corners, or that in recent decades native states have wrested nominal political control from colonizing ones, for the industrial regime hardly cares which cadres run the state as long as they understand the kinds of duties expected of them. It is remarkably protean in that way, for it can accommodate itself to almost any national system—Marxist Russia, capitalist Japan, China under a vicious dictator, Singapore under a benevolent one, messy and riven India, tidy and cohesive Norway, Jewish Israel, Muslim Egypt—and in return asks only that its priorities dominate, its markets rule, its values penetrate and its interests be defended, with 14,000 troops if necessary, or even an entire Desert Storm. Not one fully industrialized nation in the world has had a successful rebellion against it, which says something telling about the union of industrialism and the nation-state. In fact, the only places where a popular national rebellion has succeeded in the past two centuries have been in pre-industrial lands where the nation-state emerged to pave the way for the introduction of industrialism, whether in the authoritarian (Russia, Cuba, etc.) or the nationalistic (India, Kenya, etc.) mold.

5. But resistance to the industrial system, based on some grasp of moral principles and rooted in some sense of moral revulsion, is not only possible but necessary. It is true that in a general sense the Luddites were not successful either in the short-run aim of halting the detestable machinery or in the long-run task of stopping the Industrial Revolution and its multiple miseries; but that hardly matters in the retrospect of history, for what they are remembered for is that they resisted, not that they won. Some may call it foolish resistance ("blind" and "senseless" are the usual adjectives), but it was dramatic, forceful, honorable and authentic enough to have put the Luddites' issues forever on record and made the Luddites' name as indelibly a part of the language as the Puritans'.

What remains then, after so many of the details fade, is the sense of Luddism as a moral challenge, "a sort of moral earthquake," as Charlotte Brontë saw it in *Shirley*—the acting out of a genuinely felt perception of right and wrong that went down deep in the English soul. Such a challenge is mounted against large enemies and powerful forces not because there is any certainty of triumph but because somewhere in the blood, in the place inside where pain and fear and anger intersect, one is finally moved to refusal and defiance: "No more."

The ways of resisting the industrial monoculture can be as myriad as the machines against which they are aimed and as varied as the individuals carrying them out, as the many neo-Luddite manifestations around the world make clear. Some degree of withdrawal and detachment has also taken place, not alone among neo-Luddites, and there is a substantial "counterculture" of those who have taken to living simply, working in community, going back to the land, developing alternative technologies, dropping out or in general trying to create a life that does not do violence to their ethical principles.

The most successful and evident models for withdrawal today, however, are not individual but collective, most notably, at least in the United States, the Old Order Amish communities from Pennsylvania to Iowa and the traditional Indian communities found on many reservations across the country.

For more than three centuries now the Amish have withdrawn to islands mostly impervious to the industrial culture, and very successfully, too, as their lush fields, busy villages, neat farmsteads, fertile groves and gardens, and general lack of crime, poverty, anomie and alienation attest. In Indian country, too, where (despite the casino lure) the traditional customs and lifeways have remained more or less intact for centuries, a majority have always chosen to turn their backs on the industrial world and most of its attendant technologies, and they have been joined by a younger generation reasserting and in some cases revivifying those ancient tribal cultures. There could hardly be two systems more antithetical to the industrial—they are, for example, stable, communal, spiritual, participatory, oral, slow, cooperative, decentralized, animistic and biocentric—but the fact that such tribal societies have survived for so many eons, not just in North America but on every other continent as well, suggests that there is a cohesion and strength to them that is certainly more durable and likely more harmonious than anything industrialism has so far achieved.

6. Politically, resistance to industrialism must force the viability of industrial society into public consciousness and debate. If in the long run the primary success of the Luddite revolt was that it put what was called "the machine question" before the British public during the first half of the nineteenth century—and then by reputation kept it alive right into the twentieth—it could also be said that its failure was that it did not spark a true debate on that issue or even put forth the terms in which such a debate might be waged. That was a failure for which the Luddites of course cannot be blamed, since it was never part of their perceived mission to make their grievance a matter of debate, and indeed they chose machine-breaking exactly to push the issue beyond debate. But because of that failure, and the inability of subsequent critics of technology to penetrate the complacency of its beneficiaries and their chosen theorists, or to successfully call its values into question, the principles and goals of industrialism, to say nothing of the machines that embody them, have pretty much gone unchallenged in the public arena. Industrial civilization is today the water we swim in, and we seem almost as incapable of imagining what an alternative might look like, or even realizing that an alternative could exist, as fish in the ocean. The political task of "resistance" today, then—beyond the "quiet acts" of personal withdrawal Mumford urged—is to try to make the culture of industrialism and its assumptions less invisible and to put the issue of its technology on the political agenda, in industrial societies as well as their imitators. In the words of Neil Postman, a professor of communications at New York University and author of *Technopoly,* "it is necessary for a great debate" to take place in industrial society between "technology and everybody else" around all the issues of the "uncontrolled growth of technology" in recent decades. This means laying out as clearly and as fully as possible the costs and consequences of our technologies, in the near term and long, so that even those overwhelmed by the ease/comfort/speed/power of high-tech gadgetry (what Mumford called technical "bribery") are forced to understand at what price it all comes and who is paying for it. What purpose does this machine serve? What problem has become so great that it needs this solution? Is this invention nothing but, as Thoreau put it, an improved means to an unimproved end? It also means forcing some awareness of who the principal beneficiaries of the new technology are—they tend to be the large, bureaucratic, complex and secretive organizations of the industrial world—and trying to make public all the undemocratic ways they make the technological choices that so affect all the rest of us. Who are the winners, who the losers? Will this invention concentrate or disperse power, encourage or discourage self-worth? Can society at large afford it? Can the biosphere?

7. **Philosophically, resistance to industrialism must be embedded in an analysis—an ideology, perhaps—that is morally informed, carefully articulated and widely shared.** One of the failures of Luddism (if at first perhaps one of its strengths) was its formlessness, its unintentionality, its indistinctness about goals, desires, possibilities. If it is to be anything more than sporadic and martyristic, resistance could learn from the Luddite experience at least how important it is to work out some common analysis that is morally

clear about the problematic present and the desirable future, and the common strategies that stem from it.

All the elements of such an analysis, it seems to me, are in existence, scattered and still needing refinement, perhaps, but there: in Mumford and E. E. Schumacher *(Small Is Beautiful)* and Wendell Berry *(The Unsettling of America)* and Jerry Mander *(In the Absence of the Sacred)* and the Chellis Glendinning manifesto (*Utne Reader,* March/April 1990); in the writing of the Earth Firsters and the bioregionalists and deep ecologists; in the lessons and models of the Amish and the Iroquois; in the wisdom of tribal elders and the legacy of tribal experience everywhere; in the work of the long line of dissenters-from-progress and naysayers-to-technology. I think we might even be able to identify some essentials of that analysis, such as:

- Industrialism, the ethos encapsulating the values and technologies of Western civilization, is seriously endangering stable social and environmental existence on this planet, to which must be opposed the values and techniques of an organic ethos that seeks to preserve the integrity, stability and harmony of the biotic communities, and the human community within it.
- Anthropocentrism, and its expression in both humanism and monotheism, is the ruling principle of that civilization, to which must be opposed the principle of biocentrism and the spiritual identification of the human with all living species and systems.
- Globalism, and its economic and military expression, is the guiding strategy of that civilization, to which must be opposed the strategy of localism, based upon the empowerment of the coherent bioregion and the small community.
- Industrial capitalism, as an economy built upon the exploitation and degradation of the earth, is the productive and distributive enterprise of that civilization, to which must be opposed the practices of an ecological and sustainable economy built upon accommodation and commitment to the earth and following principles of conservation, stability, self-sufficiency and cooperation.

A movement of resistance starting with just those principles as the sinews of its analysis would at least have a firm and uncompromising ground on which to stand and a clear and inspirational vision of where to go. If nothing else, it would be able to live up to the task that George Grant, the Canadian philosopher, has set this way: "The darkness which envelops the western world because of its long dedication to the overcoming of chance"—by which he means the triumph of the scientific mind and its industrial constructs—"is just a fact. The job of thought at our time is to bring into the light that darkness as darkness." And at its best, it might bring into the light the dawn that is the alternative.

What to Do about Welfare?

Jack Beatty

The *Atlantic Monthly* has one of the richest sites on the entire Web, especially for college students looking to get beyond CNN Headline News. Indeed, Atlantic Unbound's current banner proudly reads, "Bound since 1857; Unbound since 1993."

Two especially valuable regular features of Atlantic Unbound are the Executive Decision (like the one here) in which a policy issue is presented as a decision-making memo from a fictitious staffer for you, the president, to act on; and the Flashbacks feature, in which the *Atlantic* collects important material from its archives on a single topic.

For more on welfare-related Web sites, see the Web work for Exploration 13f.

At the *Visions* Web Site Link to other Executive Decisions at the Atlantic Unbound.

READING PROMPT

The national welfare system, largely set up since 1940, was completely overhauled by Congress in 1996, in part because it was a system that almost no one liked. Reading between the lines in Beatty's memo, how was it ever possible to create a system that displeased so many people?

To: The President of the United States
From: D. N. Forser, Chief of Staff
Re: Welfare
Date: July 15, 1996

If welfare did not exist, our politicians would have to invent it. Where would they be without ritualistic condemnation of welfare chiselers, breeders, queens, layabouts, and so on? Only in a politics of distraction can welfare rank as such a major issue. To understand welfare's true importance, perform the following thought-experiment. Imagine welfare "fixed"—with every possible recipient of Aid for Families with Dependent Children (AFDC) in either a private- or public-sector job. Would that make any difference to you in your daily life? The streets, swept by AFDC mothers, would be cleaner. But beyond that what else would be different? Costs would be higher, not lower, to pay for jobs and day care—expenses not incurred by the current system. Crime? Who knows? The telling correlation seems to be between fatherlessness and crime, not welfare and crime, and the "make 'em work like the rest of us" welfare reforms do not address fatherlessness directly, though it would obviously help to end welfare incentives that prevent family formation. Improved morals among the poor? Possibly. Work is a form of discipline that consists of controlling one's urges for sensation or novelty and of delaying present gratifications. If indiscipline is what ails welfare recipients, then work could prove a moral tonic. Fertility? Already rates of childbirth among welfare recipients are no higher than the national average; what more progress can be made on the fertility front?

So streets might be cleaner, spending would be higher, crime might be slightly lower, the moral character of welfare recipients might be increased, and their fertility might be lowered. Is this worth all the heated controversy, all the inflated rhetoric, that has long attended welfare?

In contrast look at the bipartisan silence that has fallen on the subject of "corporate welfare"—the $75–$150 billion of taxpayer support given annually to the Gallo Winery, Ocean Spray, McDonald's, Walt Disney, the sugar industry, the mohair industry, and the rest of the recipients of more than one hundred different sorts of subsidies. "You see a lot of politicians attacking welfare queens," Colin Powell said last year, "but you see them a little reluctant to take on the welfare kings on K Street, the lobbyists who are there getting what? 'Preferences' for corporations." Powell puts his finger squarely on the truth.

Robert Reich for the Democrats and John Kasich and John McCain for the Republicans tried and failed during the past year to bring their parties around for an attack on this brand of welfare. So long as the scandalous system of campaign finance remains untouched, don't expect an attack on corporate welfare.

It is odious to treat welfare as a huge issue and forget General Powell's words.

That said, there are only two alternatives that deserve respect in the debate over welfare reform. One is the fifty-state approach. The other is the national approach. Other proposals, such as cutting off money to single teenage mothers or ending AFDC payments altogether, are ferocious political gestures that if implemented would put God knows how many children out on the streets. If that is your idea of "reform," you need a curing dose of Jonathan Swift. In any case, Mr./Ms. President, we will present only nondemagogic alternatives.

The selections are:

A: The Fifty-State Solution (Please read a memo containing a brief argument in favor of Option A.)

B: The National Solution (Please read a memo containing a brief argument in favor of Option B.)

OPTION A: THE FIFTY-STATE SOLUTION

Mr./Ms. President:

Welfare reform is properly the business of the states. Why? Because dependent poverty, though it exists across the nation, has stubbornly local roots. The forces that lead people in rural California to file for public assistance to provide for themselves and their children have to do with the cyclical, weather-contingent nature of low-wage agricultural employment. Welfare generally has a different face in rural states than in urban ones, where failures of education and school-to-work preparation and, no doubt, "culture of poverty" effects lead people to seek public assistance.

To deal with local roots, local remedies are needed. Wisconsin, Massachusetts, and Maryland, among other states, have launched locally responsive experiments in welfare reform. Tennessee has brought managed care to Medicaid—the health-insurance program for the medically indigent.

Thus far each state has had to seek waivers from the federal government to permit these experiments. Let's end the federal role altogether, Mr./Ms. President. Let's systematize policy innovation. Let the states set up their own versions of AFDC and Medicaid. Let's end welfare as a federal entitlement.

Critics charge that the states would treat children and the sick much more callously than the federal government does; that in recessions states would cut benefits, making women and children homeless. That is a libel on the states. Yes, in the days of segregation, many states could not be trusted to deal humanely with minorities among the dependent poor. But those days are over, surely. In post-racist America, real federalism would not be an instrument of white supremacy.

Let the states decide, Mr./Ms. President. The objective of welfare reform is to get recipients back to work without disrupting family formation or hurting children—and the states can find their different paths to that single overriding goal.

OPTION B: THE NATIONAL SOLUTION

Mr./Ms. President:

Everyone is unhappy with the current welfare system. Most who are on welfare want to get off, but either can't find a job or can't find one that compensates for the loss of health care and child care. Those who pay for welfare with their taxes resent those who are on the government dole. What we need is a system that encourages people to enter the workforce while still providing them with a safety net. The reform now being discussed by Congress would turn welfare from a federal entitlement into a block grant given to states, with few limits on how they spend the money. States would be free to lower their own contributions to the welfare system by as much as fifty percent. It's not hard to guess what system this would create—one of migrations to states with comparatively generous welfare benefits.

According to William Julius Wilson, who was brought in to advise you on this matter, if you really want to solve the welfare problem, you need to look back to its root: poverty. The lack of jobs, decaying inner cities, and poor schooling systems that all define poverty have created a climate in which public assistance is necessary. In our country, welfare recipients are often despised, while the poor as a group elicit sympathy. A survey conducted by the National Opinion Research Center from 1983–1991 showed that while a "substantial majority felt that too little was being spent to help the poor, only slightly more than twenty percent felt that too little was being spent on welfare programs." Your job is to persuade the public that it's not just welfare we need to reform, but our prescriptions for poverty itself. This will be a big and expensive task, but in these post–Cold War times we have the resources to accomplish it.

The proposal I'm about to suggest will be difficult to enact, unless you are able to sell it as an attack on the causes of poverty rather than simply as welfare reform. The country will need to rethink its view of poverty as the result of individual failings and realize that much of it is caused by big social and economic forces beyond an individual's control. You will have to lead a return to the attitude, prevalent during the days of the New Deal, that poverty and inequality can only be eliminated through enlightened public policy.

We cannot approach this without addressing the lack of opportunities for unskilled workers. Changes in the economic climate during the past few decades have resulted in fewer jobs for unskilled workers, and the remaining jobs pay much less in today's dollars than their counterparts of thirty years ago. Millions of low-wage jobs have been replaced by mechanization and computers, or have been lost to the international labor market. Why focus on getting people off welfare if there aren't enough jobs for them?

Following a proposal by Mickey Kaus of *The New Republic,* you should implement a neo–Works Progress Administration program that would provide a job, with pay slightly below minimum wage, for anyone who wants one (not just welfare recipients). These jobs would not compete with existing ones in the private or public sector. Instead they would provide services that have fallen through the cracks during the government belt-tightening of the eighties and nineties: repairing the infrastructure in older cities, keeping libraries open at night, maintaining public schools and parks, and cleaning streets more than once every several weeks.

Guaranteed health care and child care would necessarily be a part of this plan, as would a continuation of the earned-income tax credit for those in low-paying jobs. Otherwise, financially strained parents would face the same choice they do today: stay on welfare or look for a job that does not provide enough support for their families. Because of the below-minimum-wage pay, the WPA program would encourage people to seek other, higher-paying jobs, while providing them with basic training for jobs in the real world. You might also want to look into a transportation initiative that would increase transportation from inner-city neighborhoods, either through extra buses or carpool systems, so that people without cars could work outside their immediate area. With the above guarantees you could limit the amount of time able-bodied people are allowed to receive welfare, without consigning them to deeper poverty. People would be working for their money and answering the critiques of those who deride public assistance recipients as layabouts. A program like this, which would provide viable employment opportunities for those trapped in poverty, is much more expensive than simply handing people money—it would cost an estimated twelve billion dollars to create one million jobs. Support this plan, Mr./Ms. President. The long-term benefits of reducing poverty while giving something back to communities should outweigh the cost.

chapter 13 Explorations

13a Temp Jobs

Collaboration. Share a detailed account of short-term or temporary work you may have done. What were advantages or disadvantages of such temporary employment?

Web work. The Department of Labor for the State of Georgia provides a statewide search engine for jobs, many low tech and temporary. Atlanta Classifieds lists employment ads from the *Atlanta Journal-Constitution,* while CareerPath provides links to jobs listed in local newspapers across the country.

One writing topic. Write an essay classifying and evaluating the kinds of temporary work available to students in your community.

13b Résumé Writing and Beyond

Collaboration. Share a list of what you feel are job-oriented skills you have learned or improved on in academic courses taken over the last year.

Web work. CareerWeb specializes in professional, technical, and managerial jobs. E-span is another top employment site, as is the Virtual Job Fair. CareerMosaic allows you to search their database of 50,000 résumés. Yahoo! itself lists some 500 individual résumés at Business and Economy > Employment > Resumes > Individual Resumes. Rebecca Smith's eResume site and ProvenResumes offer help in constructing your own.

For years the most popular introduction to html (hypertext markup language—the language of Web pages) has been the NCSA *Beginner's Guide to HTML*—although increasingly students are using high-powered word-processing software such as Microsoft *Word 97,* which greatly automates and simplifies the creation of a Web page, basically allowing one to save any word-processing document in html format. Then the only step required for making this document into a Web page is somehow getting it copied to a Web server.

A quite different approach to creating a Web résumé would entail using a free Web service such as Angelfire or GeoCities, where one can create a Web page by completing a simple template.

One writing topic. Prepare a résumé (real or imaginary) either to be printed or to be displayed as a Web page (that is, in html format).

13c My Next School

Collaboration. Describe and then share your thoughts as to how much and what kind of further schooling (graduate or professional) you think you might ever acquire beyond a four-year college degree.

Web work. *U.S. News & World Report* has a truly excellent site for college students interested in all sorts of career choices, school or otherwise. For information about graduate education generally, check out Peterson's Online site; or look at Peterson's undergraduate site if you have any interest in transferring.

The College Board Online will give you information about the kind of tests you will have to take, while Test.com will help you prepare online.

If you are interested in the law, check out the Law School Admissions Council Online or LawSchool.com; if interested in business school, consult MBA College Edge or MBA Plaza.

One writing topic. Visit the Web sites of at least three schools you are thinking of attending and write an essay comparing the programs and services they offer and how they present themselves via the Web.

13d Unpaid Work: Community Service

Collaboration. Share a detailed account of a time when people in a school you attended assisted in a community project.

Web work. A good place to start is with Yahoo!'s Society and Culture: Issues and Causes > Philanthropy > Community Service and Volunteerism > Organizations.

One writing topic. Write an argument in support of or opposed to a school's having a policy that requires the completion of a service component for graduation.

13e Harlem

Collaboration. Harlem is perhaps the most famous predominantly black community in the United States and hence one of the best known ethnic communities, although there are distinctive ethnic neighborhoods in almost all older cities (this despite the notion of the United States as a melting pot). Share a detailed description of an ethnic neighborhood you know or have visited.

Web work. Here's a scattered list of Harlem sites: HomeToHarlem, a comprehensive guide to the community; HarlemLive, published by the youth of Harlem; and the Ralph Bunche School. Also check out the Greater Harlem Chamber of Commerce Web site.

The National Urban League has been dealing with problems of underemployment of black Americans since 1910.

See these links for a general overview of contemporary African American culture: AfroNet; the Black World Today; and NetNoir, the last with more of a world emphasis.

Finally, the Web site Cafe Los Negroes describes itself as "New York's black and latino hangout."

One writing topic. Write an essay on the geography of ethnic difference within your own community (either college or hometown), focusing on one or more neighborhoods and what each has to offer to those who live there.

13f Welfare Reform and Jobs

Collaboration. Make and share lists of reasons people on welfare (mostly single women with dependent children) should or should not be required to work to receive benefits.

Web work. In May 1995, the Atlantic Unbound revisited, with links to, 1971 and 1987 pieces the *Atlantic* had run on welfare reform. In March 1997, the Atlantic Unbound held an online roundtable on this topic.

The *Washington Post* has a special section on welfare reform. The Welfare Information Network, WIN, has links to practically everything imaginable connected to welfare reform. The Welfare Reform Resource Project is hosted by Regents University, an institution connected with conservative evangelist Pat Robertson. Welfare Watch calls itself a "media monitoring project" run by the highly respected Annenberg School of Communication.

One writing topic. Write an argument supporting or opposing work requirements for people receiving basic financial assistance from the government.

chapter 14

Working Less

READING QUESTIONS

- What aspects of work have you most disliked in the past or do you most dread in the future?
- How effective are you at managing your time? What aids do you use (or feel you should use) to manage your time more effectively?
- What would you do with your future if you did not need a monthly paycheck?

QUOTATIONS

A specter is haunting the world—the specter of leisure. All the great powers have conspired against it: Popes and presidents, bankers and unionists, hamburger chains and environmentalists. But, mighty as they are, they will soon fail. Machines may exist solely to produce, but humans are awakening from a dark eon of drudgery to the dawning realization that there is more to life than work. . . . Leisure is not a promise that makes us press our noses to the grindstone; it is a basic human right. To work for the vague hope of a short respite is criminal extortion. We will achieve the leisure we deserve only when we change our priorities and live the revolution of the deed. Our debt of leisure constantly mounts, but do we ask for our due? Do we realize what we're owed?

— *The Leisure Party Manifesto, an Internet publication*

Hard work never killed anybody . . . but why take chances?

— *Anonymous*

The Net is a waste of time, and that's exactly what's right about it.

— *William Gibson, science fiction novelist*

To be able to fill leisure intelligently is the last product of civilization, and at present very few people have reached this level.

— *G. B. Shaw (1930), Irish playwright and social critic, 1856–1950*

The Speed Trap *Jay Walljasper*

Jay Walljasper is editor at large of the *Utne Reader* and has written about politics for several progressive magazines, including *The Nation, Tikkun,* and *In These Times.* As the following piece indicates, Walljasper is in some ways the flip side of Newt Gingrich: Whereas Gingrich made his reputation as a "conservative" bent on creating a future largely liberated from most current practices (liberated from what some might call *tradition*), Walljasper is a "radical" who seems nostalgic for the past and opposed to most of what passes as *progress.* Welcome, then, to millennial politics—where high-tech "conservatives" are contemptuous of tradition and low-tech "progressives" are contemptuous of progress.

Walljasper's rearranging of political boundaries is apparent in his interviews with eight prominent social critics, "Rethinking the Left." When the following piece appeared originally in the *Utne Reader,* it was accompanied by a time line charting the introduction of new technologies that have helped speed up modern life, including the introduction of long-distance dialing in 1951, Carl Swanson's first TV dinner (turkey, gravy, cornbread, peas, and sweet potatoes) in 1953, and the start-up of Federal Express in 1973.

At the *Visions* Web Site Links to the Netherlands Design Institute; to Brian Eno; to the DayRunner Corporation; to the LongNow Foundation (building a 10,000-year clock).

READING PROMPT

Which of the speedy things Walljasper mentions do you experience in your life? Are there other fast-paced aspects of your life that he does not mention?

The alarm rings and you hop out of bed. Another day is off and running. A quick shower. Wake the kids and rush them through breakfast so they won't miss the bus. Down a cup of coffee. Shovel a bowl of cornflakes. Hurry out to the car, not forgetting a swift kiss on your partner's cheek. Hightail it to the freeway, making a mental note to grab some takeout Thai on the way home. (The kids' soccer practice starts at 6:15 sharp.) Weave back and forth looking for the fastest lane while the radio deejay barks out the minutes—8:33, 8:41, quarter to. Reaching work, you sprint into the building and leap up the stairs three at a time, arriving at your desk with seconds to spare. You take a couple of deep breaths, then remember that the project you didn't finish last night must be faxed to New York by 10:00. Meanwhile, you've got five voice-mail messages and seven more on e-mail, two of them marked urgent.

More and more it feels like our lives have turned into a grueling race toward a finish line we never reach. No matter how fast we go, no matter how many comforts we forgo in order to quicken our pace, there never seems to be enough time.

It wasn't supposed to turn out this way. As a kid in the 1960s I remember hearing that one of the biggest challenges of the future would be what to do with all our time. Amazing inventions were going to free up great stretches of our days for what really matters: friends, family, fun. But just the opposite has happened. We've witnessed a proliferation of dazzling time-saving innovations—jet travel, personal computers, Fed Ex, cell phones, microwaves, drive-through restaurants, home shopping networks, the World Wide Web—and yet the pace of life has been cranked to a level that would have been unimaginable three decades ago.

Curiously, there has been scant public discussion about this dramatic speed-up of society. People may complain about how busy they are and how overloaded modern life has become, but speed is still viewed as generally positive—something that will help us all enrich our lives. Journalists, business leaders, politicians, and professors feed our imaginations with visions of the new world of instantaneous communications and high-speed travel. Even many activists who are skeptical of the wonders of modern progress, the folks who patiently remind us that small is beautiful and less is more, look on speed as an undeniable asset in achieving a better society. Four-hundred-mile-an-hour trains, they assure us, will curtail pollution, and modem links across the planet will promote human rights.

Revving up the speed, in fact, is often heralded as the answer to problems caused by our overly busy lives. Swamped by the accelerating pace of work? Get a computer that's faster. Feel like your life is spinning out of control? Increase your efficiency by learning to read and write faster. No time to enjoy life? Purchase any number of products advertised on television that promise to help you make meals faster, exercise faster, and finish all your time-consuming errands faster.

Yet it seems that the faster we go, the farther we fall behind. Not only in the literal sense of not getting done what we set out to do, but at a deeper level too. I feel this keenly in my own life. Like many Americans, I've always moved at a fast clip. I can't stand small talk, waiting in line, or slow numbers on the dance floor. It has always seemed obvious to me that the faster I move, the more things I can do and the more fun and meaning my life will have. But it has gotten to the point where my days, crammed with all sorts of activities, feel like an Olympic endurance event: the everydayathon. As I race through meals, work, family time, social encounters, and the physical landscape on my way to my next appointment, I'm beginning to wonder what I've been missing, what pleasures I've been in too much of a hurry to appreciate or even notice.

I hear an invisible stopwatch ticking even when I'm supposed to be having fun. A few weeks ago I promised myself a visit to a favorite used-book store that I hadn't stopped in for a while. It was a busy day, of course, and I rushed through what I was doing and dashed over to the bookshop fully aware that I would have only a few minutes there before I needed to be going somewhere else. Heading for the travel section I bumped—literally—into a friend I hadn't seen for at least three months. He was in a hurry too, and we proceeded to have a hasty conversation without even looking at one another as we both frantically scanned the bookshelves. It must have looked highly comical—two talking heads bobbing up and down the aisle. Finally we each grabbed a book, raced to the cash register, and hollered good-bye as we sped off in opposite directions. Walking away, I felt suddenly flat, anticipation of a pleasurable pastime giving way to dulled disappointment. I had not enjoyed a meaningful conversation with my friend nor experienced the joy of browsing, and now I was carrying home a $12.50 book about London in the 1890s that I wasn't even sure I wanted.

Experiences like this are making me question the wisdom of zooming through each day. A full-throttle life seems to yield little satisfaction other than the sensation of speed itself. I've begun voicing these doubts to friends and have discovered that many of them share my dis-ease. But it's still a tricky topic to bring up in public. Speaking out against speed can get you lumped in with the Flat Earth Society as a hopelessly wrongheaded romantic who refuses to face the facts of modern life. Yet it's clear that more and more Americans desperately want to slow down. A surprising number of people I know have cut back to part-time work in their jobs or quit altogether in order to work for themselves, raise kids, go back to school, or find some other way to lead a more meaningful, less hurried life—even though it means getting by on significantly less income. And according to Harvard economist Juliet Schor, these are not isolated cases.

Schor, author of the 1991 bestseller *The Overworked American,* says her research shows that "millions of Americans are beginning to live a different kind of life, where they are trading money for time. I believe that this is one of the most important trends going on in America."

Fed up with what compressed schedules are doing to their lives, many Americans want to move out of the fast lane; 28 percent in one study said that they have recently made voluntary changes that resulted in earning less money. These people tend to be more highly educated and younger than the U.S. workforce as a whole although they are being joined by other people who are involuntarily trading paychecks for time off through layoffs and underemployment.

People want to slow down because they feel that their lives are spinning out of control, which is ironic because speed has always been promoted as way to help us achieve mastery over the world. "The major cause in the speed-up of life is not technology, but economics," says Schor. "The nature of work has changed now that bosses are demanding longer hours of work." After a long workweek, the rest of our life becomes a rat race, during which we have little choice but to hurry from activity to activity, with one eye always on the clock. Home-cooked meals give way to frozen pizzas, and Sundays turn into a hectic whirlwind of errands.

Yet there is a small but growing chorus of social critics, Schor among them, who dare to say that faster is not always better and that we must pay attention to the psychological, environmental, and political consequences of our constantly accelerating world. Environmental activist Jeremy Rifkin was one of the first to raise questions about the desirability of speed in his 1987 book, *Time Wars.* "We have quickened the pace of life only to become less patient," he wrote. "We have become more organized but less spontaneous, less joyful. We are better prepared to act on the future but less able to enjoy the present and reflect on the past.

"As the tempo of modern life has continued to accelerate, we have come to feel increasingly out of touch with the biological rhythms of the planet, unable to experience a close connection with the natural environment. The human time world is no longer joined to the incoming and outgoing tides, the rising and setting sun, and the changing seasons. Instead, humanity has created an artificial time environment punctuated by mechanical contrivances and electronic impulses."

Rifkin closed his book with an eloquent call for a new social movement to improve the quality of life and defend the environment, a movement of people from all walks of life gathering under the "Slow Is Beautiful" banner. Perhaps appropriately, progress in forging such a movement has moved forward very slowly in the decade since *Time Wars* was published, while the pace of modern life has revved up considerably thanks to breakthroughs in technology and new economic demands imposed by the globalizing economy.

IS SLOW REALLY BEAUTIFUL?

A number of these advocates of slowness gathered in Amsterdam last November for a conference hosted by the Netherlands Design Institute. Drawing an overflow crowd of designers, computer professionals, scholars, journalists, environmentalists, business leaders, and activists from around the world, the conference marked the first large-scale forum on the cultural and political implications of unmitigated speed in our ever-accelerating world.

Not everyone at the conference, which focused on the design and technological aspects of the issue, was convinced that speed poses any real problems. Some of the younger participants were appalled that anyone would advocate slowing down the pace of life.

Historian Stephen Kern, a professor at Northern Illinois University whose book *Culture of Time and Space* chronicled the soaring velocity of life between 1880 and World War I, pointed out that "new speeds have always brought out alarmists." In the 1830s, he noted, it was feared that train passengers would suffer crushed bones from traveling at speeds as high as 35 miles an hour. Kern considers the current concern about the effects of our speeded-up lives a similar kind of hysteria. "Technologies that promote speed are essentially good," he said, adding that "the historical record is that humans have never, ever opted for slowness."

Danny Hillis, who pioneered the conceptual design behind high-speed supercomputers, disagreed with Kern, warning that our obsession with speed forces us to lose sight of the future and remain trapped in the present. He recommended cultivating what he calls "a new aesthetic of slowness." To illustrate what that might look like he told a story about how Oxford University replaced the gigantic oak beams in the ceiling of one of its dining halls. When the beams began to show signs of rotting, university officials were concerned that they wouldn't be able to find enough lumber large and strong enough to replace them. But the university's forester explained to them that, when the dining hall was built 500 years ago, their predecessors had planted a grove of oak trees so that the university could replace the beams when the time came.

In that spirit, Hillis is now at work with musician Brian Eno and others on designing the world's slowest clock, which will chime just once a millennium. He hopes that at a conference 3,000 years from now, people will look back on our time and see this clock as a symbol of "the moment when they took responsibility for the future. When they stopped believing in just now."

The prominent German environmental thinker Wolfgang Sachs shares Hillis' interest in devising an aesthetic of slowness and offers his own ideas about what form it would take. "Medium speeds will be considered an accomplishment, something well done," he says. "And when you see someone going fast, you shrug your shoulders, saying, 'What's the point?'"

Sachs believes that speed is an underrecognized factor fueling environmental problems. As he puts it, "It's possible to talk about the ecological crisis as a collision between time scales—the fast time scale of modernity crashing up against the slow time scale of nature and the earth." In his view, genetic engineering, with all its potential for ecological havoc, is an example of how we interfere with natural processes in the name of speeding up evolution. Sachs' recent report *Sustainable Germany*—which maps a route to a Green society—embraces slowing down as a key environmental objective, proposing to put 100 kilometer-an-hour (60 m.p.h.) speed limits on Germany's autobahns and scrap plans for a high-speed rail network. He also recommends strengthening local economies and cultures so that people won't have to rely as heavily on long-distance travel.

"A society that lives in the fast lane can never be a sustainable society," Sachs told the conference, adding that a slower society would also be a more pleasant, elegant place to live. "In a fast-paced world we put a lot of energy into arrivals and departures and less into the experience itself. Raising kids, making friends, creating art all run counter to the demand for speed. There is growing recognition that faster speeds are not just a natural fact of the universe. It's an issue for public attention. What has not been discussed before now is: What kind of speed do we want?"

Jogi Panghaal, a designer who works with community groups in India, defines the issue as not simply whether speed is good or bad, but whether the world of the future will allow a variety of speeds. He worries that a monoculture of speed in which the whole world is expected to move at the same pace will develop globally.

India and other traditional societies of Asia, Latin America, and Africa are undergoing culture shock as the rule of Western efficiency bears down upon them. People who once lived according to the rhythms of the sun, the seasons, and nature are now buying alarm clocks, carrying pocket calendars, and feeling the pressure to move faster and faster. At the conference, Panghaal warned that inhabitants of the industrialized nations may feel this loss as much as the traditional peoples do because less modernized cultures provide inspiration for finding a slower, simpler way of living—including the two-week vacation in the Third World that has become a necessary ritual of replenishment for many of us.

Sachs and Panghaal raise the question of whether we will have any choice in determining the tempo of our lives or will we all be dragged along by the furious push of a technologically charged society. When I hear friends complain that their lives move too fast, they're not talking about a wholesale rejection of speed so much as a wish that they could spend more of their time involved in slow, contemplative activities. One can love the revved-up beat of dance music, the fast-breaking action on the basketball court, or the thrill of roller coaster rides without wanting to live one's life at that pace. A balanced life—with intervals of creative frenzy giving way to relaxed tranquility—is what people crave. Yet the pressures of work, the demands of technology, and the expectations of a fast-action society make this goal increasingly difficult to achieve.

Another speaker at the conference, Ezio Manzini, director of the Domus Academy design institute in Milan, sees hope for a more balanced approach to speed springing from the same source that fuels the acceleration of our lives: modern mastery of all that stands in our way. "This is the first time in history in which people think they can design their lives," he said.

In an age of technological marvels, we've come to expect that solutions will be found to help us overcome our problems. So if the problem now appears to be too many things coming at us too fast, we'll naturally begin looking for ways to slow down. Humans may not have opted for slowness in the past, but they have also never had to contend with constantly soaring speeds not only diminishing the quality of life, but also endangering the future of the planet. As Wolfgang Sachs declared to the audience in Amsterdam, "Slow is not only beautiful, but also necessary and reasonable."

HOW TO HASTEN SLOWLY

All this stimulating talk at a splendid conference is fine, but how do we even think about the enormous undertaking of slowing down a world that's been on a spiral of growing acceleration for more than a century and a half? Especially when the captains of the global economy dictate that speed is an essential ingredient of tomorrow's prosperity? How do we begin to apply the brakes in our lives when the world around us seems to be stomping on the gas pedal?

Right outside the theater where the conference was held, the city of Amsterdam itself seemed to offer an answer. More than almost any city in world, Amsterdam has consciously curtailed the speed of traffic, creating a delightful urban environment in which a bike rolling past at 15 miles an hour seems speedy. Strolling the narrow streets for just a few minutes, you encounter all sorts of shops, restaurants, nightclubs, parks, public squares, banks, and movie theaters—an impressive array of shopping and entertainment that would take at least an hour's worth of driving and parking to reach in most American cities. You're moving slower than in a car but experiencing much more.

Amsterdam's efforts have been widely imitated around the world by advocates of traffic calming, a burgeoning popular movement that seeks to improve safety and environmental quality by reducing the speed of cars. The spread of traffic-calming techniques like speed bumps throughout Europe, Australia, and now North America provides a sterling example of how a grassroots movement can bring about the slowing down of society.

This idea of calming could be taken out of the streets and into workplaces, government, and civic organizations. It's true that transnational corporations wield near autocratic authority in today's global economy, but a spirited worldwide campaign for shorter work hours, more vacation, and a less intense work pace might crystallize worker discontent into a potent political force that would undermine that power. Juliet Schor contends that additional leisure time, not further economic growth, will be the chief political goal of the coming age. (We've already seen the start, with women's groups and labor unions leading a successful campaign for family leave policies in American workplaces.)

But before any political movement can take hold, people need to begin thinking differently about speed and how important it really is. For 150 years we've been told (and believed) that the future will inevitably be faster than the present and that this is the best way to broaden human happiness. And speed has brought major improvements to our world. But in taking advantage of its possibilities, we have become blind to its drawbacks. While the acceleration of life that started with the first steam locomotive didn't crush our bones, it may have crushed our spirits. Our lives have grown so hurried and so hectic that we often don't take in the thrill of a sunset, the amusement of watching a youngster toddle down the sidewalk, or the good fortune of bumping into a friend at a bookstore. We can regain the joy of those things without giving up the World Wide Web, ambulance service, and airline flights to Amsterdam. Rather than accept that the world offers just one speed, we have the privilege, as Ezio Manzini says, of "designing" our lives.

How does Manzini himself do it? "Like everybody's, my life is in a hurry," he admits. "When I am at work I'm in the machine, and there is nothing I can do to move slow. But I try to be conscious that it is not a good way to live. And when I leave work, I try to switch off—slow down and do things that make me feel good, like go out to the country and relax. This is what you might call selective slowness. It's the beginning of consciousness that you can get out of the machine."

Wolfgang Sachs, who is project director of the Wuppertal Institute for Climate, Energy, and the Environment in Wuppertal, Germany, says, "It's a struggle for me to slow down, as it is with many people. But the key is to be able to dedicate yourself to the proper rhythm, geared to what you are doing, whether you are playing with a child, writing a paper, or talking to friends." One thing that keeps his life from whirling out of control is walking to work each day. Those strolls offer him 20 minutes each morning and evening when he's out of reach of the rushing insistence of the modern world.

Juliet Schor has slowed the pace of her life by setting firm limits on when she works. "My work time is limited by my childcare hours. I don't work on weekends. My life outside of work has also been simplified. I rarely drive a car. I ride my bike. I just don't do all the things that make me crazy. And my husband, who is from India and has a much calmer approach to life, has been instrumental in helping me slow down. He has taught me to just do one thing at a time."

We all have a chance to slow down. Maybe not at work or in raising kids, but someplace in our lives. It might be turning off the rapid-fire imagery of television and taking a stroll through the neighborhood. It might be scaling back the household budget and spending Saturdays fishing or gardening instead of shopping. It might be clearing a spot on your daily calendar for meditation, prayer, or just daydreaming. It might be simply deciding to do less and not squeezing in a trip to the bookstore when you don't have time for a relaxing visit.

Manzini has another suggestion. "In Italy there is something called the movement for slow food," he says. "It's a group of people who have decided to promote the idea that there are things that matter more than speed. The idea of defending the quality of something that is slow is very interesting to me."

Launched after the arrival of McDonald's in the heart of Rome, the group soon attracted 40,000 members in 40 countries.

That's how I've started the "Slow Is Beautiful" revolution in my own life—right in the kitchen, scaling back my busy schedule to find more time for cooking good meals and then sitting down to enjoy them in a festive, unrushed way with my wife, son, and friends. Even cleaning up after dinner can offer a lesson in the pleasures of slowness, as I learned a while back when our dishwasher went on the fritz. Before that I had always just tossed dirty dishes into the machine as fast as possible and hurriedly wiped the counters so that I could get on to more worthwhile activities. But when I was forced to wash dishes by hand, I discovered that although it took longer I had way more fun; I'd put some jazz or blues or zydeco on the stereo and sing along, or just daydream as I stacked dishes and glasses on the drying rack. What had been 5 or 10 minutes of drudgery, filling the dishwasher and desperately wishing I was doing something else, turned into 15 or 20 minutes of relaxation. Our dishwasher is fixed now, but I still find myself looking forward to cleaning up the kitchen. A lot of nights I wash the dishes by hand anyway, and when I load the dishwasher, now I do it slowly and without the slightest hint of displeasure.

Workers of the world, relax. You have nothing to lose but your microwaved burritos and your overstuffed Day Runners.

Rapture of the Net

Reva Basch

Reva Basch is the author of *Secrets of the Super Net Searchers*. She lives in northern California, but her spirit roams the Net. And yes, there really are sites concerned with IA (Internet addiction).

At the *Visions* Web Site Links to sites dealing with addiction.

READING PROMPT

What is it about the Web that Basch sees as potentially addictive? Does she see it as something good or bad—and how do you feel about it?

My friend Hank is approaching retirement age. He's ready to do it, his wife wants him to do it, and they have the financial security to do it. He doesn't have many hobbies, though, other than fast cars, bicycle racing, gourmet food, and fine wine—each of which carries some built-in limitations if pursued to excess. But he's a Type A, workaholic kind of guy, and he's looking for a consuming passion—something to fill the long, empty hours he foresees stretching out ahead of him once the Social Security checks start rolling in. So I introduced him to the Web.

We sat down at my computer one morning and began by checking out the hot Italian cars at www.vol.it/ferrari/. Then we zoomed over to the Tour de France (www2.letour.fr) and monitored near-real-time results from the race in progress. From there, we pedaled to www.le-sommelier.fr to check the prices on some of the great Bordeaux. Gulp. Perhaps a more modest domestic vintage instead? We hopped the Atlantic and browsed the wine list at the Virtual Vineyard (www.virtualvin.com). After more site-seeing, we headed for the recipes and restaurant reviews at www.epicurious.com/.

No wonder we'd gravitated toward the food pages. By the time we shook ourselves out of our mutual trance, it was almost dinnertime. Hank—glazed, boggled, and mildly intoxicated from his daylong Tour de Net—had his first glimpses of what life after retirement might be like. And once more, I'd watched a neophyte computerist succumb to what divers call the "rapture of the deep": Enthralled by your surroundings, a new universe unfolding in all directions, you lose track of time and the outside world.

Athletes feel something similar during peak performance. So do actors and musicians, painters and computer programmers. It's called "flow," a fugue state—almost a trance—in which you become utterly involved with your environment, whether it be the ocean, the effort, or the machine. You're in flow when you're totally immersed in a good book or caught up in a movie, where the intrusion of the "real" world comes as a shock.

The concept was promulgated by psychologist Mihaly Csikszentmihaly in the book *Flow: The Psychology of Optimal Experience*. Inevitably, the Web marketeers have begun to move in, investigating whether flow can be exploited to make consumers more susceptible to sales pitches, or whether, on the contrary, the experience of Web surfing is so riveting in its own right that unwanted messages are unconsciously blocked.

The ingenious playwright Tom Stoppard once said his technique when lecturing is "to free-associate within an infinite regression of parentheses." He was describing the same kind of intuitive, endlessly interconnected structure that gave the Web its name. The Web's hyperlinks mirror the creative leaps made by the human mind at play. Web surfing puts us back in touch with the way we explored the world when we were kids, before books and school curricula channeled our thought processes into orderly and often arbitrary sequences with definite beginnings and endings.

The Web lets ordinary mortals experience the altered state of mind where artists and virtuosos do their best work. Flow is what keeps us pointing and clicking long after we've found, or forgotten, what we went online for in the first place. Flow is like an endorphin high; it goes a long way toward explaining the addictive appeal of the Web. I suspect that Hank is on the threshold of a new obsession that might last the rest of his life. Welcome to retirement, pal. Don't worry about it; just go with the flow.

To Work Is Human, to Slack Divine

Hal Niedzviecki

Hal Niedzviecki is co-editor and founder of *Broken Pencil* magazine, the guide to independent culture in Canada, and co-organizer of Canzine, the annual festival of alternative publications in Canada. He is also the author of *Smell It*, available online from Coach House Books.

At the *Visions* Web Site Links to sites dealing with slackers (covered in Exploration 14e at the end of this chapter).

READING PROMPT

There is exaggeration and humor in Niedzviecki's piece, but a serious message as well. What is that message, and to what extent do you share the writer's concerns?

For several years now, I have relied on stupid jobs to pay my way through the world. This isn't because I am a stupid person. On the contrary, stupid jobs are a way to avoid the brain-numbing idiocy of full-time employment. They are the next best thing to having no job at all. They keep you sane—and smart.

I'm lazy sometimes. I don't always feel like working. On the stupid job, you're allowed to be lazy. All you have to do is show up. Hey, that's about as much of an imposition on my life as I'm ready to accept.

Understanding your stupid job is the key to wading your way through the muck of the work week and dealing with such portentous concepts as the Youth Employment Crisis and the Transformation of the Workplace.

"Out of Work: Is There Hope for Canada's Youth?" blurted the October 1997 issue of *Canadian Living*. My answer? There is more hope than ever. I'm not talking about ineffectual governments and their well-intentioned "partners," the beneficial corporations, all banding together to "create" jobs. What kind of jobs do you think these corporations are going to create? Jobs that are interesting, challenging, and resplendent with possibilities? Hardly. These are going to be stupid jobs. Bring me your college graduates, your aspiring business mavens, your literature lovers and we will find them rote employment for which servility and docility are the best training.

But hope is something different. Hope is the process whereby entire generations learn to undervalue their work, squirm out of the trap of meaningless employment, work less, consume less, and actually figure out how to enjoy life. I hope I'm right about this, because the reality of the underemployed, overeducated people of Canada is that the stupid job is their future.

As the middle-aged continue to occupy all the "real" jobs, as the universities continue to hike tuition (forcing students to work and study part time), as the government continues to shore up employment numbers with make-work and "retraining," there will be more stupid jobs than ever. And these stupid jobs won't be reserved for the uneducated and poor. The growth of the stupid job is already producing a crop of middle-class youngsters whose education and upbringing have, somehow, given way to (supposedly) stalled prospects and uncertain incomes.

Many young people think their stupid jobs are temporary. Most of them are right—in a way. Many will move on from being, as I have been, an usher, a security guard, a delivery boy, a data coordinator, a publishing intern. They will get marginally better jobs, but what they have learned from their stupid jobs will stay with them forever. I hope.

If I'm right, they will learn that you must approach stupid jobs—and, by extension, all jobs—with willing stupidity. Set your mind free. Using it isn't necessary, and it can be an impediment. While your body runs the maze and finds the cheese, let your mind go where it will.

Look at it this way: You're trading material wealth and luxury for freedom and creativity. While you may have less money to buy things, you will have more time to think up ways to achieve your goals without buying stupid things.

In the stupid-job universe, time isn't quantifiable. You're making so many dollars an hour, but the on-job perks include daydreams, poems scribbled on napkins, novels read in utility closets, and long conversations about the sexual stamina of Barney Rubble. How much is an idea worth? An image? A moment of tranquility? A bad joke? The key here is to embrace the culture of anti-work.

For a couple of years, I was hired as a security guard at a big craft show in Toronto. The people who worked with me were all university graduates (or students) with artistic pretensions. We loved to tell jokes on our walkie-talkies. There was a lot of pot smoking. We used code words over the radio. Whenever something had to be done, it was difficult to track one of us down. Many of us were outside in the parking lot getting high.

We worked 15-hour days. The pay was low, but the hours amassed. I didn't have to explain my stupid-jobs philosophy to anyone there. They were way ahead of me. Most of the craft show security guards were painters and musicians. But others had no skill, no craft. This group deserves special mention in the stupid-jobs pantheon. They are urban creatures, aberrant socialites well-versed in anarchist thought, the best punk bands in Saskatchewan, and what's on cable at 3:30 a.m. They can't imagine working 9 to 5, they have strange ideas, and they probably deserve paychecks just for being their loquacious selves.

Passivity is the difference between near slavery and partial freedom. It's a mental distinction. Your body is still in the same place for the same amount of time (unless you're unsupervised), but your mind is figuring things out. Personally, I'd take low-level servitude over a promotion that means I'll be working late the rest of my life. You want me to work weekends? You better give me the rest of the week off.

Meanwhile, it's not like my life is all that great. I may claim to have determined the best way to live, but I remain—like so many other would-be social engineers—caught in the trap of my own contradictions. Every year at the end of the craft show, the worst offenders were barred from working there again. I didn't get banned. I'm still a little embarrassed about it.

My father's plight is a familiar one. He started working at 13. He's 55 now. His employer of 12 years recently forced him to take early retirement. On his last day of work, I helped him clean out his office. The sight of him stealing staplers, blank disks, and Post-its is something I'll never forget.

But the acquisition of more stuff is not what he needs to put a life of hard work behind him. I wish that he could look back on his years of labor and think fondly of all the hours he decided not to work, those hours he spent reading a good book behind the closed door of his office, or skipping off early to take the piano lessons he never got around to.

Despite his decades of labor and my years of being barely employed (and the five degrees we have between us), we have both ended up at the same place. He feels cheated. I don't.

The Simple Life | *Jackson Lears*

T. J. Jackson Lears is a noted historian, specializing in consumption and advertising in U.S. life; in 1994 he published *Fables of Abundance: A Cultural History of Advertising in America.* The following essay began as a review of David Shi's 1985 book, *The Simple Life: Plain Living and High Thinking in American Culture.*

At the *Visions* Web Site Link to the American Studies Crossroads Project.

READING PROMPT

The "simple life," according to Lears, seems to be not so simple—at least not in terms of its historical role in U.S. culture. Explain.

Not long ago, as I drove across the cold prairie and flipped the radio dial, I heard a resentful male voice singing that "a country boy can survive." The voice told about a friend who'd gone to the city and prospered for a while before being laid off, mugged and killed when he took to packing a gun to defend himself. The song went on to say that a country boy didn't need that kind of trouble: he could bring down a buck at fifty yards, set a trout line—hell, he could even make his own wine. Country folk could survive. The ideal of the self-sufficient simple life was reduced to the stuff of survivalist fantasy.

There was a time when the simple life kept more polite company. In the late 1970s, L. L. Bean was, temporarily at least, the clothier of choice among Washington's upwardly mobile; in the mid-1980s there are no doubt many unworn pairs of work boots moldering in the closets of refurbished town houses behind Capitol Hill. The change of fashion embodies a change of mind. Simplicity chic, for all its silliness, occasionally reflected an awareness that the pursuit of wealth required some constraints, if only to minimize environmental health hazards. That awareness has dissipated as the era of limits has yielded to the decade of the entrepreneur. The will to growth is legitimized by intellectuals such as anthropologist Mary Douglas, who belittles campaigns against pollution by comparing them to primitive purity rites.

David Shi's fine book *[The Simple Life]* makes clear that the ideal of the simple life so popular in the 1960s and 1970s involved more than a passing yen for trail mix and alfalfa sprouts. It was in fact only the most recent expression of a complex intellectual tradition which has shaped American culture for centuries. As Shi acknowledges, the simple life has often been less a life of rural simplicity than of "stylish rusticity," but it has also occasionally led to a serious critique of private gain at the public expense. If many of Shi's sources for this tradition are familiar, he does a masterful job of bringing them together in a balanced, sensitive and comprehensive account of a major theme in American cultural history.

Shi begins at Plymouth Rock, steering clear of the unregenerate colonists at Jamestown. The Puritan ministry fretted continually about their communities' "declension" from high standards of frugality. Even the most pious Puritan's dedication to simplicity was sorely tried by the results of his work ethic; diligence brought luxury and the temptations to indolence. (The Quakers encountered the same fatal dialectic.) The growth of glittering fortunes provoked increasing shrillness in the Puritans' characteristic rhetorical form, the jeremiad. During the Revolutionary era, the jeremiad became politicized: in the colonists' imagination, luxury was a peculiarly British disease, and colonial leaders defined an emerging sense of nationhood by their own Spartan simplicity. As the infant republic required its citizens to elevate "public virtue" over private gain, a scorn for display became evidence of patriotism.

But in the early nineteenth century, the preoccupation with frugal living gave way to material success as the measure of public virtue. A superabundance of natural resources and entrepreneurial opportunities promoted the steady privatization of the protestant republican ethic. By midcentury, the ideal of republican simplicity was confined to the domestic novels of Sarah

Joseph Hale and the esthetic pronouncements of Andrew Jackson Downing, who urged "sincerity" in residential design.

Yet at about the same time a more romantic version of simplicity surfaced in the writings of Henry David Thoreau and the Concord transcendentalists. The wilderness that had nurtured entrepreneurship also bred the idea that this was "nature's nation," where one could establish Emerson's "original relation with the universe." A longing for what the essayist John Burroughs called "direct and immediate contact with things, life with the false trappings torn away," intensified during the second half of the nineteenth century. A craving for immediate experience led William James into the Adirondacks and Theodore Roosevelt up San Juan Hill; by 1910 it had energized numerous back-to-the-land movements and handicraft revivals. Some simple-lifers dug turnips, split wood and sat by the fire reading Burroughs's latest piece in *Century.* A few heard the call of the wild and followed John Muir into Yosemite. Most merely sent their sons to the nearest dude ranch to toughen up.

In both its serious and superficial guises the simple life remains an organizing principle for agrarian communards, an impetus for the decentralist critiques of Paul Goodman, Lewis Mumford and E. F. Schumacher and a cultural resource for conservationists and ecologists. Ancient and international in its origins, the gospel of simplicity has resonated with particular force in nature's nation. Yet Shi's story is, all told, a sad one of failed utopias and frustrated hopes. Often beneficiaries of industrial capitalism, the devotees of simplicity were doomed to be its countercultural conscience, heeded (if at all) only during periods of "economic decline, political crisis, or moral excitement."

Shi can hardly be blamed for not separating all the strands in this richly textured tradition. He might have explored more carefully the split between solitude and community that runs from Puritanism to the present. He might also have devoted more attention to the shift in the meaning of simplicity that occurred around the mid-nineteenth century, when romantic spontaneity appeared alongside Calvinist discipline and mass production began to create the clutter that made simplicity seem all the more desirable.

Shi is right, though, to stress a certain continuity. Secularized Calvinism persisted in a variety of forms, notably in the functionalist ethics and esthetics congenial to many educated professionals in the twentieth century. Combining distaste for ornament with devotion to systematic achievement, this functionalism found its characteristic embodiment in the academics' apartment, done in "institutional brown and library gray," which Norman Mailer satirized in *Armies of the Night.* Functionalism lay behind Thorstein Veblen's critique of "conspicuous consumption" and accounts for Veblen's (otherwise inexplicable) popularity among liberals and radicals for the last several decades. The legitimate heirs of this utilitarian brand of simplicity have been some of the purist leaders of the consumer movement in their less attractive moments—during the 1930s, for example, when many consumer activists seemed more concerned with eliminating waste than with ending injustice.

In consumer movements and elsewhere, the elite social background of the simple-lifers has been evident. As Shi observes, from William Penn to *The Nation*'s own E. L. Godkin, the ideologues of simplicity sometimes prescribed constraints for the lower orders that they were unwilling to accept for themselves. But the difficulty ran deeper: by presenting social problems as individual moral failings, the jeremiad implicitly reaffirmed the structural status quo. The most recent example was Jimmy Carter's "crisis of confidence" speech of July 1979, in which (as Shi notes) he attacked self-seeking hedonism without even a nod toward the corporate institutions that promote it. But at that moment, the idiom of "declension" was exhausted. Carter's plea for public virtue evoked universal derision as a dour evangelical's attempt to turn a people of plenty into "a nation of monks." Reagan effectively used the optimistic side of the jeremiad, reaffirming America's sacred purpose in terms of private abundance and military might. Carter became the skeleton at the feast set by a resurgent monied class.

Filled with apocalyptic claims and counterclaims, the debate over the simple life has touched some of the most sensitive nerves in the American polity. In our public discourse, at least, we are still a Protestant nation, sometimes fretfully examining ourselves for evidence of moral decline, sometimes claiming absolute assurance of salvation and damning all doubters. The ideal of simplicity has attracted the doubters—not only moralists like Carter but skeptics who question whether our enthusiasts of abundance have provided the pleasure and freedom they promise. Talk of serenity through simplicity has often concealed a vein of anxious longing for a world elsewhere. Often the anxiety has been peculiarly male. (In republican ideology, effeminacy was handmaiden to luxury and as great a threat to public virtue.) In the mid-nineteenth century, manly commitments to civic duty made room for dreams of eternal boyishness; countless novelists climbed aboard Huck Finn's raft and fled domestic entrapment. But behind male fantasies of escape, the simple life has contained a more inclusive hope as well. Edward Bok, editor of *Ladies' Home Journal,* articulated it in 1901 when he claimed that America might become a nation of "real people, where each man and woman is measured by his or her own true worth . . . where lives are happiest because they are lived simplest: where the air is clear and where people look you in the eye, and where the clothes you wear do not signify."

The horror of sham, the reverence for the "real," the dream of transparent sincerity in social relations—these emotions have animated much of the modern passion for the simple life. Their most recent incarnation is in the criticism of consumer culture voiced by Herbert Marcuse and Ivan Illich (among others), particularly in the notion that mass marketing strategies create "false needs" for commodities undreamt of in simpler societies. A number of criticisms could be brought against this view, but the most serious problem is that there is not a substratum of "true need" we can use as a bench mark; all human needs are mediated by cultural and social circumstance. There is a certain pathos in Bok's dream of a world where "the clothes you wear do not signify." In any organized society, clothes convey meaning; it is a commonplace of semiotics that clothing constitutes a kind of language. The irony is that even as Bok preached simplicity, the magazine he edited became a major vehicle for national advertisements that encrusted clothes and other commodities with new layers of social significance.

Still, the ideal of the simple life has scientific, political and philosophical dimensions worth pondering. About the science I am ill qualified to write, but many such as Rachel Carson and Michael Brown have made the case plainly and are not to be lightly dismissed by the fanciful analogies of Mary Douglas. Politically, the simple life has democratic implications: it encourages basic competence and discourages dependence on technical elites. To know how to house and warm yourself, "to be in direct and personal contact with the sources of your material life" (as Burroughs said) is not necessarily to succumb to survivalism. The decision to "do it yourself" can imply a refusal to slip into rapt contemplation of technological miracles peddled by experts, a reassertion of the value of ordinary intelligence and judgment.

This does not require a resurrection of obsessive Victorian work habits. On the contrary: since the mid-nineteenth century, devotees of simplicity have often sought not to reject pleasure but to enhance it. They have fled a driven bourgeois society where people labor as anxiously at leisure as they do on the job. One of the great contributions of Marcuse and the Frankfurt school was to suggest the ways that leisure could become mechanized and performance-oriented under advanced capitalism; Thoreau had sensed the beginnings of the problem a century earlier. The most thoughtful advocates of the simple life have never merely preached asceticism. They have realized that the standard of living is a subjective value, and that affluence comes in many guises. One way to attain it is by reducing wants and working less, seeking to lengthen that "Sabbath of the affections and the soul" envisioned by Thoreau.

The ultimate significance of this tradition is that it encourages a prolonged encounter with the nonhuman world. We have heard so much about the "sentimentality" of "environmental extremists" we forget that it is at least as easy to sentimentalize human accomplishments by placing them at the center of the universe. In everyday life, it is easy to forget the cosmic indifference of the natural world. But one can still find it, and there are few stronger antidotes to human self-importance. The startled stare of a wild animal can still bring the realization: I have come to this place, and here is this Other—this silent passion, this strange fragility, this fate outside our fate and yet bound to it.

chapter 14 Explorations

14a TV

Collaboration. Not counting sleeping, nothing occupies more of people's leisure time than TV. Share a brief account of the amount and kinds of television you now watch on a weekly basis. How does this compare with the amount you watched at other stages in your life?

Web work. *TV Guide* achieved huge success for years by bringing the new world of television to the more traditional magazine format. Check out the TV Guide Web site and see how it is trying to continue that success online; also look at the site of a competitor, TotalTV. Meanwhile, TeeVee offers a critical look at the boob tube.

Multichannel News and TV Rundown both report on the business side of the cable industry. The Museum of Television and Radio provides a historical overview.

One writing topic. Write an essay on reasons why television is likely to become more or less popular over the next few years.

14b Sleep and Related Somnolent Activities

Collaboration. Share accounts of some of the best and worst experiences with sleep or napping you have ever had.

Web work. SleepNet has "Everything you wanted to know about sleep disorders but were too tired to ask," including a section to check your biorhythms and a sleep newsletter entitled *The SnoozePaper.* Also check out Yahoo!'s page on napping at Health > Medicine > Sleep Medicine > Napping.

To study dreams, begin with the Association for the Study of Dreams, and be sure to check out its gallery of dream-associated artwork and its reprints of dream articles. Other sites to visit include Dream Central, Dream Wave, and Dr. Dream, with the motto " You will spend about 21 years of your life asleep. During 8 of them, you will be Dreaming! Wake UP!" Meanwhile, Dream Gate offers a monthly online History of Dreaming class.

Many sites offer help with interpreting your dreams, including Dream Lover Inc., which for your assistance has an extensive dictionary of symbols. Dream Reader is one of several commercial sites offering dream interpretation—seen by some as just a small step removed from psychic hot lines, in which the Web abounds.

Aristotle's short essay "On Dreams" is available online.

One writing topic. Argue either position: that most of us too often ignore the insight that comes to us in our dreams, or that people who base their decisions on their dreams are acting foolishly.

14c Web Surfing

Collaboration. Share accounts of different ways you use the Web to fill up leisure time.

Web work. Finding interesting and unusual sites on the Web has become a cottage industry. Yahoo! began essentially as a site listing other sites and still collects its best entertainment sites together as "cool links" at Entertainment > Cool Links. Yahoo! also lists best new picks of the day; the pick for February 6, 1998, was The East India Company, founded by Elizabeth I almost four hundred years earlier, in 1600, to explore commercial interests in India. There is also a site called Cool Site of the Day that actually began the craze. For the more impetuous surfer, there is the Cool Site of the Minute site.

One writing topic. Write an argument, pro or con, on the idea that Web surfing is really a highly educational experience, akin to browsing in a library.

14d Beach Life

Collaboration. Most people have an image of the ideal place they would like to retire to—or move to and not have to work. Write and share an account of such a place.

Web work. For many, beach life represents the epitome of leisure, and Pathfinder hosts an incredible exhibition of photos from the archives of *Life* magazine: Life Is a Beach. Sons of the Beach Sandcastle page has photos of their creations from South Padre Island. Meanwhile, Stephen Leatherman of the Laboratory for Coastal Research at the University of Maryland produces an annual ranking of the best beaches in the United States.

The best all-around surfing site may be Surflink. *Yahoo Internet Life* (YIL) has a feature site, Surfin URL, with links to, among other places, a story on Dick Dale, the king of the surf guitar. Mike Wheeler's homepage for Brian Wilson is perhaps the closest thing on the Web to an official Beach Boys site.

There's a lone beach volleyball Web site—beach volleyball fans probably have better things to do with their time than create Web pages. Of course, beach life means beach music: Check out the e.traveller and 21st Century Adventures e-zine.

For many, leisure life is best represented by tropical island life; one site for images and stories is *Islands* magazine.

One writing topic. Write an essay, serious or otherwise, on the leisure activities available in your hometown or in some other town in which you have lived.

14e Slackers

Collaboration. Share a description of someone you know, either personally (no names please) or indirectly (real or fictional), who was widely criticized for underachieving.

Web work. The term *slackers* seems generally to be applied somewhat despairingly to underachieving, unmotivated, vaguely amoral young adults (post–college years); the term *Generation X* to high-achieving (at least in financial terms), career-oriented, vaguely amoral young adults. The two terms, in other words, could be opposites except for the fact that both are used to refer in an unflattering way to people in their twenties not leading a traditional (married with children) life.

The slacker concept has been popularized through film, especially Richard Linklater's 1991 feature *Slacker.* The Visions Web site has an alphabetical list of other films that portray slacker life.

There are lots of electronic magazines (e-zines) that attempt to capture the hedonistic slacker/generation X spirit. *Urban Desires* and *Generation Next* are two. Princeton University's alternative magazine, *The Princeton Review,* goes against the grain in offering young readers solid and progressive news. *Gen-X Necessities,* meanwhile, is trying to make a buck by providing all the links to Web stuff that any X-er could ever want.

Pathfinder's Alt.Culture is an amazing site: just select autopilot and watch a cultural history of pop culture of the 1990s move across your screen. Those who find the 1990s a bit much might want to check out the 1980s server.

One writing topic. Write an argument, serious or satiric, in favor of underachievement in general.

14f Early Retirement

Collaboration. Write and share an account of someone you know who is retired from work.

Web work. The AARP (American Association of Retired Persons) remains one of the most effective political forces in our national life, lobbying the government to take better care of all of us in our old age—or, depending on your point of view, making all young people work extra hard to pay for governmental services for the elderly.

Visit Social Security Online and the Social Security Educational Center, located at Miami University. Americans for Tax Reform makes the conservative case for privatizing social security; the National Committee to Preserve Social Security and Medicare offers the counter argument, as does the AARP.

The Dolans, consisting of Ken and Daria, who bill themselves as the "First Family of Finance," and Retire Web are two Web sites that will help you decide if you are well equipped to plan for your early retirement.

Good online magazines for seniors include *Grand Times* and *Third Age*, which bills itself as "the Web for grown-ups," while the University of Pittsburgh sponsors a site, Generations Together.

Meanwhile, Yahoo! has links to some 80 planned retirement communities at Business and Economy > Companies > Real Estate > Planned Communities > Retirement Communities.

One writing topic. How do you envision the ideal way to spend your retirement? One way to organize your response is by considering two or more different possibilities before selecting the one you prefer.

chapter 15

Women, Work, and Technology

READING QUESTIONS

- Do you feel there are any important differences between men's and women's ways of relating to computers?
- Do you find any common messages or themes in the writings of women who have been successful in working with computers? If so, are there lessons here for men as well?
- Are you personally more comfortable with words or with numbers? What do you feel schools could do to help strengthen your weaker area?

QUOTATIONS

Mathematics expressions give us another way of thinking about relationships that would otherwise be unavailable to us.

— *Sheila Tobias (1987), coauthor of* Breaking the Science Barrier (1992)

Fear of mathematics is the result and not the cause of . . . negative experience in mathematics.

— *Sheila Tobias (1978)*

Stand firm in your refusal to remain unconscious during algebra. In real life, I assure you, there is no such thing as algebra.

— *Fran Lebowitz (1977), American humorist and social commentator*

Women have been more systematically excluded from doing serious science than from performing any other social activity except, perhaps, from frontline warfare.

— *Sandra Harding (1986), professor of social science and education*

It is not hard work which is dreary; it is superficial work.

— *Edith Hamilton, classical scholar, 1867–1963*

Women and the Sexed Machine: An Online Exchange

Pamela McCorduck and Karen Coyle

The following exchange comes from a regular feature called Brain Tennis in the Web publication *Wired,* in which two experts engage in a running email debate on a prominent issue. The Brain Tennis series began in June 1996 with a debate on journalistic objectivity. The following exchange between Pamela McCorduck and Karen Coyle took place between August 26 and September 5, 1996, and can be read online. A complete list of Brain Tennis debates is available online.

Here's how the editors at *Wired* introduced the issue: "Are women underrepresented in technology fields solely because of gender-biased cultural forces, or because the existing technology itself is male-centric? If the latter, what would female digital technology look like—and would its existence widen the gender and technology gap even further?"

Pamela McCorduck is the author of five books on computers, including, with Nancy Ramsey, *The Futures of Women: Scenarios for the 21st Century* (1996). Karen Coyle, a librarian and chair of the Berkeley, California, chapter of Computer Professionals for Social Responsibility, takes the position that "there is nothing inherently masculine about computers" and that all of us are too influenced by gender stereotypes fostered by society generally. Coyle has a homepage.

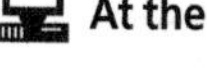 **At the *Visions* Web Site** Link to WITI (Women in Technology International).

READING PROMPT

The online exchange is a peculiarly contemporary form of exchange. To what extent is it like and unlike an oral argument? Like and unlike a traditional essay?

PAMELA MCCORDUCK, POST NO. 1

Long ago I published a book, *The Universal Machine,* where in one chapter, I deconstructed the June 1983 issue of *PC Magazine.* I had some fun with the ads, drawing a comparison between them and Edwin Abbott's little classic, *Flatland,* about an impoverished, two-dimensional world.

"What a shock it must be for the woman (or any other adult) who takes the first tentative step toward computing by buying a magazine such as this," I wrote. "Fresh from a world of ambiguities, texture, difficulties that evade solution, systems that have depth as well as breadth, nuance and indirection, zest, she encounters a fundamentalism so arid and simplistic in its rules, so petty and cramped in its ambitions, so stubbornly resistant to large concerns. . . .Micro indeed."

I typed those words on my old IBM PC, whose manual, though written in my native language, was incomprehensible to me: how to use the machine I learned by asking my friends. When I moved on later to a Macintosh (whose interface had originated at the old Xerox PARC, a joint project between Alan Kay and Adele Goldberg, among others) I began to see that the computer didn't have to be so opaque and intractable. It could actually be a tool to achieve interesting ends, instead of an obstacle to them, if only some thought were given to its human design.

Other things have changed, too. A recent *Times–Mirror* study claims that among users under 35, men and women use the computer in equal numbers, though for different purposes. Moreover, women have begun to move into positions of significant power: Besides the names everyone knows from the media, there are also distinguished computer scientists, like Mary Shaw, formerly chief scientist of the Software Institute at Carnegie Mellon, and Anita Jones, responsible for R&D at the Pentagon, with a budget of some US$40 billion.

In other words, women cope very well with this technology—some of them even love it. But coping is not the same as finding it so well-fitting and intuitive that it simply fades away and lets you get on with the important things. This is not only possible, it is overdue.

All technology is embedded in its culture—and shapes and is shaped by it. This is not a negligible fact. The atmosphere of arrested development that still permeates the computer magazines goes a long way toward explaining why expectations for this technology continue to be so dismally low. I praise the Macintosh, but it's technology of 20 years past. I cannot for the life of me think why the designers of Windows 95 haven't been put in the public stocks, so we can pelt them with the tomatoes and offal that they richly deserve.

KAREN COYLE, POST NO. 2

If you look at the vision of computing that is presented in the popular magazines that we find on every newsstand, you will come away with the impression that computers are mainly used by men. These magazines are astonishingly male. Women are a small minority of the staff writers and appear only occasionally in the advertisements. Those ads are rife with "guy" imagery—lots of sports cars, basketballs, and superheroes.

The fact is that $\frac{2}{3}$ of the people sitting at work in front of a computer are women, and more than $\frac{1}{3}$ of professional programmers are women. In the household, men and women are equal users of the home computer, though he is more likely to use it for games and online access and she to do word processing and spreadsheets.

Pamela, you seem to blame the machine itself. Yet when women do excel in the computer arena, their accomplishments often go unnoticed. Ada Lovelace, a.k.a. Lady Byron, advised Babbage on the invention of his Difference Engine and, in a letter to him, wrote the first articulation of what today we call an "algorithm." Admiral Grace Hopper worked with the earliest computers, and was a key player in the development of the first assembler language.

Like the early days of the automobile, some cultural force is working to define computers as the realm of men, even though women are just as capable of using these machines. It's like the way we divide up chores in the home: Men carry out the garbage and mow the lawn; women cook and remember family birthdays. There is no logic to it, but we seem to have a hard time questioning this gender-based division.

Considering that the computer is just a machine, I find it interesting that we have it invested with such a strong gender image. John Dvorak once called the IBM AT "a man's computer, built by men, for men." There are people who find the command-line interface and completely unadorned ASCII of DOS to be inherently "masculine." We interpret "user-friendly" to mean "easy," "easy" to imply "feminine," and thus Dvorak himself at one point refers to the Macintosh's "girlish GUI."

With this, we've just entered into the world of math-is-hard Barbie, where we only expect women to undertake things that we have defined as easy. Men, however, are challenged to prove their masculinity by taking on a world where the "grep" command is a perfectly reasonable way to search for something in a file.

So, are men genetically inclined to "grep," while women need a little icon of binoculars to perform the same function? Is one of these more masculine than the other? If there is anything that is keeping women and computers from a happy relationship, it isn't the machine itself, which is capable of a wide range of uses, but the culture and society that surround the machine. Women are more hindered by their lack of leisure time and expendable income than they are by the technology of the computer.

PAMELA MCCORDUCK, POST NO. 3

I have no quarrel with the machine itself, Karen; no issue with the yin and yang of all those zeros and ones. I protest what gets shoveled on top, so bad you'd think it's intentional. It's the Rule-Crazed Patriarch of the Arbitrary: DO IT RIGHT THE FIRST TIME, OR ELSE. Please, sir? What's right?

If I accidentally leave off a dot com when I'm mailing, it takes hours for the mailer daemon to inform me (and I must figure out what I did wrong). No online system I use can tolerate synonyms or obvious typos. Technology that could manage those modest tasks is at least 30 years old. Where is it?

Men complain about these things too, and so they should, because we're onto two different styles of problem solving, "hard mastery" and "soft mastery." Many—but not all—men say they prefer hard mastery, and many—but not all—women are soft masters. Systems now are moving from hard to softer (even guys are ready for better), but they're not nearly good enough. I blame that not on the machine, but on the culture that produces it.

Current research suggests that, for whatever reasons, most women operate differently from most men: their ethics are more situational than rule-based (Gilligan), their language differs (Tannen), they even do business differently (Hampden-Turner). Surely the operating systems and interfaces women would design—always exquisite cultural signatures—would also be different.

But larger cultural issues keep women out of engineering; or down, if they persist in entering. In the halls at the Women in Technology meetings, get an earful of American engineering culture, the 1950s captured in amber. Comical, maybe—but the price we all pay for this cultural fossilization is our present computer systems.

KAREN COYLE, POST NO. 4

"Hard mastery," "soft mastery"—Pamela, the words "hard" and "soft" are red flags for me, they have so much cultural baggage.

The problem I have with math-is-hard Barbie is not what she says—because math IS hard, and that's the joy of it, with its intricate patterns and hidden surprises. It's the implication that because it is hard, women should not attempt it.

This same cultural attitude keeps women away from computers. Sure, early computers were hard to use. There wasn't much friendliness you could pack into a machine with 2K of memory and a plain ASCII display. But were they really harder for women than for men? Of course not.

No, the problem was that women were essentially forbidden from trying. Those stories in the halls at WITI [Women in Technology International] are about women breaking the rules, going where no woman has gone before, not because she wasn't capable, but because she wasn't supposed to try.

Yes, computers are getting "softer." They are moving rapidly away from their origins as tools for scientists and engineers toward being a general consumer product. And it's false for us to see this as a change from "masculine" to "feminine" unless we are clearly able to distinguish that from "for men" and "for women." The new computer interfaces are designed for kids, grandparents, and all the humanists among us who don't have the word "algorithm" in their daily vocabulary.

In each of these categories, we should find women and men in equal numbers.

The nature of computing will not change until women master the computer as their tool and start shaping it into the machine they want it to be. And for this to happen, we have to embrace all of the wonderful complexity that is computing.

We have to take on something "hard."

PAMELA MCCORDUCK, POST NO. 5

"Hard" and "soft" mastery refers not to difficult and easy, Karen, but to different styles of problem solving. Mathematicians and computer scientists (and writers and painters) use both approaches, though personality, not sex, makes them prefer one to the other.

We agree: Many women ($\frac{2}{3}$ of users by your survey; half of users under 35 by mine) use computers every day. I've already named female superstars whose mastery is second to none, and I could name more. Then what's the problem?

The problem is that software is still shaped by the treehouse attitude that pervades science and engineering. In my new book, *The Futures of Women,* an imaginary woman engineer describes her professional life—actually a composite of interviews I conducted. *Fortune* magazine found that monologue so bizarre (or comical) that it was reprinted in the body of the book review.

Except it's not funny. To have to out-boy the boyz just to practice your profession falsifies who you are. It does women no good to "master the computer as their tool," as you insist, if their professors, employers, and even their clients can't recognize or accept solutions beyond narrow expectations.

Far from shirking difficulties, or evading complexity, these women do what no man does: They're bicultural. They play the boys' game well (often outstandingly), and then they return home. Some even adopt the dominant culture as their own. Recently, I was part of a group exploring scenarios of future warfare, a group where women were in the distinct minority. When I finally, reluctantly (did it have to be me?), mentioned how women might affect future wars, one colleague, a distinguished female computer scientist, said to me privately: My God, I'm so brainwashed I can't even see the obvious anymore.

What should computing obviously be?

KAREN COYLE, POST NO. 6

Right, women can use computers, and men aren't going to give them credit for what they accomplish with them. This seems obvious.

But it's just as obvious that the average computing engineer (white, male, around 30–35) isn't going to create software for women or for anyone culturally different from himself. So we have no choice but to get involved to create our own version of computing.

What does it mean to create software for women? What would computers be if women dominated the computing field?

Women today still are burdened with keeping the world running on a mundane level. We haven't seen a society where women can put their time into abstract thinking or theory. Women invent practical technologies, like agriculture and weaving. And probably they would initially do the same with software.

I can imagine a "Home Suite" that has tools for family finances and for family members' schedules. It would help you keep an archive of important documents (mortgages, legal records) and would notify you when your child is due for her tetanus booster. The language on the screen would not intimidate first-time users, and the metaphors would be from the everyday life of women.

I can imagine an "Office Suite" for the thousands of new women-run businesses. The functions would be the same as other office software, because accounting for a woman's business is the same as accounting for a man's business. But there would be something—something I can't express in words—about the look and feel that would be more *Cosmo* than *Sports Illustrated*. I don't know what it is, but I'll know it when I see it.

The basic technology of this software will be the same as that which is used to write megadeath games and boring spreadsheet programs. The same technology that men use.

PAMELA MCCORDUCK, POST NO. 7

I like those ideas, Karen. My own specs for good computing are somewhat more general, and all are possible right now:

Let me talk to and touch my information, in addition to typing text and manipulating icons. In the beginning, narrow domains for these more natural means of communication are better than none at all.

Give me tools that can be customized easily (or that learn my needs themselves by watching me). Ease my learning curve, with payoffs well before we reach the peak. One of those tools should archive my work in human terms so that I—or it—can retrieve something effortlessly, and regardless of changing platforms.

Good computing should teach me from my mistakes instead of only punishing me for them.

It should build up an internal model of me so that it, too, can learn from my mistakes and compensate for them. Excellent—not merely good—technology will co-evolve with me.

Offer me tools to improvise and simulate with ease in many different domains. "With ease" means a simple set of shared concepts understood by me and my machine; it means a set of simple (and memorable) commands that can cross contexts, changing meanings as appropriate. Spare me from superfluous upgrades that I don't need and don't want to pay for. (Let me choose the features I want, or simply bill me for the ones I use.)

When fail it must, good technology will degrade gracefully, not leaving me in the lurch. It self-diagnoses before it goes completely.

"The best thing about computers is they can grow things from seed," says artist Brian Eno. Good technology will give me many opportunities to grow things from seed, to generate rather than simply reproduce, to be surprised and inspired by what we accomplish together.

KAREN COYLE, POST NO. 8

Great visions, Pamela. I guess I'm just more cynical, or maybe this is one of those "bad" days.

I'd be happy with a computer that wasn't designed to put me out of a job even though it can only do a fraction of what I really do.

I'd like to live in a world where the word "power" isn't used to sell a machine with a skill set less than that of a gnat.

Life would be better if we'd set aside all claims of artificial intelligence until we've gotten some grasp of what real intelligence is (and, no, Marvin Minsky should not be allowed to designate himself as the definition of intelligence).

I'd like to see computers designed as servants, not masters.

I want to live long enough to see a time when computers are considered no more sexy than toasters (but I don't think I'll make it).

I want someone to calculate pi out far enough to figure out the *#&^ pattern so we can do something more interesting with our supercomputers.

In my perfect world, schoolchildren feel sorry for computers because they are so stupid. They know that their dog is a more advanced being than the microprocessor.

The most important thing for people to learn is that the computer is just a machine. In itself it has no power. It is being used today to disadvantage some people while it makes others insanely rich. All of these decisions are being made by people, not by computers. But we're being told that it's the computer that has the power, and we're supposed to turn our anger against the machine, not the people behind the machine.

I say it's time to look behind the curtain, at the little man. It's time to point and laugh. The computer has no clothes.

Oh my god, I sound like a cross between Dorothy and a Trotskyite. I have mixed my metaphors and am melting.

Facing the Big Chill in Science

Ann Gibbons; Constance Holden; Jocelyn Kaiser

The following article on the plight of programs designed to encourage more women and minorities to enter the sciences was first published in the March 29, 1996, issue of *Science,* which bills itself as the "global weekly of research" and which is published by the American Association for the Advancement of Science.

At the *Visions* Web Site Link to Science Online.

READING PROMPT

Compare the authors' observations about females in science courses with your own experience.

While Sue Rosser was head of women's programs at the National Science Foundation (NSF) last year [1995], she got a series of concerned telephone calls. "Are you still in business?" worried voices asked. "Have your programs been cut?" Rosser reassured them that the programs were alive and well, but the calls are a sign of growing nervousness in the academic community over the vulnerability of programs for women and minorities in science. And Rosser worries that the very perception that these programs are in trouble may stunt their effectiveness. "There is a good bit of fear about federal budget cuts. There's been this chilling effect even though there have been no cuts yet," she says.

Like Rosser, many educators and policy-makers worry that efforts to foster women and minorities in science are endangered in today's economic and political climate. A backlash against affirmative action has sent a strong message that minorities and women may no longer get a helping hand from government and academia. Many programs to encourage greater diversity in science are under review; some may be in jeopardy, and others are likely to be recast to be sex- and race-blind. At the same time, Congress is slicing federal science budgets, and in many fields research jobs are scarcer than ever before. Meanwhile, women and minorities are still seriously underrepresented in many areas, and in an era of belt-tightening, "men will be more likely to do better compared to women and minorities," says NSF Deputy Director Anne C. Petersen. "There's a lot of smoke, and if our alarms are not going off, then there's something wrong with our detectors," says Georgia Institute of Technology sociologist Mary Frank Fox, who recently surveyed 3,800 doctoral students about their career prospects in science and engineering.

Students, among the first to spot smoke signals, are already avoiding what they consider unfriendly territory. Applications from underrepresented minorities to the University of California (UC) dropped last fall in the wake of a resolution by the Board of Regents to end preferences based on race and sex. Although solid national figures are scarce, the National Action Council for Minorities in Engineering reports a 7% drop in minority freshmen enrolling in engineering in 1994, with the largest losses among African Americans. Those numbers are particularly disturbing because they come on the heels of a tremendous increase—60%—in minority enrollment from 1986 to 1993.

In this climate, educators say they must work harder than ever to convince minorities that it's still worthwhile to pursue careers in science. "There has been a chilling effect," says Freeman Hrabowski, president of the University of Maryland, Baltimore County (UMBC), and founder of the highly successful Meyerhoff Scholars program to increase the number of black scientists. "Many minorities believe that most universities don't want them there."

The tough environment presents a major challenge to educators, who predict that only programs with a proven track record and strong institutional support will survive. But the veterans of 2 decades of efforts to improve diversity in science are already employing new strategies for braving the political and economic chill. They are focusing on systemic reform of kindergarten through grade 12 (K–12) science and math education, and opening their programs to students of all colors, while making sure that those who need help most—minorities and women—reap the greatest gains. "We are seeing a backlash against affirmative action, but I do feel that we are closer than we have ever been to achieving equity in mathematics and science education in schools [at the K–12 level]," says Beatriz Chu Clewell, principal research associate at the Urban Institute.

These changes come at a time when some politicians and educators question whether women and minorities still deserve favored treatment in science and engineering. The answer is yes if you look at the numbers, says Catherine Gaddy, executive director of the nonprofit Commission on Professionals in Science and Technology, which analyzes human resource data on scientists. Although more women are earning science degrees, the percentage of degrees earned by underrepresented minorities has changed little, especially in the natural sciences and engineering. "My biggest concern is that the percentage of minorities earning degrees in science and engineering isn't going anywhere—and for African Americans it's just hovering," says Gaddy. Minorities earned just 10.7% of all science bachelor's degrees in 1992—compared to 10.2% of such degrees in 1981. And although women are closing the gender gap, they are still relatively scarce in many natural sciences, and they are rare at the top, in the ranks of full professors.

BACKLASH BLUES

Closing those gaps is likely to get even tougher if Congress outlaws federal affirmative action programs, as proposed by Senator Bob Dole (RKS) in a bill introduced last summer. And a recent Supreme Court ruling sets narrow limits on any programs that make distinctions based on race or sex and is forcing a review of all such programs in the federal government. It is still unclear what all this will mean to programs at NSF and the National Institutes of Health. But if these and other government programs are reduced or eliminated, the impact could be devastating because the government is by far the biggest sponsor of efforts to increase diversity in science. "A small change in government spending will make a big difference," says Ted Greenwood, program officer at the Alfred P. Sloan Foundation, which spends about $3 million a year on minority programs. Foundations do sponsor scholarships and academic programs, but "the government is the 600-pound gorilla here," says Greenwood. Indeed, even the $350 million spent since 1988 on science education and diversity programs by the largest philanthropic organization, the Howard Hughes Medical Institute (HHMI), is minimal compared to the more than $2 billion spent by the government over the past 20 years. "It's unrealistic to think the private sector can be a replacement for cuts in the federal budget," says HHMI President Purnell W. Choppin.

Few such cuts have actually been made yet, given the long-drawn-out funding cycle and the wrangling over the federal budget. Yet shifting priorities are already taking their toll. For example, NSF pulled the plug on one of its programs, Research Careers for Minority Scholars (RCMS), which provided $6 million a year for research by minority undergrads. One of the programs left out in the cold is at the City University of New York (CUNY), which in September will lose $200,000 a year, money that program director Neville Parker says he used "to nurture students and keep them in the pipeline."

The NSF canceled RCMS and used the money to refocus its efforts on K–12 education and to support another undergrad program called the Alliance for Minority Participation (AMP), which spends $25 million a year on alliances of colleges that offer research opportunities for minority undergrads, says Luther Williams, assistant director for education and human resources at the NSF. But Williams doesn't want institutions to get too dependent on that funding either: "My intent was that AMP would catalyze these programs . . . that the institutions would take them over, and the NSF would not fund them in perpetuity."

Meanwhile, at Kenyon College in Ohio, economist Susan Palmer is struggling to find funding for COSEN, an 8-year-old consortium of eight colleges in Ohio and the Carolinas (including Duke University) that has a strong track record of helping minorities and women earn degrees in science. But it is losing its funding from the Pew Science Program in Undergraduate Education, which has given it $1.3 million. Like NSF, Pew is shifting resources to K–12 education and prodding such consortiums to become self-sufficient after almost a decade of funding. But the cuts come at a time when new money is hard to find. "There's no way [our] institution can contribute that kind of money," sighs Palmer. "I've been looking for money, but I've been discouraged by the general climate against affirmative action."

These two are just warning signs that the once-abundant programs for women and minorities may be shrinking. "I think for sure there will be a reduction of targeted programs," predicts NSF Deputy Director Petersen. As the NSF reviews its efforts, it is trying to see how to get the same results without targeting women and minorities. But, Petersen adds, "the worry is if you lower the profile of these programs—if you never talk about women and minorities—will they have the same kind of potency?"

Poor prospects. These potential cuts in agency funding for diversity programs coincide with major economic trends that make it harder to land a traditional research job (*Science,* 6 October 1995, p. 123). A new study by Fox and Georgia State University economist Paula Stephan identified three major factors that may hamper students' career prospects: government deficits, which may prompt cuts in federal research funding; the end of the Cold War, which curtails defense spending; and the lifting of the mandatory age for retirement, which slows turnover at universities.

The result of those changes is that many of the 3,800 science and engineering doctoral students surveyed for Fox's study were glum about their prospects for finding university research positions, particularly in physics and chemistry. And women and minorities, who already drop out of science in greater numbers than men do, are the most easily discouraged, because they start out with uncertainty about whether they "belong," says sociologist Elaine Seymour of the University of Colorado, Boulder. The result, says Fox, is that the economic and political changes are "a closing of the door, so minorities and women don't have access to careers in science."

Even before the jobs are gone, the perception of a tough market may deter students, says Rosser, who now directs a women's studies center at the University of Florida, Gainesville. Already, the number of freshmen who enrolled in engineering last fall is "way down" for both men and women, says Linda Sax, associate director of the Higher Education Research Institution at UC Los Angeles, which surveyed 240,000 freshmen at 473 colleges. Ironically, there are still jobs in engineering.

STRATEGIC PLANNING

Given the economic uncertainty, policy-makers are changing their tactics even before the ax falls on their programs. Heeding Williams's warning, Parker is seeking funds from the CUNY administration and other new sources. "Every time you get money from the government, you better figure out how not to depend on it," he says. Commitment from institutions themselves is critical. "I didn't agree to do this job unless there was institutional commitment behind it," says Parker.

But even where the highest echelon of an institution turns its back on affirmative action, administrators can still take steps to increase diversity, as shown by Berkeley Chancellor Chang-Lin Tien. In the face of the regents' order to end preferential treatment in admissions, Tien has begun a program called the Berkeley Pledge, which sends recruiters to high schools to find minorities who qualify for admission, especially in science and math.

At other institutions, the answer is to broaden the pool. At UMBC, for example, Hrabowski launched the Meyerhoff Scholars program in 1989, which has helped 48 African Americans earn B.S. degrees. But now the highly successful program is reconfiguring to ensure its survival. This fall it will open its doors to students of all colors—if they can demonstrate that they are committed to helping underrepresented groups in the sciences, such as by working with inner-city kids. "Our most important goal is to increase the number of successful minorities in science, and it's not just minorities who can help," says Hrabowski.

The same tactic is being used in programs that assist women. New NSF-sponsored programs now enlist both male and female faculty members in sessions on how to be more effective in teaching and working with women, says Rosser.

And no matter what the government does, nothing is stopping institutions from taking matters into their own hands. At Tulane University, President Eamon Kelly is dangling a juicy carrot in front of departments that hire minorities—$750,000 from its general fund to hire and support qualified new minority faculty members. This is not a new approach, but it's unusual in these times.

FUTURE STOCK

While Kelly works at the faculty level, many educators are looking to the future, which means today's K–12 students. In El Paso,Texas, where 80% of public school students are Hispanic and 5% are black, a diverse group is working to boost the performance of students in grades K–12 and college. The El Paso Collaborative for Academic Excellence—led by the mayor, a county judge, a nun, and other educators—seeks to have all students master math and a deep understanding of science, rather than tracking only "gifted" kids into these courses. "There are just not enough kids making it through high school to get into science and math in college," says Susana Navarro, an educational psychologist at the University of Texas, El Paso, and executive director of the collaboration.

The El Paso project is one of 25 urban areas to receive an Urban Systemic Initiative Grant from NSF. The bulk of the students in these cities, including New York City, Detroit, New Orleans, and Chicago, belong to minorities, but the initiatives are not affirmative action programs because they include every student in the system. So they should be immune to the anti–affirmative action backlash, yet fill the pipeline with better prepared minorities and girls.

So far, this systemic reform is getting high marks from people like Kati Haycock, who directs the education trust at the American Association for Higher Education. Colleges that join these initiatives tap into a pool of better prepared minorities and women, she says. "CUNY had the best prepared freshman class in decades," Haycock, referring to the College Preparatory Initiative, a project run by the New York City Public Schools and CUNY. "Pushing students—and schools—toward higher level classes is having a payoff in higher education as well."

In fact, when institutions make structural changes to improve the way they teach minorities and women, almost all students seem to benefit. In New York City, the number of minorities passing college-prep science courses more than doubled—and the number of whites passing these classes rose by 34%. A 2-year study at the University of Southern California found the same result: When faculty members changed their teaching methods and curricula to boost the retention of women in science, men's grades also improved. All this suggests a key strategy to thaw the ice surrounding diversity programs, says Rosser: Get the message out that in many cases, what's good for women and minorities is good for everyone in science.

What about Us Grils? *Kathe Davis*

Kathe Davis here pursues the interesting connections between feminism and science fiction, seeing both as governed by ideas as to how best to organize society. For more on sci-fi sites see the Web work for Exploration 17e.

At the *Visions* Web Site Links to National Organization for Women (NOW); to other feminist organizations.

READING PROMPT

Why is Davis more interested in stories about the future than about the past?

Once, long ago and far away (i.e., Long Beach in 1969), I saw written on a wall among other graffiti "For a good time, call the grils at 789–1010." Some Mrs. Grundy of the women's room had made the correction to "girls." Below that someone else had pencilled "What about us grils?"

The pathos of the question stayed with me. I imagined well-meaning but inept aliens living undercover and hoping for a chance to get themselves integrated into Southern California culture or mutants, inconspicuous but definitely other, waiting like the aliens for a chance at inclusion—eager and wistful, new beings wanting someone to reach out. What about them?

In *Hemingway's Genders* (1994), Nancy Comley and Robert Scholes's recent book arguing that Papa's views on gender were more complex, and troubled, than has generally been acknowledged, the authors quote a Gautier character who, after a period of cross-dressing, loses patience with both available options: "In truth, neither of the two sexes are mine; I have not the imbecile submission, the timidity, or the littleness of women; I have not the vices, the disgusting intemperance, or the brutal propensities of men: I belong to a third, distinct sex, which as yet has no name." Maybe she was a gril.

That was in 1835. In the intervening 160 years, great numbers of women have arrived at similar conclusions, and many of those expressed their dissatisfaction by joining organized movements to free women from their littleness and men's disgusting intemperance. How well they fared is a matter of debate. Men also found their prescribed role too confining, at least occasionally. But on the whole the gender in power has little incentive to alter the arrangements that put them there. So men's movements there were none, until very recently. Now that women have gained enough power for men not merely to object but to begin demanding it back (it didn't take much), the Iron Johns seem more concerned to reclaim than discard their brutal propensities (fearing, doubtless, imbecile submission). So what's a poor gril to do?

She can read SF and take heart. Gender has always been an issue in science fiction. Not coincidentally, the genre is just about as old as "The Woman Question" (and Gautier's gril). The monster Mary Shelley created to embody anxieties about technological overreaching grew also from her own childbearing traumas, and since then it has remained true that "a fascination with the process of reproduction dominates science fiction," as Robin Roberts claims in *A New Species: Gender and Science in Science Fiction* (1993).

Notwithstanding that beginning and that preoccupation, the genre historically has been notoriously a male domain, sexist when it deigned to deal with women at all and overwhelmingly male also in readers, writers, and values.

For some of us grils that was part of its charm. Really. Not long after my pleased father brought home *Galaxy*, Vol. 1, No. 1, I began competing with him for possession of each new issue and ransacking his old *Amazing Stories* and *Astoundings*. I even read the editorials and took vicarious pride in the rigorous insistence on "hard" SF scientifically up-to-the-minute and in accord with what is known to be known, in Samuel Delaney's phrase. I felt thus aligned with obviously preferable "male" values, with rationality and scientific power and progress. Discussing the merits of a Robert Sheckley story with Dad, I felt superior to the giggling silly girls at school ("airhead" and "fluff chick" were terms not yet coined) and somewhat less threatened by the fate of my irrational, entrapped mother. SF provided grist for "serious" conversation with boys in high school biology and, indeed, college philosophy classes. It was not until graduate school and the 1970s that I learned that my male-identification was a surrender to divide-and-conquer tactics and that sisterhood is powerful.

By that time a new sort of SF was available, and it has been getting bigger and mostly better ever since. But I want to maintain that old SF, even the "hardest," even the overtly sexist, also contributed importantly to what would become my feminism. For one thing, making myself at home in a male realm was empowering and was somehow possible in SF's speculative openness as it was not in the closed societies of westerns and detective fiction (notwithstanding Nancy Drew).

More significantly, SF encouraged and developed tendencies of mind that carried over into feminism: the conviction that today's culture can be—indeed, almost certainly will be—different tomorrow and the habit of thinking about that change.

For SF and feminism are at a profound level engaged in a shared endeavor, whether or not a female character is in sight. Wait a minute—"feminism"? What do we mean by this much-misunderstood term?

My bumper sticker says "Feminism is the radical notion that women are people." SF provides a handy way to talk about women as not human, or, vice versa, it has been observed at least since the mid-1970s. Robin Roberts goes so far as to say that, "in science fiction, the Other always represents the feminine." But SF can present that Other sympathetically, as Roberts and Ursula Le Guin and Susan Gubar have all noticed.

More formally, feminism simply means "advocacy of the claims and rights of women" (Concise OED). Opinions as to what those claims and rights are or ought to be obviously vary widely, which is why we use the plural, "feminisms." But feminisms of every stripe require change.

And SF is the literature of change, whether you call it "cognitive estrangement" (Darko Suvin) or "the What If" (Joanna Russ). SF shows us how, if we don't change, we'll end up where we're headed. It shows us what change might bring, what might happen if a perpetual fog eliminated technology, or if technology eliminated human effort. The parallel universe is not merely a favorite device of SF; it is the nature of SF. In a sense, the whole field of SF is a collection of parallel universes. Training in imagining alternatives is not sufficient to make one a feminist, but it is necessary.

Since the future is where change can happen, it is only logical to expect feminists to utilize SF, the fiction of the future, to depict changed possibilities for women: new psyches, roles, societies, planets. So they have done, as this issue richly documents.

Even in the bad old days (or the Golden Age, depending on your perspective), SF stories weren't about technological change only. They often portrayed conflicting assumptions in confrontation. I remember a small Asian-looking people who conduct war by killing themselves (as if Gandhi on acid orchestrated the war in Vietnam—off-planet). I remember a crash-landed pilot who figures out that his kindly treatment by a giant alien is due to his being regarded as a pet, like his own cat. I remember "Kindergarten," in which the arrogant human race learns that with space flight it has just entered school. These are, if not quite feminist stories, certainly stories that lead one toward feminist territory.

For the change most feminisms urge is not simple reversal. If women soldiers replace men, it's still war. Women have now demonstrated that they can adopt a male mode, complete with suits and briefcases, and "make it" in the male world. What they have not done is answer the question of why they would want to do so. The male world as presently constituted is pretty hostile to human life, not to speak of the soul. It is not a world habitable twenty-four hours a day. Men have traditionally relied on women to provide the comfort, the retreat, the nightly leave, the R&R, that enable them to go back into the fray.

Now that women have themselves entered the public fray in large numbers, they're less available to provide the service and solace for men—and are noticing that no one's providing it for them. Thus the whole competitive system, which never acknowledged its secret dependence on the noncompetitive support of women, is in serious crisis.

By any index—the distribution of wealth, the number and condition of the poor, health care, the support of education, the condition of the infrastructure, the health of cities and the environment, crime rates, the peace of mind of its citizens—we in the United States, the richest and most powerful country in the world, are not doing particularly well. And we were supposed to be the future.

New futures are needed. We, a nation founded on the claim of a vision, are notably short of visionaries. The only citizens who engage in the activity for a living are science fiction writers. This makes them, to the discerning eye, a singularly valuable resource. Since women's perceptions and perspectives have been neglected in public discourse, women's visions in particular are in short supply.

What is needed is for women to bring to bear what they know as women, what they have learned from not being recognized players in the public power game. What is needed is not just for women to become players but for the game to change. Women need a new game, but so do men. So do children. So does the planet. Feminism is not only a woman's movement.

But if it's a new game we are making up as we go along, we don't know what it's supposed to look like. We don't have any models. SF can imagine them. So far from being an anomaly, feminist SF is consonant with what has always been SF's most deeply humane impulse: imagining humanly preferable alternatives to business as usual. It is SF writers, long practiced in speculation and—well, extrapolation, who can conduct the necessary thought experiments, think through the alternatives to current gender behaviors, relations, and consequences.

Science itself is moving toward more "feminine" values. Evelyn Fox Keller and Donna Haraway, among others, have alerted us to the profoundly gendered character of science as a whole, not simply at the level of the (still overwhelming) exclusion of women from the practice of science, but in its underlying assumptions and values. But the advent of relativity, the uncertainty principle, quantum mechanics, and such subsequent developments as chaos theory, have moved science from certitude to guaranteeing the impossibility of certitude, the impossibility of "objectivity" in the old sense. It was a woman, Judith Merril, who observed that in becoming "softer," "SF is catching up with science." This issue will help you count the ways.

Time in the Balance *Arlie Russell Hochschild*

Arlie Russell Hochschild is a professor of sociology at the University of California, Berkeley, and the author of *The Second Shift* and *The Time Bind: When Work Becomes Home and Home Becomes Work,* from which this article is adapted. Hochschild appeared on the PBS *NewsHour* on July 31, 1997 to discuss her latest work.

At the *Visions* Web Site Link to the Families and Work Institute.

READING PROMPT

At the center of the important issue raised by Hochschild is the balance between the benefits and the liabilities of working at home—and the question whether that balance is any different for men and women. What do you think?

We've gotten ourselves into a time bind. Feeling that we are always late, having no free time, trying to adapt as best we can to the confines of our time prisons—these are all symptoms of what has become a national way of life. There are several reasons for this. Over the past two decades, global competition and inflation have lowered the buying power of the male wage. In response, many women have gone to work in order to maintain the family income. But the legacy of patriarchy has given cultural shape to this economic story. As women have joined men at work, they have absorbed the views of an older, male-oriented work world—its views of "an honest day's work"—at a much faster rate than men have absorbed their share of domestic work and culture. One reason women have changed more than men is that the world of "male" work seems more honorable and valuable than the "female" world of home and children.

There is another factor too—we are increasingly anxious about our "culture of care." Where do we turn when we are down and out? This is a question we ask, understandably, even when we are up and in. With recent welfare "reform," the government is cutting off aid to women and the poor. With the growth of the benefitless contingency labor force, many corporations are doing the same thing to middle-class men. Meanwhile, modern families have grown more ambiguous: It's a little less clear than it once was who's supposed to take care of whom, how much and for how long. So we've grown more anxious. Given our tradition of individualism, many of us feel alone in this anxiety, and insofar as work is our rock, we cling to it.

In this context, the idea of cutting back the workday is an idea that seems to have died, gone to heaven and become an angel of an idea. We dream about it, but it's something we'd never really expect to do. No matter how a movement for work–family balance is structured, in the long run, no such balance will ever take hold if the social conditions that would make it possible—men who are willing to share parenting and housework, supportive communities, and policy-makers and elected officials who are prepared to demand family-friendly reforms—remain out of reach. It is by helping to foster these conditions that a broad social movement could have its greatest effect.

The moment has come to adjust the old workplace to the new work force. As history has shown us, the only effective way to bring about basic change is through collective action. Such a movement would have to face certain fundamental issues. As a start, since the corporation absorbs increasing amounts of family time, it is the corporation that we most need to change. Research on large companies indicates that it is hardly prudent to rely on company executives as our architects of time. Whatever their stated goals, whatever they believe they are doing, they are likely to exacerbate, not relieve, the time bind of their workers.

Therefore, a time movement would need to find its center outside the corporation, however important it may be to cooperate with advocates of family-friendly policies inside the company. The struggle for the eight-hour day that began in the nineteenth century and triumphed in the thirties was spearheaded mainly by unionized male workers. A new time movement would have to be made up of a wider range of stakeholders and the organizations that represent them. Male and female workers, labor unions, child advocates and feminists, as well as work-family balance advocates and even the leaders of some progressive companies, would act as the vanguard of such a campaign.

Supporters of the eight-hour day strove to expand the leisure time of workers but said little about families per se. Perhaps this was because most unionized workers at that time were men not responsible for the direct care of children. But now that most mothers are on the job, work time is inextricably linked to family life. A new time movement would need to focus far more on the nature of this linkage. On the other hand, corporations that do provide family-friendly programs tend to associate these programs only with middle-class women, leaving out middle-class men as well as the working class and poor of both genders. Clearly, although women would be a significant constituency of a time movement, men have just as much to gain. Male workers, who often average longer hours than women and whose presence is often sorely missed at home, need a time movement at least as much as women do.

As my study of Amerco (a pseudonym for a Fortune 500 company that offers options for shorter hours that few workers take advantage of) has shown, however, even when the jobs of working parents are secure, pay a sufficient wage and provide family-friendly programs, many working parents are still reluctant to spend more time at home. American fathers spend less time at home than mothers do, expand their work hours when children are small and, if Amerco is any indication, are reluctant to take paternity leaves. We know from previous research that many men have found a haven at work. This isn't news. The news is that growing numbers of working women are leery of spending more time at home as well. They feel guilty and stressed out by long hours at work, but they feel ambivalent about cutting back on those hours.

Women fear losing their place at work; having such a place has become a source of security, pride and a powerful sense of being valued. As a survey conducted by Bright Horizons (a Boston-based company that runs on-site daycare centers in twenty-three states) indicates, women are just as likely to feel appreciated as men at the workplace; as likely as men to feel underappreciated at home; and even more likely than men to have friends at work. To cut back on work hours means risking loosening ties to a world that, tension-filled as it is, offers insurance against even greater tension and uncertainty at home. For a substantial number of time-bound working parents, the stripped-down home and the community-denuded neighborhood are simply losing out to the pull of the workplace.

Many women are thus joining men in a flight from the "inner city" of home to the "suburbs" of the workplace. In doing so, they have absorbed the views of an older, male-oriented work world about what a "real career" and "full commitment to the job" really mean. Women now make up nearly half the labor force. The vast majority of them need and want to be there. There is definitely no going back. But women have entered the workplace on "male" terms. It would be less problematic for women to adopt a male model of work—to finally enjoy privileges formerly reserved for men—if the male model of work were one of balance. But it is not.

All this is unsettling news, in part because the children of working parents are being left to adjust to or resist the time bind—and all of its attendant consequences—more or less on their own. It is unsettling because while children remain precious to their parents, the "market value" of the world in which they are growing up has declined drastically. One need not compare their childhoods with a perfect childhood in a mythical past to conclude that our society needs to face up to a serious problem.

But the role of a movement for the reform of work time should not be limited to encouraging companies to offer policies allowing shorter or more flexible hours. As my research has shown, such policies may serve as little more than fig leaves concealing a long-hour work culture. A time movement would also need to challenge the premises of that work culture. It would ask: Are workers judged mainly on the excellence of their performance, or mainly on the amount of time they are present at the workplace? Is there a culture of trust that allows workers to pinch-hit for one another as needs arise? Is there job security? The answers to these questions are crucial, for shorter hours can have little appeal as long as employees fear that the long hours they now work may disappear entirely.

To start with, a time movement should press to restructure corporate incentives. For example, the Commerce Department could be pressured to broaden the criteria for receipt of its coveted Malcolm Baldrige National Quality Award (annually given to companies for outstanding achievement in meeting the standards of "performance excellence") to include the successful implementation of family-friendly programs as measured by the number of employees who actually use them. How many working parents at a given company report that they have enough time for their families? How many go to P.T.A. meetings? How many volunteer in the schools? These could be signs of a company's success in establishing a work–family balance.

Then, too, a time movement must not shy away from opening a national dialogue on the most difficult and frightening aspect of our time bind: the need for emotional investment in family life in an era of familial divestiture and deregulation. It would have to force a public reckoning with private ways out of the time bind—emotional asceticism, the love affair with capitalism, the repeatedly postponed plans of the potential self—that seem only to worsen the situation.

Finally, a time movement would need to compel us to face the issue of gender. In the early stages of the women's movement many feminists, myself included, pushed for restructuring work life to allow for shorter hours, more flexible jobs and restructuring of home life so that men could get in on the action. But over the years, this part of the women's movement seems to have surrendered the initiative to feminists more concerned with helping women break through the corporate glass ceiling into long-hour careers. A time movement would have to bring us all back to the question of how women can truly become men's equals in a more child-oriented and civic-minded society.

A time movement would need to tackle a number of other tough questions as well. How many hours a day, a week, a year, should people work? How can we press for better work environments without inadvertently making them havens from life at home? How can both partners in a relationship achieve a stable and compatible understanding of work–family balance? In an era of growing income inequality, how can more time be made available to the working poor as well as to the better off?

Sweden, a global competitor long held up as a model of work–family balance, allows 360 days of parental leave, thirty at 85 percent pay and 210 at 75 percent—and 78 percent of fathers take part of it. In addition, Sweden offers up to 120 days of leave a year at 75 percent of income for the care of sick children.

Travelers in Sweden quickly sense that they are in a child-friendly environment. (Even trains have children's play areas with little slides, crawl spaces and tables.) Swedish family policy specifies that children have the right to be looked after properly while their parents work. The government subsidizes childcare, maintaining high standards for secure and stimulating environments at childcare centers nationwide.

In the nineties, Sweden has become a model in a more unexpected way as well. Pressured by more conservative members of the European Union, Sweden has taken tentative steps to cut back on family benefits for working parents. In response, grassroots protest groups have sprung up across the country. The Children's Lobby, established in 1991, is fighting cuts in children's benefits, as is Support Stockings, a group formed by women from all the main political parties. It has even threatened to create a separate party if politicians don't work harder to support family-friendly issues. In opinion polls, a third of Swedish women indicated that they might vote for such a party.

Any successful movement for social change begins with a vision of life as it could be, with the notion that something potential could become real. So let's imagine a mother picking up her daughter at childcare twice a week at 3 PM. instead of 6 PM. Picture a father working half-day Fridays and volunteering at his child's center. Let's imagine P.T.A. meetings to which a large majority of the parents come, and libraries where working parents can afford to devote their spare time to reading or literacy programs, or community gardens in which they and their children have the leisure to grow vegetables. Picture too the voting booths in which parents choose candidates who make flexible worktime possible.

But vision alone will not be enough. A time movement will not succeed without changes in many of the underlying social conditions that make it necessary. The rising power of global capitalism, the relative decline of labor unions and the erosion of civil society will all test the resolve of such a movement. Yet it should not be forgotten that such trends not only tighten the time bind we live with but highlight the urgent need for a way to gain release from it. Job scarcity can make people "work scared" (and thus work longer hours), but it could also allow corporations and unions to look at ways to share more lower-hour jobs. Under the right political and social conditions, the growth of technology, which is extending the "anywhere, anytime" workplace into the home, might help people balance work and family, even as it squeezes non-worktime even more.

Finally, I believe that the rising number of women in the labor force—and their partners—are a growing constituency for a time movement. This is especially true for those in the middle ranks of the corporate world. It is these workers whose potential selves—if not yet their real ones—are clamoring for more time at home. At a hypothetical meeting of time activists, a unionized auto worker who wants to cut down on overtime in order to give hours back to laid-off comrades may yet join together with an upper-middle-class, nonunion working mom who wants to job-share. Both could find common cause in their children. The most ardent constituency for a solution to the time bind are those too young as yet to speak up.

chapter 15 Explorations

15a Equal Access to Technology

Collaboration. Make and share a list of factors that currently limit (or until recently have limited) your access to computer technology.

Web work. Home and school access to the Internet is about to undergo a major transformation. In the late 1990s, the Internet was still something of a hobbyist's world, where individual and small institutions generally had to buy an expensive piece of equipment and then do a lot of the tweaking themselves. But at some point in the future (precisely when, no one knows), obtaining Internet service will be as foolproof and easy and (maybe) as relatively inexpensive as obtaining a dial tone for local phone service. Just how this will happen among the war of competing vendors is also a great mystery and a legitimate object of public debate. There are two basic concerns here: the kind of equipment one will have to buy to get online (the phone, so to speak) and the quality of the connection to the Internet (the phone line).

One emerging and related technology aims to offer individuals cheaper access to the Web by using the least computer power possible (called a *thin client*). Such is the highly advertised Web TV (see the link from Philips Magnavox's homepage), which works as an add-on to one's TV.

The other issue involves cost of connecting to the Web through what is called an Internet service provider (ISP). To check out current rates, try CNET's comparison of national ISPs, or see Yahoo!'s more general list, at Computers and Internet > Internet > Information and Documentation > Reviews > Internet Service Providers.

Of course there are a number of new competing technologies for connecting home computers to the Internet, all of which offer much faster modes of transmission (or fancier stuff, like video clips, at the same slow speeds) and the ability to send and receive data without tying up one's phone line. Just how affordable any of them will be is an entirely different issue, but one can assume that the companies will charge as much as they can or as much as government regulators will allow them to.

The High Bandwidth Web page collects information about all these services.

One writing topic. Write an essay (serious or otherwise) describing the technological conveniences of the home of the future.

15b Single-Gender Education

Collaboration. Make and share a list of the major reasons for and against single-gender high schools.

Web work. Peterson's has been a leader in publishing guides to schools and colleges. Its site can give links to traditional all-boy and all-girl primary and secondary schools, most with boarding arrangements.

The search page of *U.S. News & World Report*'s college site allows you to locate the short list of single-sex U.S. colleges. Visit any of the Web sites of these schools and see what they have to say about the their single-sex admissions policy.

The Citadel and Virginia Military Institute have been two schools visibly in the news recently in connection with single-sex education, mainly because of their resistance to admitting women into their corps of cadets. Check out their homepages, especially the admissions sections, to see what they say about women applicants.

U.S. News & World Report offered a short debate on the issue of single-sex education following the Supreme Court decision against the Citadel and VMI, plus links to related articles. The Court decision, written by Justice Ruth Bader Ginsburg, can be read online.

A lengthy report on "Women's Colleges in the United States: History, Issues, and Challenges" by Irene Harwarth, Mindi Maline, and Elizabeth DeBra is available on the Web.

Two highly publicized Washington, D.C., events, both held there in the late 1990s, were the male-only Million Man March (see *Time* magazine's coverage) and the Promise Keepers rally. Yahoo! has a list of sites related to the latter organization, including a section on opposing views, at Society and Culture > Religion > Faiths and Practices > Christianity > Organizations > Men > Promise Keepers.

One writing topic. Write an argument in favor of or opposed to single-sex high schools.

15c Caregivers, Male and Female

Collaboration. Share a detailed account of a male you know (personally or indirectly) who routinely or occasionally performs such domestic chores as feeding, cleaning, and clothing other people.

Web work. *Fathering* magazine is a rare publishing venture into the world of domestic magazines dominated by women and women's issues. The National Center for Fathering has as its motto "Strengthening Families . . . and America . . . by Strengthening Fathers." Forrest Seymour's "A Father's Journal" is a testament to one person's commitment to family responsibility.

Parents Place and Parent Soup both cover parents' issues without regard to gender.

One writing topic. Write an essay on changing notions of fatherhood, motherhood, or parenthood.

15d Girl Talk

Collaboration. Make and share three lists of magazines: those that seem to appeal largely to men, those that seem to appeal largely to women, and those that seem to appeal equally to both men and women.

Web work. Browse through the Web sites of these unabashedly fun and feminist e-zines: gURL, the Australian e-zine; geekgirl; PlanetSleep, which has been dubbed "a virtual pajama party"; Riotgrrl; and Nrrd Girl.

One writing topic. Write an essay characterizing the images of the ideal woman reader of one or more print or electronic magazines aimed at a largely female market.

15e Math, a Guy Thing?

Collaboration. For years there has been widespread concern about female students having more problems than their male counterparts in math and science courses. Share detailed but anonymous accounts of your greatest success or failure in a math or science course, and discuss whether or not these narratives display any hints as to their authors' gender.

Web work. Of interest are the "new" math curriculum, as developed by the National Council of Teachers of Mathematics, and Bill Quick's site opposing the "new" math. The latest controversy started in California, where plans for a new math curriculum were introduced in September 1997; here is the introduction to this report. A group of Palo Alto parents formed HOLD to oppose these changes.

Girls to the Fourth Power is a research program at Stanford University to help girls achieve in high school algebra. The Web site for Voices of Girls in Science, Mathematics, and Technology reports on a three-year study funded by the National Science Foundation.

Canadian site MathMania invites students to work on ongoing math problems in an open environment. Cut-the-knot is a site founded by Alexander Bogomolny to answer basic but fun math questions.

Before leaving all this math, try taking Dilbert's daily mental workout.

While on the subject of math as a guy thing, see if you can see what, if anything, makes the Woman Motorist or the Women's Automotive Help Center especially women-centered. Or, for that matter, what's especially "guylike," if anything, about the movies Scott Meyer describes in *The Guys' Guide to Guys' Videos?*

One writing topic. Write a critique in support of or in opposition to using more group assignments and collaborative work generally in all areas of the high school and/or college curriculum, possibly including math and science classes.

15f Affirmative Action

Collaboration. Share a detailed account of a time when you gained or lost an advantage in a competitive situation because of the special treatment of a supervisor.

Web work. Affirmative action remains a controversial issue in the United States. In the late 1990s much attention focused on a ballot initiative in California to prohibit the state from basing any hiring, school admissions, or business preferences on a person's gender or ethnic identity. Yahoo! offers a page (Society and Culture > Affirmative Action) for general affirmative action links, and another page (headlines.yahoo.com > Full Coverage > US > Proposition 209) just for the latest news on the California Civil Rights Initiative.

The *Atlantic Unbound* has numerous articles on general issues relating to affirmative action, including Nicholas Lehmann's Executive Decision memo "Is Affirmative Action Unfair?" and a reprint of McGeorge Bundy's 1977 article "Who Gets Ahead in America?" as well as important later articles on race relations in the United States by Stanley Fish, Juan Williams, and others.

Two feminist sites visible in their support of affirmative action include the Feminist Majority and the National Organization for Women.

One writing topic. Write an essay on whether or not, or under what conditions, you believe schools should be allowed to consider biographical or historical information about students in determining admissions to highly competitive programs.

part VI

Future Thoughts

The final three chapters act as a reprise or reworking of the major themes of the entire anthology—focusing in on the basic questions of just what kind of world we want to live in, how we can go about creating that world, and what role technology can play in our efforts. The chapters in Part I, Objects of Desire, look at such issues from the specific point of view of tools, toys, and gadgets; the selections in this final part all tend to address a similar question of quantity versus quality but in more general fashion, often speaking directly about big ideas like *the past* and *the future* (as the Tofflers do in their essay, "Getting Set for the Coming Millennium").

Chapter 16, High Tech/Low Tech, plays off two readings brimming with optimism about high-tech solutions to our problems (Bill Gates's "The Information Age" and Alvin and Heidi Toffler's "Getting Set for the Coming Millennium") with two pieces that celebrate, if anything, the struggles and joys of small triumphs and of generally getting along: the broad-ranging ruminations of poet-farmer-philosopher Wendell Berry, in "Feminism, the Body, and the Machine," and Tracy Baxter's personal account ("Into the Outdoors") of taking inner-city youth on an expedition into the bayous of southern Louisiana. The exchange started in these four pieces is intended to leave us with even more questions about just what fulfillment means for humans—issues taken up in the chapter-end Explorations that deal with the difficult matter of animal rights and the growth of what can be seen as the opposite of high-tech training: namely, wilderness education.

Chapter 17, "Utopia/Dystopia," also involves the pairing of readings. Christopher Columbus describes the Earthly Paradise, a place where the shape of the world is more in line with our desires; Tim Appelo evokes a dark vision of the future as presented in popular Hollywood films. There has always been an element of pessimism in utopian writings, playing reality off against a more glorious vision, while the word *utopia* itself was coined by Sir Thomas More from the Greek for "nowhere." *Dystopia*, meanwhile, refers to literary works that depict the future strictly as ruin and chaos—as in Ridley

Scott's classic 1982 film *Blade Runner,* a work that seems especially to haunt the thoughts of people who work with high technology. Explorations 17a and 17b take us into the strange worlds of intentional communities (where everything is supposed to be planned for the best) and of cults (where we assume that, despite intentions, everything works out for the worst).

Chapter 18, Global Concerns, looks at the extent to which U.S. culture, with its fascination with invention, gadgetry, and consumption, really does seem to be taking over the world. Obviously, whether this is for better or for worse remains the big unanswered question. High technology really does seem to be making us into one world (or, as Bob Wehling suggests, one huge market, in this case for Procter & Gamble products), sometimes at an alarmingly fast rate—and perhaps to the detriment at times of various local (and lower-tech) interests. These are the issues raised by the first three readings: David Ignatius's review of William Greider's take on the new global economy, Newt Gingrinch's vision of the future of the world as the United States writ large, and Vandana Shiva's take on how the women's movement and the ecology movement can be combined to resist what Shiva sees as destructive tendencies in global economic expansion.

What these three readings suggest is how difficult it is for many of us to imagine a future life other than one like the present, only better: that is, a life in which I have all that I want, and so does everyone else. Hence, it is often difficult to imagine anything else for the world writ large: We envision that the future of the planet and the human race is for everyone to become like us, especially to share our high level of material comfort, a key part of our high standard of living. (And who doesn't want more comfort?) Yet at the same time, it is difficult to imagine Spaceship Earth (Buckminster Fuller's term) not choking on its own gaseous fumes if the billions of people in India and China—to name just two countries—start to approach our current U.S. levels of consumption. Or can the automotive companies really design nonpolluting cars that run on sunshine? Must we then have a different vision, not for other people's future, but for our own? Might our survival, in other words, depend more than ever before not on our strength or even our intelligence (both dwarfed by our technology) but on the one power so far reserved strictly for humans alone: that of the imagination?

Explorations for Chapter 18 are quite varied, ranging from a look at the huge market for personal-care products and the battle over sneakers and worker rights to a look at efforts to reinvigorate barter, the oldest form of human exchange. Exploration 18f looks at images available on the Web of the simple life, and as such serves as a transition to the Thoreau reading, "The Wild," that serves as a coda to the entire anthology. Here in this final piece we have Thoreau's vision, not of utopia—life perfected by or for humans—but of life totally untouched by technology, hence life essentially without any human interaction except that of the author: that is, Thoreau's sense of what it is like to stare directly into the wild.

chapter 16

High Tech/Low Tech

READING QUESTIONS

- What arguments can be made for or against the notion that the government should support certain kinds of work (like running a small family farm) for the good of society as a whole, even if such work is not economically "efficient"?
- How important do you think it is that all or most manual labor be eliminated in the future—or, conversely, that manual labor be better rewarded and more highly regarded?
- How does each of the authors here—and how do you—feel about progress and the passage of time? Does the future promise to be better, worse, or more of the same—and, whatever your answer, do you see that as good or bad?

QUOTATIONS

Do not dwell in the past, do not dream of the future, concentrate the mind on the present moment.

— *Buddha*

Unspoilt by progress.

— *Advertising slogan for a British ale*

The sanitary and mechanical age we are now entering makes up for the mercy it grants to our sense of smell by the ferocity with which it assails our sense of hearing. As usual, what we call "Progress" is the exchange of one Nuisance for another Nuisance.

— *Havelock Ellis (1912), British psychologist, 1859–1939*

Vorsprung durch Technik. ("Progress through technology.")

— *Advertising slogan for Audi cars (1986)*

I have seen the future and it works.

— *American journalist Lincoln Steffens describing a visit to the new Soviet Union in 1919*

The development of society and culture depends upon a changing balance, maintained between those who innovate and those who conserve the status quo. Relentless, unchecked, and untested innovation would be a nightmare. . . . If repetition and rigidity are the dark side of the conservative coin, loyalty and stability are its bright side.

—*Judith Groch (1969), American author*

Change is inevitable, except from vending machines.

—*Anonymous*

The Information Age | *Bill Gates*

Bill Gates, the cofounder of Microsoft Corporation (along with the lesser-known Paul Allen, featured in Donald Katz's piece in Chapter 9), has come to epitomize the entire personal computer industry and hence to play an increasingly public role as the spokesperson for the benefits (to all the rest of us) for the computer technology he continues to shape.

At the *Visions* Web Site Links to *The Road Ahead,* Gates's 1995 book; to *Time* magazine's January 1997 cover story; to the official Bill Gates homepage at Microsoft.com.

READING PROMPT

A person's attitude about time—the relative values of the past, present, and future—is at the core of his or her attitude about technology: explain how Gates's belief (for better or worse) is clearly stamped in his forehead.

This is an exciting time in the Information Age. It is the very beginning. Almost everywhere I go, whether to speak to a group or to have dinner with friends, questions come up about how information technology will change our lives. People want to understand how it will make the future different. Will it make our lives better or worse.

I've already said I'm optimistic, and I'm optimistic about the impact of technology. It will enhance leisure time and enrich culture by expanding the distribution of information. It will help relieve pressures on urban areas by enabling individuals to work from home or remote-site offices. It will relieve pressure on natural resources because increasing numbers of products will be able to take the form of bits rather than of manufactured goods. It will give us more control over our lives and allow experiences and products to be custom tailored to our interests. Citizens of the information society will enjoy new opportunities for productivity, learning, and entertainment. Countries that move boldly and in concert with each other will enjoy economic rewards. Whole new markets will emerge, and a myriad of new opportunities for employment will be created.

When measured by decades, the economy is always in upheaval. For the past hundred years, every generation has found more efficient ways of getting work done, and the cumulative benefits have been enormous. The average person today enjoys a better life than the nobility did a few centuries ago. It would be great to have a king's land, but what about his lice? Medical advances alone have greatly increased life spans and improved standards of living.

In the first part of the twentieth century, Henry Ford *was* the automotive industry, but your car is superior to anything he ever drove. It's safer, more reliable, and it certainly has a better sound system. This pattern of improvement isn't going to change. Advancing productivity propels societies forward, and it's only a matter of time before the average person in a developed country will be "richer" in many ways than anyone is today.

Just because I am optimistic doesn't mean I don't have concerns about what is going to happen to all of us. Major changes always involve tradeoffs, and the benefits of the information society will carry costs. Societies are going to be asked to make hard choices about the universal availability of technology, investment in education, regulation, and the balance between individual privacy and community security. We'll confront tough new problems, only a few of which we can foresee. In some business sectors, dislocations will create a need for worker retraining. The availability of virtually free communications and computing will alter the relationships of nations and of socioeconomic groups within nations. The power and versatility of digital technology will raise new concerns about individual privacy, commercial confidentiality, and national security. There are equity issues that will have to be addressed because the information society should serve all of its citizens, not only the technically sophisticated and economically privileged. I don't necessarily have the solutions to the many issues and problems we'll confront, but . . . now is a good time for a broad debate to begin. . . .

The presence of advanced communications systems promises to make countries more alike and reduce the importance of national boundaries. The fax machine, the portable videocamera, and Cable News Network are among the forces that brought about the end of communist regimes and the Cold War because they enabled news to pass both ways through the Iron Curtain. Most sites on the World Wide Web are in English so far, which confers economic and entertainment benefits on people around the world who speak English. English-speaking people will enjoy this advantage until a great deal more content is posted in a variety of languages—or until software does a first-rate job of translating text on the fly.

The new access to information can draw people together by increasing their understanding of other cultures. But commercial satellite broadcasts to countries such as China and Iran offer citizens glimpses of the outside world that are not necessarily sanctioned by their governments. Some governments are afraid that such exposure will cause discontent and worse, a "revolution of expectations" when disenfranchised people get enough information about another lifestyle to contrast it with their own. Within individual societies, the balance between traditional and modern experiences is bound to shift as people use the network to expose themselves to a greater range of possibilities. Some cultures may feel under assault as people pay greater attention to global issues and cultures and less to their traditional local ones.

"The fact that the same ad can appeal to someone in a New York apartment and on an Iowa farm and in an African village does not prove these situations are alike," commented Bill McKibben, a critic of what he sees as television's tendency to override local diversity with homogenized common experiences. "It is merely evidence that the people living in them have a few feelings in common, and it is these barest, most minimal commonalities that are the content of the global village."

American popular culture is so potent that outside the United States some countries now try to ration exposure to it. They hope to guarantee the viability of domestic content producers by permitting only a certain number of hours of foreign television to be aired each week, for instance. In Europe, the availability of satellite and cable-delivered programming has made it harder for governments to control what people watch. The Internet is going to break down boundaries and may promote a world culture, or at least a greater sharing of cultural activities and values. But the network will also make it easy for people who are deeply involved in their own ethnic communities at home or abroad to reach out to other people who share their preoccupations no matter where they are. This may strengthen cultural diversity and counter the tendency toward a single, homogenized world culture. It's hard to predict what the net effect will be—a strengthening or a weakening of local cultural values.

Feminism, the Body, and the Machine

Wendell Berry

Wendell Berry is poet, novelist, essayist, and moral critic, the author of some thirty two books, all representing a vision of modern life that has been shaped almost entirely by his own personal relationship with the rough farmland of northern Kentucky.

At the *Visions* Web Site Links to more by and about Berry, including a short excerpt from his 1973 book, *The Unsettling of America: Culture and Agriculture* (San Francisco: Sierra Club Books) and a poem to an editor at the business magazine *Inc.*

READING PROMPT

Explain how Berry is difficult to pin down politically. Which of his ideas would traditionally be considered "conservative," which "liberal"?

Some time ago *Harper's* reprinted a short essay of mine in which I gave some of my reasons for refusing to buy a computer. Until that time, the vast numbers of people who disagree with my writings had mostly ignored them. An unusual number of people, however, neglected to ignore my insensitivity to the wonders of computer enhancement. Some of us, it seems, would be better off if we would just realize that this is already the best of all possible worlds, and is going to get even better if we will just buy the right equipment.

Harper's published only five of the letters the editors received in response to my essay, and they published only negative letters. But of the twenty letters received by the *Harper's* editors, who forwarded copies to me, three were favorable. This I look upon as extremely gratifying. If these letters may be taken as a fair sample, then one in seven of *Harper's* readers agreed with me. If I had guessed beforehand, I would have guessed that my supporters would have been fewer than one in a thousand. And so I suppose, after further reflection, that my surprise at the intensity of the attacks on me is mistaken. There are more of us than I thought. Maybe there is even a "significant number" of us.

Only one of the negative letters seemed to me to have much intelligence in it. That one was from R. N. Neff of Arlington, Virginia, who scored a direct hit: "Not to be obtuse, but willing to bare my illiterate soul for all to see, is there indeed a 'work demonstrably better than Dante's' . . . which was written on a Royal standard typewriter?" I like this retort so well that I am tempted to count it a favorable response, raising the total to four. The rest of the negative replies, like the five published ones, were more feeling than intelligent. Some of them, indeed, might be fairly described as exclamatory.

One of the letter writers described me as "a fool" and "doubly a fool," but fortunately misspelled my name, leaving me a speck of hope that I am not the "Wendell Barry" he was talking about. Two others accused me of self-righteousness, by which they seem to have meant that they think they are righter than I think I am. And another accused me of being more concerned about my own moral purity than with "any ecological effect," thereby making the sort of razor-sharp philosophical distinction that could cause a person to be elected president.

But most of my attackers deal in feelings either feminist or technological, or both. The feelings expressed seem to be representative of what the state of public feeling currently permits to be felt, and of what public rhetoric currently permits to be said. The feelings, that is, are similar enough, from one letter to another, to be thought representative, and as representative letters they have an interest greater than the quarrel that occasioned them.

Without exception, the feminist letters accuse me of exploiting my wife, and they do not scruple to allow the most insulting implications of their indictment to fall upon my wife. They fail entirely to see that my essay does not give any support to their accusation—or if they see it, they do not care: My essay, in fact, does not characterize my wife beyond saying that she types my manuscripts and tells me what she thinks about them. It does not say what her motives are, how much work she does, or whether or how she is paid. Aside from saying that she is my wife and that I value the help she gives me with my work, it says nothing about our marriage. It says nothing about our economy.

There is no way, then, to escape the conclusion that my wife and I are subjected in these letters to a condemnation by category. My offense is that I am a man who receives some help from his wife; my wife's offense is that she is a woman who does some work for her husband—which work, according to her critics and mine, makes her a drudge, exploited by a conventional subservience. And my detractors have, as I say, no evidence to support any of this. Their accusation rests on a syllogism of the flimsiest sort: my wife helps me in my work, some wives who have helped their husbands in their work have been exploited, therefore my wife is exploited.

This, of course, outrages justice to about the same extent that it insults intelligence. Any respectable system of justice exists in part as a protection against such accusations. In a just society nobody is expected to plead guilty to a general indictment, because in a just society nobody can be convicted on a general indictment. What is required for a just conviction is a particular accusation that can be *proved.* My accusers have made no such accusation against me.

That feminists or any other advocates of human liberty and dignity should resort to insult and injustice is regrettable. It is equally regrettable that all of the feminist attacks on my essay implicitly deny the validity of two decent and probably necessary possibilities: marriage as a state of mutual help, and the household as an economy.

Marriage, in what is evidently its most popular version, is now on the one hand an intimate "relationship" involving (ideally) two successful careerists in the same bed, and on the other hand a sort of private political system in which rights and interests must be constantly asserted and defended. Marriage, in other words, has now taken the form of divorce: a prolonged and impassioned negotiation as to how things shall be divided. During their understandably temporary association, the "married" couple will typically consume a large quantity of merchandise and a large portion of each other.

The modern household is the place where the consumptive couple do their consuming. Nothing productive is done there. Such work as is done there is done at the expense of the resident couple or family, and to the profit of suppliers of energy and household technology. For entertainment, the inmates consume television or purchase other consumable diversion elsewhere.

There are, however, still some married couples who understand themselves as belonging to their marriage, to each other, and to their children. What they have they have in common, and so, to them, helping each other does not seem merely to damage their ability to compete against each other. To them, "mine" is not so powerful or necessary a pronoun as "ours."

This sort of marriage usually has at its heart a household that is to some extent productive. The couple, that is, makes around itself a household economy that involves the work of both wife and husband, that gives them a measure of economic independence and self-protection, a measure of self-employment, a measure of freedom, as well as a common ground and a common satisfaction.

Such a household economy may employ the disciplines and skills of housewifery, of carpentry and other trades of building and maintenance, of gardening and other branches of subsistence agriculture, and even of woodlot management and woodcutting. It may also involve a "cottage industry" of some kind, such as a small literary enterprise.

It is obvious how much skill and industry either partner may put into such a household and what a good economic result such work may have, and yet it is a kind of work now frequently held in contempt. Men in general were the first to hold it in contempt as they departed from it for the sake of the professional salary or the hourly wage, and now it is held in contempt by such feminists as those who attacked my essay. Thus farm wives who help to run the kind of household economy that I have described are apt to be asked by feminists, and with great condescension, "But what do you *do*?" By this they invariably mean that there is something better to do than to make one's marriage and household, and by better they invariably mean "employment outside the home."

I know that I am in dangerous territory, and so I had better be plain: what I have to say about marriage and household I mean to apply to men as much as to women. I do not believe that there is anything better to do than to make one's marriage and household, whether one is a man or a woman. I do not believe that "employment outside the home" is as valuable or important or satisfying as employment at home, for either men or women. It is clear to me from my experience as a teacher, for example, that children need an ordinary daily association with *both* parents. They need to see their parents at work; they need, at first, to play at the work they see their parents doing, and then they need to work with their parents. It does not matter so much that this working together should be what is called "quality time," but it matters a great deal that the work done should have the dignity of economic value.

I should say too that I understand how fortunate I have been in being able to do an appreciable part of my work at home. I know that in many marriages both husband and wife are now finding it necessary to work away from home. This issue, of course, is troubled by the question of what is meant by "necessary," but it is true that a family living that not so long ago was ordinarily supplied by one job now routinely requires two or more. My interest is not to quarrel with individuals, men or women, who work away from home, but rather to ask why we should consider this general working away from home to be a desirable state of things, either for people or for marriage, for our society or for our country.

If I had written in my essay that my wife worked as a typist and editor for a publisher, doing the same work that she does for me, no feminists, I daresay, would have written to *Harper's* to attack me for exploiting her—even though, for all they knew, I might have forced her to do such work in order to keep me in gambling money. It would have been assumed as a matter of course that if she had a job away from home she was a "liberated woman," possessed of a dignity that no home could confer upon her.

As I have said before, I understand that one cannot construct an adequate public defense of a private life. Anything that I might say here about my marriage would be immediately (and rightly) suspect on the ground that it would be only my testimony. But for the sake of argument, let us suppose that whatever work my wife does, as a member of our marriage and household, she does both as a full economic partner and as her own boss, and let us suppose that the economy we have is adequate to our needs. Why, granting that supposition, should anyone assume that my wife would increase her freedom or dignity or satisfaction by becoming the employee of a boss, who would be in turn also a corporate underling and in no sense a partner?

Why would any woman who would refuse, properly, to take the marital vow of obedience (on the ground, presumably, that subservience to a mere human being is beneath human dignity) then regard as "liberating" a job that puts her under the authority of a boss (man or woman) whose authority specifically requires and expects obedience? It is easy enough to see why women came to object to the role of Blondie, a mostly decorative custodian of a degraded, consumptive modern household, preoccupied with clothes, shopping, gossip, and outwitting her husband. But are we to assume that one may fittingly cease to be Blondie by becoming Dagwood? Is the life of a corporate underling—even acknowledging that corporate underlings are well paid—an acceptable end to our quest for human dignity and worth? It is clear enough by now that one does not cease to be an underling by reaching "the top." Corporate life is composed only of lower underlings and higher underlings. This is invariably revealed when the time comes for accepting responsibility for something unpleasant, such as the Exxon fiasco in Prince William Sound, for which certain lower underlings are blamed but no higher underling is responsible. The underlings at the top, like telephone operators, have authority and power, but no responsibility.

And the oppressiveness of some of this office work defies belief. Edward Mendelson (in the *New Republic,* February 22, 1988) speaks of "the office worker whose computer keystrokes are monitored by the central computer in the personnel office, and who will be fired if the keystrokes-per-minute figure doesn't match the corporate quota." (Mr. Mendelson does not say what form of drudgery this worker is saved from.) And what are we to say of the diversely skilled country housewife who now bores the same six holes day after day on an assembly line? What higher form of womanhood or humanity is she evolving toward?

How, I am asking, can women improve themselves by submitting to the same specialization, degradation, trivialization, and tyrannization of work that men have submitted to? And that question is made legitimate by another: How have men improved themselves by submitting to it? The answer is that men have not, and women cannot, improve themselves by submitting to it.

Women have complained, justly, about the behavior of "macho" men. But despite their he-man pretensions and their captivation by masculine heroes of sports, war, and the Old West, most men are now entirely accustomed to obeying and currying the favor of their bosses. Because of this, of course, they hate their jobs—they mutter, "Thank God it's Friday" and "Pretty good for Monday"—but they do as they are told. They are more compliant than most housewives have been. Their characters combine feudal submissiveness with modern helplessness. They have accepted almost without protest, and often with relief, their dispossession of any usable property and, with that, their loss of economic independence and their consequent subordination to bosses. They have submitted to the destruction of the household economy and thus of the household, to the loss of home employment and self-employment, to the disintegration of their families and communities, to the desecration and pillage of their country, and they have continued abjectly to believe, obey, and vote for the people who have most eagerly abetted this ruin and who have most profited from it. These men, moreover, are helpless to do anything for themselves or anyone else without money, and so for money they do whatever they are told. They know that their ability to be useful is precisely defined by their willingness to be somebody else's tool. Is it any wonder that they talk tough and worship athletes and cowboys? Is it any wonder that some of them are violent?

It is clear that women cannot justly be excluded from the daily fracas by which the industrial economy divides the spoils of society and nature, but their inclusion is a poor justice and no reason for applause. The enterprise is as devastating with women in it as it was before. There is no sign that women are exerting a "civilizing influence" upon it. To have an equal part in our juggernaut of national vandalism is to be a vandal. To call this vandalism "liberation" is to prolong, and even ratify, a dangerous confusion that was once principally masculine.

A broader, deeper criticism is necessary. The problem is not just the exploitation of women by men. A greater problem is that women and men alike are consenting to an economy that exploits women and men and everything else.

Another decent possibility my critics implicitly deny is that of work as a gift. Not one of them supposed that my wife may be a consulting engineer who helps me in her spare time out of the goodness of her heart; instead they suppose that she is "a household drudge." But what appears to infuriate them the most is their supposition that she works for nothing. They assume—and this is the orthodox assumption of the industrial economy—that the only help worth giving is not given at all, but sold. Love, friendship, neighborliness, compassion, duty—what are they? We are realists. We will be most happy to receive your check.

My wish simply is to live my life as fully as I can. In both our work and our leisure, I think, we should be so employed. And in our time this means that we must save ourselves from the products that we are asked to buy in order, ultimately, to replace ourselves.

The danger most immediately to be feared in "technological progress" is the degradation and obsolescence of the body. Implicit in the technological revolution from the beginning has been a new version of an old dualism, one always destructive, and now more destructive than ever. For many centuries there have been people who looked upon the body, as upon the natural world, as an encumbrance of the soul, and so have hated the body, as they have hated the natural world, and longed to be free of it. They have seen the body as intolerably imperfect by spiritual standards. More recently, since the beginning of the technological revolution, more and more people have looked upon the body, along with the rest of the natural creation, as intolerably imperfect by mechanical standards. They see the body as an encumbrance of the mind—the mind, that is, as reduced to a set of mechanical ideas that can be implemented in machines—and so they hate it and long to be free of it. The body has limits that the machine does not have; therefore, remove the body from the machine so that the machine can continue as an unlimited idea.

It is odd that simply because of its "sexual freedom" our time should be considered extraordinarily physical. In fact, our "sexual revolution" is mostly an industrial phenomenon, in which the body is used as an idea of pleasure or a pleasure machine with the aim of "freeing" natural pleasure from natural consequence. Like any other industrial enterprise, industrial sexuality seeks to conquer nature by exploiting it and ignoring the consequences, by denying any connection between nature and spirit or body and soul, and by evading social responsibility. The spiritual, physical, and economic costs of this "freedom" are immense, and are characteristically belittled or ignored. The diseases of sexual irresponsibility are regarded as a technological problem and an affront to liberty. Industrial sex, characteristically, establishes its freeness and goodness by an industrial accounting, dutifully toting up numbers of "sexual partners," orgasms, and so on, with the inevitable industrial implication that the body is somehow a limit on the idea of sex, which will be a great deal more abundant as soon as it can be done by robots.

This hatred of the body and of the body's life in the natural world, always inherent in the technological revolution (and sometimes explicitly and vengefully so), is of concern to an artist because art, like sexual love, is of the body. Like sexual love, art is of the mind and spirit also, but it is made with the body and it appeals to the senses. To reduce or shortcut the intimacy of the body's involvement in the making of a work of art (that is, of any artifice, anything made by art) inevitably risks reducing the work of art and the art itself. In addition to the reasons I gave previously, which I still believe are good reasons, I am not going to use a computer because I don't want to diminish or distort my bodily involvement in my work. I don't want to deny myself the pleasure of bodily involvement in my work, for that pleasure seems to me to be the sign of an indispensable integrity.

At first glance, writing may seem not nearly so much an art of the body as, say, dancing or gardening or carpentry. And yet language is the most intimately physical of all the artistic means. We have it palpably in our mouths; it is our *langue,* our tongue. Writing it, we shape it with our hands. Reading aloud what we have written—as we must do, if we are writing carefully—our language passes in at the eyes, out at the mouth, in at the ears; the words are immersed and steeped in the senses of the body before they make sense in the mind. They cannot make sense in the mind until they have made sense in the body. Does shaping one's words with one's own hand impart character and quality to them, as does speaking them with one's own tongue to the satisfaction of one's own ear? There is no way to prove that it does. On the other hand, there is no way to prove that it does not, and I believe that it does.

The act of writing language down is not so insistently tangible an act as the act of building a house or playing the violin. But to the extent that it is tangible, I love the tangibility of it. The computer apologists, it seems to me, have greatly underrated the value of the handwritten manuscript as an artifact. I don't mean that a writer should be a fine calligrapher and write for exhibition, but rather that handwriting has a valuable influence on the work so written. I am certainly no calligrapher, but my handwritten pages have a homemade, handmade look to them that both pleases me in itself and suggests the possibility of ready correction. It looks hospitable to improvement. As the longhand is transformed into typescript and then into galley proofs and the printed page, it seems increasingly to resist improvement. More and more spunk is required to mar the clean, final-looking lines of type. I have the notion—again not provable—that the longer I keep a piece of work in longhand, the better it will be.

To me, also, there is a significant difference between ready correction and easy correction. Much is made of the ease of correction in computer work, owing to the insubstantiality of the light-image on the screen; one presses a button and the old version disappears, to be replaced by the new. But because of the substantiality of paper and the consequent difficulty involved, one does not handwrite or typewrite a new page every time a correction is made.

A handwritten or typewritten page therefore is usually to some degree a palimpsest; it contains parts and relics of its own history—erasures, passages crossed out, interlineations—suggesting that there is something to go back to as well as something to go forward to. The bright-text on the computer screen, by contrast, is an artifact typical of what can only be called the industrial present, a present absolute. A computer destroys the sense of historical succession, just as do other forms of mechanization. The well-crafted table or cabinet embodies the memory of (because it embodies respect for) the tree it was made of and the forest in which the tree stood. The work of certain potters embodies the memory that the clay was dug from the earth. Certain farms contain hospitably the remnants and reminders of the forest or prairie that preceded them. It is possible even for towns and cities to remember farms and forests or prairies. All good human work remembers its history. The best writing, even when printed, is full of intimations that it is the present version of earlier versions of itself, and that its maker inherited the work and the ways of earlier makers. It thus keeps, even in print, a suggestion of the quality of the handwritten page; it is a palimpsest.

Something of this undoubtedly carries over into industrial products. The plastic Clorox jug has the shape and a loop for the forefinger that recalls the stoneware jug that went before it. But something vital is missing. It embodies no memory of its source or sources in the earth or of any human hand involved in its shaping. Or look at a large factory or a power plant or an airport, and see if you can imagine—even if you know—what was there before: In such things the materials of the world have entered a kind of orphanhood.

It would be uncharitable and foolish of me to suggest that nothing good will ever be written on a computer. Some of my best friends have computers. I have only said that a computer cannot help you to write better, and I stand by that. (In fact, I know a publisher who says that under the influence of computers—or of the immaculate copy that computers produce—many writers are now writing worse.) But I do say that in using computers writers are flirting with a radical separation of mind and body, the elimination of the work of the body from the work of the mind. The text on the computer screen, and the computer printout too, has a sterile, untouched, factory-made look, like that of a plastic whistle or a new car. The body does not do work like that. The body characterizes everything it touches. What it makes it traces over with the marks of its pulses and breathings, its excitements, hesitations, flaws, and mistakes. On its good work, it leaves the marks of skill, care, and love persisting through hesitations, flaws, and mistakes. And to those of us who love and honor the life of the body in this world, these marks are precious things, necessities of life.

But writing is of the body in yet another way. It is preeminently a walker's art. It can be done on foot and at large. The beauty of its traditional equipment is simplicity. And cheapness. Going off to the woods, I take a pencil and some paper (any paper—a small notebook, an old envelope, a piece of a feed sack), and I am as well equipped for my work as the president of IBM. I am also free, for the time being at least, of everything that IBM is hooked to. My thoughts will not be coming to me from the power structure of the power grid, but from another direction and way entirely. My mind is free to go with my feet.

Some of my critics were happy to say that my refusal to use a computer would not do any good. I have argued, and am convinced, that it will at least do me some good, and that it may involve me in the preservation of some cultural goods. But what they meant was real, practical, public good. They meant that the materials and energy I save by not buying a computer will not be "significant." They meant that no individual's restraint in the use of technology or energy will be "significant." That is true.

But each one of us, by "insignificant" individual abuse of the world, contributes to a general abuse that is devastating. And if I were one of thousands or millions of people who could afford a piece of equipment, even one for which they had a conceivable "need," and yet did not buy it, that would be "significant." Why, then, should I hesitate for even a moment to be one, even the first one, of that "significant" number? Thoreau gave the definitive reply to the folly of "significant numbers" a long time ago: Why should anybody wait to do what is right until everybody does it? It is not "significant" to love your own children or to eat your own dinner, either. But normal humans will not wait to love or eat until it is mandated by an act of Congress.

One of my correspondents asked where one is to draw the line. That question returns me to the bewilderment I mentioned earlier: I am unsure where the line ought to be drawn, or how to draw it. But it is an intelligent question, worth losing some sleep over.

I know how to draw the line only where it is easy to draw. It is easy—it is even a luxury—to deny oneself the use of a television set, and I zealously practice that form of self-denial. Every time I see television (at other people's houses), I am more inclined to congratulate myself on my deprivation. I have no doubt, as I have said, that I am better off without a computer. I joyfully deny myself a motorboat, a camping van, an off-road vehicle, and every other kind of recreational machinery. I have, and want, no "second home." I suffer very comfortably the lack of colas, TV dinners, and other counterfeit foods and beverages.

I am, however, still in bondage to the automobile industry and the energy companies, which have nothing to recommend them except our dependence on them. I still fly on airplanes, which have nothing to recommend them but speed; they are inconvenient, uncomfortable, undependable, ugly, stinky, and scary. I still cut my wood with a chainsaw, which has nothing to recommend it but speed, and has all the faults of an airplane, except it does not fly.

It is plain to me that the line ought to be drawn without fail wherever it can be drawn easily. And it ought to be easy (though many do not find it so) to refuse to buy what one does not need. If you are already solving your problem with the equipment you have—a pencil, say—why solve it with something more expensive and more damaging? If you don't have a problem, why pay for a solution? If you love the freedom and elegance of simple tools, why encumber yourself with something complicated?

And yet, if we are ever again to have a world fit and pleasant for little children, we are surely going to have to draw the line where it is not easily drawn. We are going to have to learn to give up things that we have learned (in only a few years, after all) to "need." I am not an optimist; I am afraid that I won't live long enough to escape my bondage to the machines. Nevertheless, on every day left to me I will search my mind and circumstances for the means of escape. And I am not without hope. I knew a man who, in the age of chainsaws, went right on cutting his wood with a handsaw and an axe. He was a healthier and a saner man than I am. I shall let his memory trouble my thoughts.

Getting Set for the Coming Millennium

Alvin Toffler and Heidi Toffler

The Tofflers practically invented the term *futurist* with the publication of *Future Shock* in 1970. The Tofflers' most recent thinking is most easily accessed through online interviews, including two with *Micro-Times*, "Surfing the Third Wave" and "Perspectives for a Changing World." There is also a Hotwired interview, "Shock Wave (Anti) Warrior."

Almost by definition, futurists are optimists, and increasingly the Tofflers have found themselves aligned with political interests trying to free private businesses from government controls so as to allow the expansion of technologies and goods in the United States and throughout the world. Such a political position is in many ways deeply radical (in the sense of advocating change at the root), in that it evinces contempt for many traditional practices while expressing near unbounded optimism about the society that unconstrained technology will evolve into; even so, it is still almost always labeled *conservative.*

At the *Visions* Web Site Links to sites celebrating the steam locomotive—that rare aspect of industrial culture still capable of arousing nostalgia.

READING PROMPT

The Tofflers are, if nothing else, generalizers on a grand scale. What do you find most and least appealing about their approach?

A new civilization is emerging, but many people are trying to resuscitate the old industrial society instead of easing the transition to the new.

American politics is presented by our media as a continuing gladiatorial contest between two political parties. Yet Americans are increasingly alienated, bored, and angry at both the media and the politicians. Party politics seem to most people a kind of shadow-play, insincere, costly, and corrupt. Increasingly, they ask: Does it matter who wins?

The answer is Yes—but not for the reasons we are told.

In *The Third Wave,* we wrote:

> The most important political development of our time is the emergence in our midst of two basic camps, one committed to Second Wave civilization, the other to Third. One is tenaciously dedicated to preserving the core institutions of industrial mass society—the nuclear family, the mass education system, the giant corporation, the mass trade union, the centralized nation-state, and the politics of pseudorepresentative government. The other recognizes that today's most urgent problems, from energy, war, and poverty to ecological degradation and the breakdown of familial relationships, can no longer be solved within the framework of an industrial civilization.

LOBBYING FOR THE PAST

The reason the public does not, even now, recognize the crucial importance of this cleavage is that much of what the press reports on is, in fact, the politics-as-usual conflict between different Second Wave groups over the spoils of the old system. But despite their differences, these groups quickly coalesce to oppose Third Wave initiatives.

This is the reason why, in 1984, when Gary Hart campaigned for the Democratic Party presidential nomination and won the New Hampshire primary by calling for "new thinking," the Second Wave barons in the Democratic Party united to stop him and nominated solid, safe, Second Wave Walter Mondale instead.

It is why, more recently, Second Wave Naderites and Buchananites found common cause against the North American Free Trade Agreement (NAFTA).

It is why, when Congress passed an "infrastructure bill" in 1991, $150 billion was allocated to roads, highways, bridges, and fixing potholes—providing profits to Second Wave companies and jobs for Second Wave unions—while a trivial $1 billion was allocated to help build the much-touted electronic superhighway. Necessary as they may be, roads and highways are part of the Second Wave infrastructure; digital networks are the heart of the Third

Wave infrastructure. The point here is not whether or not the government should subsidize the digital network, but the imbalance of Second and Third Wave forces in Washington.

The imbalance is why Vice President Gore—with one toe wet in the Third Wave—has been unable, despite his efforts, to "re-invent" the government along Third Wave lines. Centralized bureaucracy is the quintessential form of organization of Second Wave societies. Even as advanced corporations, driven by competition, are desperately trying to dismantle their bureaucracies and invent new Third Wave forms of management, government agencies, blocked by Second Wave civil service unions, have managed to stay largely unreformed, un-re-engineered, un-re-invented—to retain, in short, their Second Wave structure.

Second Wave elites fight to retain or reinstate an unsustainable past because they gained wealth and power from applying Second Wave principles, and the shift to a new way of life challenges that wealth and power. But it isn't only elites. Millions of middle-class and poor Americans also resist the transition to the Third Wave out of an often justified fear that they will be left behind, lose their jobs, and slide further down the economic and social slope.

To understand the vast inertial power of Second Wave forces in America, however, we need to look beyond the old muscle-based industries and their workers and unions. The Second Wave sector is backed up by those elements of Wall Street that service it. It is further supported by intellectuals and academics, often tenured, who live off grants from the foundations, trade associations, and lobbies that serve it.

Their task is to collect supportive data and hammer out the ideological argument and slogans used by Second Wave forces—for example, the idea that the information-intensive service industries are "unproductive" or that service workers are doomed to "sling hamburgers" or that the economy must revolve around manufacturing.

SECOND WAVE RHETORIC

From Hart in the '80s to Gore in the '90s, the party's core constituencies make it impossible for the Democratic Party to act on what its most forward thinking leaders say. The party thus finds itself still trapped by its anachronistic blue-collar image of reality.

The failure of the Democrats to make themselves the party of the future (as, indeed, it once was) throws the door wide open for their adversaries. The Republicans, less rooted in the old industrial Northeast, thus have an opportunity to position themselves as the party of the Third Wave—although their recent presidents have signally failed to seize this opportunity. And the Republicans, too, rely on knee-jerk Second Wave rhetoric.

Republicans are basically right when they call for broad scale deregulation because businesses now need all the flexibility possible to survive the global competition. The Republicans are basically right in calling for privatization of government operations because governments, lacking competition, don't generally run things well. The Republicans are basically right when they urge us to take maximum advantage of the dynamism and creativity that market economies make possible. But they, too, remain prisoners of Second Wave economics. For example, even the free-market economists on whom Republicans rely have failed, as yet, to come to terms with the new role and inexhaustibility of knowledge.

Moreover, Republicans tend to play down the potentially immense social dislocations that are likely to flow from any change as profound as the Third Wave. For example, as skills become obsolete overnight, large numbers of the middle class, including highly trained people, may well find themselves thrown out of work. California defense scientists and engineers are a chastening case in point. Free-marketism and trickle-downism twisted into rigid theological dogma are an inadequate response. A party facing the future should be warning of problems to come and suggesting preventive change. For example, today's media revolution will bring enormous benefits to the emerging Third Wave economy. But TV shopping and other electronic services could well slash the number of entry-level jobs in the traditional retail sector, precisely the place under-educated young people can get their start.

If free markets and democracy are to survive the great and turbulent transitions to come, politics must become anticipatory and preventive. Yet asking our political parties to think beyond the next election is hard and thankless work.

Instead, both parties are busy mainlining nostalgia into their constituents' veins. The Democrats, for example, until recent years, spoke of "reindustrializing" or "restoring" American industry to its period of greatness in the 1950s (in reality an impossible return to the Second Wave mass-production economy). The Republicans, meanwhile, in a kind of mirror image, appeal to nostalgia in their rhetoric about culture and values, as though one could return to the values and morality of the 1950s—a time before universal television, before the birth-control pill, before commercial jet aviation, before satellites and home computers—without also returning to the mass industrial society of the Second Wave.

TOMORROW'S CONSTITUENCY

However powerful Second Wave forces may seem today, their future is diminishing. At the start of the industrial era, First Wave forces dominated society and political life. Rural elites seemed destined to dominate forever. Yet they did not. Had they, in fact, done so, the Industrial Revolution would never have succeeded in transforming the world.

Today the world is changing again, and the overwhelming majority of Americans are neither farmers nor factory workers. They are, instead, engaged in one or another form of knowledge work. America's fastest growing and most important industries are information-intensive. The Third Wave sector includes not only high-flying computer and electronics firms or biotech start-ups. It embraces advanced information-driven manufacturing in every industry. It includes the increasingly data-drenched services—finance, software, entertainment, the media, advanced communications, medical services, consulting, training and education—in short, all the industries based on mind-work rather than muscle-work. The people in this sector will soon be the dominant constituency in American politics.

Unlike the "masses" during the Industrial Age, the rising Third Wave constituency is highly diverse. It is de-massified. It is composed of individuals who prize their differences. Its very heterogeneity contributes to its lack of political awareness. It is far harder to unify than the masses of the past.

Thus the Third Wave constituency has yet to develop its own think tanks and political ideology. It has not systematically marshaled support from academia. Its various associations and lobbies in Washington are still comparatively new and less well connected. And except for one issue—NAFTA—in which the Second Wavers were defeated, the new constituency has few notches on its legislative belt.

KEY ISSUES FOR THIRD WAVERS

Yet there are key issues on which this broad constituency-to-come can agree. To start with: Liberation. Liberation from all the old Second Wave rules, regulations, taxes, and laws laid in place to serve the smokestack barons and bureaucrats of the past. These arrangements, no doubt sensible when Second Wave industry was the heart of the American economy, today obstruct Third Wave development.

For example, depreciation tax schedules, lobbied into being by the old manufacturing interests, presuppose that machines and products last for many years. Yet in the fast-changing high-tech industries, and particularly the computer industry, their usefulness is measured in months or weeks. The result is a tax bias against high tech. Research and development deductions, too, favor big old, Second Wave companies over the dynamic start-ups on which the Third Wave sector depends. The tax treatment of intangibles means that a company with a lot of obsolete sewing machines may well be favored over a software firm that has very little in the way of physical assets. Yet changing such rules will take a bitter political fight against the Second Wave firms that benefit from them.

Companies in the Third Wave sector have special characteristics. They tend to be young—both in age and in the age of their work force. Work units in them tend to be small, compared with those in Second Wave firms. They tend to invest more than average in research and development, training, education, and human resources. Ferocious competition forces them to innovate continuously—and that means short product life cycles and often implies a rapid turnover of people, too, and administrative practices. Their key assets are symbols inside the skulls of their people. Should these firms and industries be expected to play the game according to rules that penalize them for precisely these characteristics?

Much of the Third Wave sector is engaged in providing a dazzling, ever-changing array of services. Instead of decrying the rise of the service sector, continually attacking it as a source of low productivity, low wages, and low performance, shouldn't it be expressly supported and expanded—or at least freed of old shackles? America needs more, not less, service-sector employment to improve the quality of life of its people. That means jobs for everyone from the electronics repairman and the recycler to health-care providers, people to help the elderly, police, firefighters, and—yes—even jobs for child care and domestic workers, desperately needed in millions of two-job homes. A Third Wave economic policy should not pick "winners and losers," but it should clear away the obstacles to the professionalization and development of the services needed to make life in America less stressed out, frustrating, and impersonal. But no political party as yet has even begun to think this way.

Despite this political lag, the Third Wave constituency is growing in power every day. It increasingly expresses itself outside the conventional political parties because neither party has so far noticed its existence. Thus it is Third Wavers who fill the ranks of the ever more numerous and potent grassroots organizations around the country. It is Third Wavers who dominate the new electronic communities springing up around the Internet. And it is the same people who are busy de-massifying the Second Wave media and creating an interactive alternative to it. Traditional party politicians who ignore these new realities will be swept aside like the Members of Parliament in nineteenth-century England who imagined their rural "rotten borough" seats in Parliament were permanently secure.

The Third Wave forces in America have yet to find their voice. The political party that gives it to them will dominate the American future. And when that happens, a new, dramatically different America will rise from the ruins of the late twentieth century.

PRINCIPLES FOR A THIRD WAVE AGENDA

We are living through the birth pangs of a new civilization whose institutions are not yet in place. A fundamental skill needed by policy makers, politicians, and politically active citizens today—if they really want to know what they are doing—is, therefore, the ability to distinguish between proposals designed to keep the tottering Second Wave system on life-support from those that ease the growth of the next, Third Wave civilization.

Here are some ways to tell:

1. Does It Resemble a Factory?

The factory became the central symbol of industrial society. It became, in fact, a model for most other Second Wave institutions. Yet the factory, as we have known it, is fading into the past. Factories embody such principles as standardization, centralization, maximization, concentration, and bureaucratization. Third Wave production is post-factory production based on new principles. It occurs in facilities that bear little resemblance to factories. In fact, an increasing amount is done in homes and offices, cars and planes.

The easiest and quickest way to spot a Second Wave proposal, whether in Congress or in a corporation, is to see whether it is still, consciously or not, based on the factory model.

America's schools, for example, still operate like factories, subjecting the raw material (children) to standardized instruction and routine inspection. A question to ask of any proposed educational innovation is simply this: Is it intended to make the factory run more efficiently—or is it designed, as it should be, to get rid of the factory system and replace it with individualized, customized education? A similar question could be asked of health legislation, of welfare legislation, and of every proposal to reorganize the federal bureaucracy. America needs new institutions built on post-bureaucratic, post-factory models.

2. Does It Massify Society?

People who ran those factories in the brute-force economy of the past liked large numbers of predictable, interchangeable, don't-ask-why workers for their assembly lines. And as mass production, mass distribution, mass education, mass media, and mass entertainment spread through the society, the Second Wave also created the "masses."

Third Wave economies, by contrast, will require (and tend to reward) a radically different kind of worker—one who thinks, questions, innovates, and takes entrepreneurial risk. Workers who are not easily interchangeable. Put differently, it will favor individuality (which is not necessarily the same as individualism).

The new brain-force economy tends to generate social diversity. Computerized, customized production makes possible highly diverse material lifestyles. Just check the local Wal-Mart with its 110,000 different products. Or check the wide choice of coffees now offered by Starbucks, as against those sold in America only a few years ago. But it isn't just things.

Much more important, the Third Wave also de-massifies culture, values, and morality. De-massified media carry many different, often competing, messages into the culture. There are not only more different kinds of work but also more different kinds of leisure, styles of art, political movements. There are also more different religious belief systems. And, in multiethnic America, more different national, linguistic, and racial groups as well.

Second Wavers want to retain or restore the mass society. Third Wavers want to figure out how to make de-massification work for us.

3. How Many Eggs in the Basket?

The diversity and complexity of a Third Wave society blow the circuits of highly centralized organizations. Concentrating power at the top was, and still is, a classic Second Wave way to try to solve problems. But while centralization is sometimes needed, today's lopsided over-centralization puts too many decisional eggs in one basket. The result is "decision overload." Thus in Washington today Congress and the White House are racing, trying to make too many decisions about too many fast-changing, complex things that they know less and less about.

Third Wave organizations, by contrast, push as many decisions as possible down from the top and out to the periphery. Companies are hurrying to "empower" employees, not out of altruism, but because the people on the bottom often have better information and can respond faster to both crises and opportunities than the big shots on top.

Putting eggs in many baskets, instead of all in one, is hardly a new idea, but it is one that Second Wavers hate.

4. Is It Vertical or Virtual?

Second Wave organizations accumulate more and more functions over time and get fat. Third Wave organizations, instead of adding functions, subtract or sub-contract them to stay slim. As a result, they outrace the dinosaurs when the Ice Age approaches.

Second Wave organizations find it hard to suppress the impulse toward "vertical integration"—the idea that, to make a car, you also have to mine the iron ore, ship it to the steel mill, make steel, and ship it to the auto plant. Third Wave companies, by contrast, contract out as many of their tasks as possible, often to smaller, more specialized high-tech companies and even to individuals who can do the work faster, better, cheaper. Carried to its limit, the corporation is deliberately hollowed out, its staff reduced to a minimum, its activities carried out at dispersed locations, the organization itself becoming what Oliver Williamson of Berkeley has called a "nexus of contracts." Charles Handy at the London Business School has argued that these "minimalist, partly unseen organizations" are now the "linchpins of our world." . . .

5. Does It Empower the Home?

Before the Industrial Revolution, the family was large, and life revolved around the home. The home was a place where work took place, where the sick were tended, and where the children were educated. It was where the family entertained itself. It was the place the elderly were cared for. In First Wave societies, the large, extended family was the center of the social universe.

The decline of the family as a powerful institution did not begin with Dr. Spock or *Playboy* magazine. It began when the Industrial Revolution stripped most of the functions out of the family. Work shifted to the factory or office. The sick went off to hospitals, kids to schools, couples to the movie theater. The elderly went into nursing homes. What remained when all these tasks were exteriorized was the "nuclear family," held together less by the functions its members performed together than by all too easily snapped psychological bonds.

The Third Wave re-empowers the family and the home. It restores many of the lost functions that once made it so central to society. An estimated 30 million Americans now do some part of their work back in the home, often using PCs, faxes, and other Third Wave technologies. Many parents are choosing to home-school their kids, but the real change will come when computers-cum-television hit the household and are incorporated into the educational process. As to the sick? More and more medical functions from pregnancy testing to taking one's blood pressure—tasks once done in hospitals or doctors' offices—are migrating back to the home.

All this points to a stronger, not weaker, home and a stronger role for families—but families of many diverse types, some nuclear, some extended and multigenerational, some composed of remarrieds, some big, some small or childless, some giving birth to children while young, others in maturity. This diversity of family structure reflects the diversity we find in the economy and culture as the Second Wave mass society de-massifies.

RIDING THE WAVE OF CHANGE

America is where the future usually happens first, and if we are suffering from the crash of our old institutions, we are also pioneering a new civilization. That means living with high uncertainty. It means expecting disequilibria and upset. And it means no one has the full and final truth about where we are going—or even where we should go.

We need to feel our way, leaving no group behind, as we create the future in our midst. These few criteria can help us separate policies rooted in the Second Wave past from those that can help ease the way to our Third Wave future. The danger of any list of criteria, however, is that some people will be tempted to apply them literally, mechanically, even fanatically. And that is the opposite of what is required.

Toleration for error, ambiguity, and above all diversity, backed up by a sense of humor and proportion, are survival necessities as we pack our kit for the amazing trip into the next millennium. Get ready for what could be the most exciting ride in history.

Into the Outdoors *Tracy Baxter*

Tracy Baxter's piece recounts some of the work of the Sierra Club's ICO (Inner City Outing) program, which was started in 1976 and which, as Baxter relates, took almost 14,000 inner-city children to the outdoors for a wilderness experience in 1996. At the center of this effort is the goal of overcoming the public perception that the Sierra Club's efforts and conservation as a whole are principally a concern of the wealthy and privileged. Those without wealth, so the argument goes, do not have the luxury for such concerns, and instead are more interested in supporting industrial expansion and economic growth.

At the *Visions* Web Site Links to the Sierra Club; to guides to Louisiana swamps.

READING PROMPT

Buried deep in Baxter's tale is a message about the future—how would you describe that message?

This is not the New Orleans I encountered ten years ago, when a primordial heat thickened perspiration into a gummy body sheath. The last 48 hours have seen temperatures no higher than 63, and the rain's so heavy it wraps around Kate's car like an endless car wash.

Our first stop is the St. Thomas public housing projects in central New Orleans. We pick up the Taylors—Dexter, Lucretia, and Nikia—and pack them into the backseat. Dexter is disappointed to hear that some kids won't be joining us because of the weather.

Our drive across the Bonnet Carre Spillway and beyond is a 40-mile race with storms. Clabbering gray haze continually eclipses feeble patches of pale blue sky. "We need this to clear up so we can make it to Turtle Cove, you guys," Kate says.

"Dexter, you've got to wish hard for the clouds to go away." The girls snicker, Nikia demurs, "Miss Kate, I can't even stop a toilet from running." But Dexter isn't quite old enough to scoff at a sincere wish. He murmurs along with Kate for reprieve.

Since 1976, the Sierra Club's Inner City Outing program has brought wilderness experiences to children much more familiar with concrete compounds than open space. Nearly 14,000 kids last year took to the hills, or to the shore, or to somewhere in between, led by volunteers like our driver, Kate

Mytron. Kate's a veteran of the anti-war and civil-rights movements, dedicated to "sharing the world." But the kids aren't the only ones who profit from the experience. "It's so incredibly rewarding, rediscovering what it feels like to see frogs for the first time," says one leader. "Days after a trip, the volunteers still talk about the good time we had."

Off the Manchac exit ramp we slow down to look out for other members of our party. We roll up to Middendorf's parking lot for a pit stop but are shooed away by an approaching restaurant employee. He gestures broadly to a power line downed by whipping winds. Nature denies nature's call.

A quarter of a mile ahead we see a cluster of cars parked near a shack with a sign audaciously dubbing it "The Manchac Yacht Club." Nearby wait a passel of kids ages 7 to 14: John Kuss and his little brother, Clinton; the Webster sisters, Ronata and Danielle, and Shemeka Billo. The ICO leaders—Rogerwene Washington; Paul Bergeron and his teen son, Chad; Kate; and the director of the Turtle Cove research station, Bob Hastings—all confer earnestly over the weather. The kids don't seem to be in a hurry either way. The clothes-snapping wind amuses those who brave it, while those who stay in the cars watch the spectacle of all that fluttering clothing; that, and a tumbling flurry of frantic sparrows overhead.

I feel winds intent on swatting us into Lake Pontchartrain. Kate detects a lull. Damned if she isn't right. Within 20 minutes we're packing a boat, tightening fiery orange life jackets, and chugging into newly quiescent, open water.

Bob's call for help in manning the boat is irresistible to all but our youngest: Clinton hasn't made this trip before, and he's not about to spend his time crowded around the steering console. He crawls over sleeping bags, peering into our faces as if to confirm that the speed and the spray do indeed combine to make this one truly glorious moment. He clambers to his feet, claps his thigh, and points excitedly into the distance. "Land ho!" he crows. "Land ho!"

We soon disembark at the Manchac Wildlife Area research station, which is administered by Southeastern Louisiana University. The green and white building, our home for the next 24 hours, was once a private hunting and fishing lodge. Its three levels have been converted into a lab, sleeping quarters and dining hall, and a classroom.

Students from SLU come here for field research in aquatic biology, much like we'll be doing, on our more modest scale. Rogerwene passes out pens and paper to catalog the flora we find traversing the first leg of the boardwalk. No sooner has she reminded the kids to be careful than we have our first horseplay casualty on the slippery boards of the canoe house. Danielle, a chatty imp, springs into the air and lands against Nikia, bringing the taller girl down flat. Adroitly stepping over her prone friend, Danielle innocently pronounces the obvious—"Ooh, she fell"—and skips off to join the others. I suspect Nikia is more surprised than hurt as she tenderly rubs the back of her head. Nor is she heartbroken when told to return to the station for rest, just to be on the safe side. Turtle Cove is old news to Nikia, and her gangly limbs and carefully placed twists of hair announce the birth of a teen with an associated disdain of kid stuff.

The others, though, are eager to explore. Sadly, the kids aren't seeing the original denizens of this water world. Logging companies, in a locust-like orgy of destruction, felled most of the cypress trees between the late 1890s and the 1950s. The few cypress here now, courtesy of an ongoing SLU restoration effort, are survivors of still another swamp bane, the red-fanged nutria, a rodent with an endless yen for saplings. A broad-leafed aquatic plant called bullstongue seems to dominate the marsh like a vast submerged herd of insolent bovines, but the kids' inspection turns up lavender iris, spiderwort, misty Spanish moss, and sundry colors, shapes, and fragrances—not to mention a panoply of bugs. Bob, the Turtle Cove director, calls us over to see what looks like a shivering mound of brown sugar on a piece of driftwood. When he tells the kids that the marooned ants are taking turns underwater so that they'll all have a chance to breathe, Danielle sees an excellent opportunity to work up a scare. "Ooooh, y'all," she coos. "Them's fire ants." But Lucretia, bringing into play her junior naturalist skills, disagrees. "No, they not. Them's sweet, sweet. I ain't afraid uh dem."

Plop! Shemeka's in the drink. She's climbed onto the lookout platform without incident, but in coming down takes a spill off the ladder, landing harmlessly, though noisily, into a foot of water. She wears the "oops" face, a blend of merriment and sheepishness as Rogerwene leads her back into the station to dry off. Inside, Bob has just fished one of several juvenile alligators out of a blue holding tank, offering its pale, banded underside for the touch. The boys tentatively run their fingers up its belly. Lucretia gives the reptile a more deliberate rub and watches its stubby legs slowly kick out. Bob tells them how scientists are testing transmitters to track alligator hatchlings' movements in the wild. But as the boys paddle their fingers in the shallow water, toothsome digits tinglingly close to those toothy maws, their minds are on what the creatures eat. "They've been raised in captivity so they eat mostly pellets," is Bob's mildly deflating answer. John, a chunky blond, still craves danger. Hands dripping, he flicks tank water in my face. "Gotcha!" he yells, and runs off.

Kate has set out sliced fruit and a bowl of chips on a long table in the station's dining hall upstairs. Kids pass back and forth, grabbing a tidbit, then returning to their card and board games, but Shemeka stands very carefully placing a corner of each tortilla wedge in the middle of a red-flecked, shiny orange mousse, and chews in silent delight. It's a New Orleans favorite, Kate says: melted Velveeta and a can of Rotel diced tomatoes.

Between bites of goop, Paul tells me what's apparent to any visitor to the Crescent City—people here are big. In fact, a well-publicized report has just declared New Orleanians the fastest growing population in the United States—by weight, that is. This is no surprise. Any people who routinely eat bushels of the obscenely ugly crawdad (jubilantly exhorting the novice to "twist dey tail and suck dey head") can be expected to put away far too much of more appealing fare. And they do, provided it's fried, prepared with roux, or densely seasoned.

But obesity is New Orleans' more benign badge of distinction. Violent crime, the likes of which seem to come straight out of the reject pile of potboiler fiction, is another story. In 1995, the city's murder rate averaged one per day. Who's responsible? Several of the more infamous cases involved peace officers who protected and served their own shady interests: a cop slaughtering witnesses during a restaurant heist, or murdering a woman who fingered a dirty colleague. But even a squeaky-clean police force would have trouble capping Big Easy lawlessness. "It's so brazen and so random," Kate says. "A thirteen-year-old boy was shot in the back of the head because he didn't get off a pay phone quick enough to suit a fourteen-year-old with a gun." Juveniles, in fact, are responsible for many of the nastiest offenses. "Teens think they're invincible, anyway. And I guess the heartlessness is another facet of that attitude," says Kate.

The dip bowl is emptying fast. I scrape the bottom with a chip and find, to my horror, that it is quite tasty. "Should I make some more?" I ask. "Yea," says Danielle. "But this time, hit it with a little uh dis heah." She hands over the green container of Tony Chachere creole seasoning she's found in the kitchen. "That'll make it good, yea."

That there is no time between snacks and dinner makes no difference to anyone. The kids take seats and the adults pass out plates with gooey squares of lasagna and piles of spinach salad with buttermilk dressing.

Inner city outing programs, including this one, frequently partner up with children's agencies ranging from the YMCA to pediatric AIDS care groups. Volunteers for ICO know that some of the kids are shouldering enormous burdens but are still sometimes taken aback by the particulars. Sven Thesen, a New Orleans leader, told me about a boy whose incessant need for attention was driving everyone crazy. "He was constantly acting out. We thought, 'never again,' but then we found out later that both his parents were in jail and that he had been abused and put in child porn."

A troubled background can lead to the unruliness that prompted one boy on an ICO trip to creep out and start up the boat, but harsh circumstances can also imbue kids with a love of structure and cooperation. Displays of affection are open and frequent. "Sometimes you can't go out 15 feet without someone reaching for your hand or giving you a hug. Even the boys," says Sven. "They start off being cool, but they come around quickly."

The animated and easygoing nature of the children around our table belies the Dickensian loss and peril they've known. They're chatting, smiling, and digging into their plates—all except one. Paul notices Lucretia's silence first. "What's the matter, Lucretia? You don't like lasagna?" Eyes down, she shakes her head gloomily. "Well, do you like spaghetti?" She nods enthusiastically. I detect a small bid for special treatment in this hair-splitting. Evidently, Paul does too, but he decides not to make anything of it. "Well, that's precious," he says evenly, and continues eating. No longer in the spotlight, Lucretia picks up her fork and works on her meal.

The late-night boat ride in search of red-eyed alligators has been canceled—all right by me. Bob says that the high water would make maneuverability difficult. The girls and I will walk the quarter mile around the boardwalk by flashlight instead.

I am the only shrinking violet, inching along, careful not to drift too far left or right, calling out for the group to slow down. Nothing would be more mortifying to me than an involuntary dip with the residents of the marsh, an apprehension common to Turtle Cove's urban newcomers.

The trepidation a few of the new kids feel out of the city is foreign to some of the ICO leaders. "One kid said she was afraid of trees," an exasperated leader told me. But another was able to put herself in the kids' shoes: "Their experience and survival skills don't always match my values." Keeping those differences in mind helps explain why some kids appear uninterested in or resistant to some activities.

Bob has watched many kids evolve from fearful to exuberant after a few outings. "At first they think they'll see lions and tigers and bears. When they've assured themselves that there's no significant danger, they want to look and see. We don't teach them how. They teach themselves."

Danielle suddenly breaks out from behind, giving me one alarmingly unsteady moment along the swamp's edge, and scampers up to Paul with a request. Soon she's riding on his shoulders and chatting as brightly as though it were high noon at the playground. The clouds have cheated us out of starglow to light the way, yet—but for our voices and the bump and shuffle of feet—we still have the gift of silence. No speeding cars, no popping that could be the report of gunfire, and, by specific restriction, no personal radios. It's enough, really, to make you want to sing.

"Stop—in the name of love," Danielle pleads with a high, light tremolo, "bee-fore you break my heart." The other girls earnestly coo the refrain, "Think it over, think it over."

Someone calls out a dare to touch the weather station, a windmill-like apparatus connected to the boardwalk by a beam five feet long and no more than six inches wide. Ronata and Nikia are tonight's daredevils. A lance of light in front and behind her, Nikia scoots out, touches the structure, and scoots back to cheers. Ronata, whose round face has been placid all day, dashes out, scores, and returns in seconds. Caught up in the clapping and congratulations, she steps off the boardwalk and into the swamp.

I awaken at six a.m. to shrieking, as piercing as the hoots accompanying last night's slide show of Turtle Cove's greatest moments, but with a note of theatrical terror. The girls, sleeping together on the bottom of a bunk bed, are now up and swiping at their arms, ants taking hot nips on their skin, lured into the bed by a forgotten lollipop. "I like to died, " Ronata exclaims. Danielle sweetly disavows any knowledge of who left the candy out.

A mammoth breakfast of hotcakes and strawberries fuels our last outing to the flooded backyard. By the time I arrive downstairs, the kids have collected a menagerie of water critters in a pail by dragging long-handled dip nets through the turbid, shallow water. Danielle walks primly up to the bucket with a look of immense satisfaction. In her hand is a piece of flashing quicksilver scarcely longer than a grain of rice. She drops the minnow in with the diving beetles, grass shrimp, and quartersized crabs, takes up her net, and rejoins Nikia in pulling through the muck and examining the catch.

Bob has unhitched the pirogues from the dock and Dexter and Lucretia quickly claim one of the flat-bottomed canoes. Evidently Lucretia has forgotten her somber mood of last night. With rolled-up jeans soaked well past her knees, she is an arm-flailing engine of gaiety as she pushes the vessel off. I notice Shemeka, a petite girl with dreamy looks, alone in a pirogue and hop in behind her. My modest paddling skills would strand us in open water, but here in the cove, aided by Shemeka's faint but insistent stroking, I maneuver us out of vegetation nearly as fast as I steer us in. We manage to pull out ahead of the Kuss boys: an obvious invitation to a race. Chad gives the signal, and we're off in a contest where the object seems to be who can go the slowest while expending the most energy. Shemeka tightens her shoulders to stroke quickly, determined to pull us through to victory, and a fortuitous piece of driftwood blocking the boys' path gives us a win by inches. Puffing hard, Shemeka grins proudly.

Though the sun has yet to make an appearance, a pore-clogging, enervating heat has been creeping into the air. Catching her breath and suddenly looking fatigued, Shemeka whispers, "I want to stop now"—the first sentence I've heard from her. Shemeka is HIV positive, and I wonder if the medications she takes to boost her immune system don't sometimes tucker her out. As Chad helps Shemeka out of the boat, I realize that neither the drugs nor the virus have drained her spirit. She sits on the boardwalk, watching her friends and smiling, wan, but without a trace of wistfulness.

An hour's play discharges the kids' energy. Back up in the classroom they calmly review their "Look, See, and Touch" work sheets from yesterday's excursion, play math games, and even manage to hold still as I sketch them. The contented quietude lingers on the ride back to the Yacht Club, the spray from the boat eliciting only the most meager of squeals from the girls.

Shemeka, our passenger on the way home, dozes off nearly immediately after buckling in. I ask Kate what she thought of our trip. Her reply echoes the comments on the children's handouts, where each of the letters in the words Turtle Cove starts off sentences describing our day: the rain wasn't fun but everything else was just fine.

Despite being cooped up more than we'd planned, the kids enjoyed each other's company and ate heartily—something some of them might not do consistently. I'm disquieted by the thought. When my grandma, the daughter of Louisiana sharecroppers, speaks of her childhood, she often mentions the fresh fruits and vegetables her mother would put up and the filling meals they made. My late mother talked about the many eggs she sacrificed in pursuit of the perfect butter cake. Physical hunger was never the issue for them, only the desire for life without the fetters of an intangible social order. When my mom arrived in California that opportunity seemed more possible to find. With the support of a community made up of migrants like her, and night school, she moved us out of the projects in three years. Now, as the decline of northern cities prompts some black folk to follow a south star back home, I wonder what the recourse is for those who never left.

How much ICO helps is a matter of perspective. The trips do not alter the fact that the Tailors, an extended family of ten, live on less than $400 a month and an allotment of food stamps. But the outings do add to the sense of well-being that keeps the Taylor kids on academic honor rolls. Kate tells me about two teens, Walter and John, five-year veterans of New Orleans ICO, who were recently invited by the Club to hike at its Claire Toppan Lodge in the Sierra Nevada. They might one day evolve into environmental powerhouses, or they might not. No one in the ICO program presumes to perform miracles, but as Kate recalls her family's trips to the beach and the woods, and I think about a church trip when I drank with an elated thirst, from a mountain stream until I like to burst, it's honest to say that some events change your life by encouraging you to seek out your place in the world. Maybe some of the kids will remember the trips to Turtle Cove as explorations that helped them decide where to go.

chapter 16 Explorations

16a Pumping Gas

Collaboration. New Jersey and Oregon are two states that do something many may find un-American: prohibit people from pumping their own gas at service stations. Make and share a list of reasons in support of and opposed to this unusual law.

Web work. Jim Potts maintains Primarily Petroliana: The Gas Station & Auto Service Memorabilia Web site, with links to other Web sites and information about museums and restored service stations across the country. Vic's Place offers oil and gas collectibles. To see what's happening today, visit the homepage of a major oil company such as Exxon or, on a smaller scale, the homepage of Jiffy Lube.

One writing topic. Pumping gas is obviously not the best work most people can imagine, but it may not be the worst work either. Write an essay defining what you consider to be the basic components of a good job or an acceptable one, apart from the obvious fact that it pays a living wage. That is, what kind of work is most worth fighting to protect?

16b PCs

Collaboration. Personal computers are clearly the weaving looms of the late twentieth century, the technology over which people are prepared to discuss the future of work and a whole lot more (such things as personal associations). Make and share two lists: of ways that PCs have changed your life for the better and, where applicable, for the worse. You may also want to create a third list of important aspects of your life that, so far at least, have remained largely untouched by PCs.

Web work. Dave's Guide for Buying a Home Computer is supposed to be a helpful service to would-be consumers but can easily be viewed as a parody of an industry and technology out of control. It seems to have been constructed as part of a college course. Check out the continually updated homepage of Bill Gates's book *The Road Ahead.* Visit the homepages of major sellers of home computers, such as Compaq or Gateway.

One writing topic. Write an essay on one particular aspect of your life that either you have gladly turned over to computers (or look forward to turning over to computers someday soon) or you are determined you will never do via a computer.

16c The Unabomber, or Crimes and the National Press

Collaboration. The Unabomber attracted far more national interest as the unknown, raving, menacing terrorist (the original Luddites destroyed machinery, not people) than as the captured, diminutive, disheveled Ted Kaczynski. Indeed, some portion of the U.S. public always seems ready to transform a suspect or a convicted criminal into some sort of outlaw folk hero, at times because people believe this person has been falsely charged and at other times because this person, even if guilty, represents some value higher than law-abidingness. Describe and share with others brief accounts of one or more "outlaws" that you or others have found worthy of sympathy.

Web work. CNN, in conjunction with *Time* magazine, has a well-researched site on the Unabomber, as does MSNBC.

The PBS series *The American Experience* did a show focusing on three great crimes of the century: Prohibition-era crime in Detroit; the fixing of TV game shows; and the most famous murder trial before O. J. Simpson's, the 1908 murder of architect Stanford White.

Check out the Court TV homepage to see what trials are currently in the news.

One writing topic. Write an essay on the news coverage of the Unabomber or of any criminal trial with a national news interest. If you have access to the Web, check out how the Web sites of the major news companies are covering the trial versus the other coverage you can find on the less commercial, less orderly parts of the Web.

16d World's Fairs and Expositions

Collaboration. World's fairs and expositions have been key events in the celebration of technology and progress over the last two centuries, going back at least as far as the Great Exhibition in London in 1851, best known for its central building, the Crystal Palace. For your parents, their parents, and even their grandparents, traveling to such an event remained a memorable occasion in their entire lives: the one chance, before television and international global communications, of coming face to face with the future.

Share descriptions of a momentous trip you have taken in your life, to see something in person that you could not see at home. Was it to an exposition—and hence to see something man-made and high tech—or to a historic or natural site?

Web work. Yahoo! has a page on world's fairs and exhibitions, arranged historically and starting with links to the Crystal Palace in 1851 and going to the world's fair planned for Calgary in 2005. Yahoo! also has a page of links for Disney's ongoing celebration of progress and technology.

There's a rich Web site on the St. Louis World's Fair of 1904, held as part of the centennial celebration of the Louisiana Purchase. (This same fair served as the backdrop for Vincent Minnelli's heartwarming 1944 musical *Meet Me in St. Louis.*)

One writing topic. One might argue that a variation of the old saying, seeing (in person) is believing, has formed the basis of the great popularity of exhibitions celebrating technology and progress: For generations now, people have traveled, often long distances, in order to be dazzled by the latest technological advances. Write an essay on changes, if any, in how people generally or you personally come face to face with the future. That is, where is it that you are most likely to get glimpses of new technology and of the future lifestyle it may be supporting?

16e Animal Rights

Collaboration. For years, animal rights has remained a fringe political issue, perhaps a little difficult for many people struggling with their own immediate problems to identify with, and with its advocates often portrayed as impractical idealists, even zealots. Looking at the animal rights controversy from the perspective of technology may help bring a new focus on these issues. The key issue here, and the starting point for a collaborative discussion, is the question of what restrictions, if any, you feel we should observe in using animals to help us develop technological innovations aimed at enhancing human life.

Web work. The most visible group leading the fight against what it sees as the mistreatment of animals has been the activist organization PETA, or People for the Ethical Treatment of Animals; PETA's style and programs are quite different from those of the older, more mainstream ASPCA, the American Society for the Prevention of Cruelty to Animals.

Yahoo! lists dozens of other animal rights organizations at Society and Culture > Animal Rights > Organizations and far fewer oppositional groups at Society and Culture > Animal Rights > Opposing Views > Organizations.

Check out VIVI—Vegetarians International Voice for Animals. Or, for general vegetarian information, see the Crazy Vegetarian, which calls itself "the leader of the meat-free world."

One writing topic. What restrictions, if any, should there be on humans' treating animals as mechanical natural processes subject to technological innovation to improve the quality of human life?

16f Wilderness Education, or Life without Electricity

Collaboration. Share a detailed description of the time when you felt yourself to be in most direct, immediate touch with wilderness—the world untouched by humanity; that is, a time when the lights went out.

Web work. Outward Bound is the best-known wilderness education program. Check out the Yahoo! page for wilderness education at Business and Economy > Companies > Outdoors > Wilderness Education. Bill Borrie of the University of Montana maintains links to all wilderness sites on the Web, including a section on education.

Anyone interested in blending wilderness education with the latest in high technology might want to check out the homepage of the U.S. Army Green Berets.

Meanwhile, one way of defining a wilderness experience is the absence of electricity, including all battery- and generator-powered appliances. The Electric Power Research Institute is an initiative of the electric power industry, and its homepage offers what it calls a "road map" for the next century.

One writing topic. What is it that students or people generally are expected to learn in wilderness education—in not just learning about wilderness but experiencing it? Write an argument for or against the position that some significant wilderness experience (life without electricity) should be a required part of high school education or college education?

chapter 17

Utopia/Dystopia

READING QUESTIONS

- What role does technology play in your picture of utopia?
- What role does technology play in your picture of dystopia?
- What role does technology play in your memory of your own childhood? In the childhood you would like to provide to future children?

QUOTATIONS

Imagination is the one weapon in the war against reality.

— *Jules de Gaultier, early twentieth-century French author*

Tomorrow, and tomorrow, and tomorrow,
Creeps in this petty pace from day to day,
To the last syllable of recorded time;
And all our yesterdays have lighted fools
The way to dusty death.

— *William Shakespeare,* Macbeth

President Reagan remembers the future and imagines the past.

— *Senator Eugene McCarthy (1990)*

What's the use of worrying? It's silly to worry, isn't it? You're gone today and here tomorrow.

— *Groucho Marx, American humorist and film star, 1895–1977*

Time present and time past
Are both perhaps present in time future,
And time future contained in time past.

— *T. S. Eliot (1936), British/American poet, 1888–1965*

If you want a picture of the future, imagine a boot stamping on a human face—for ever.

— *George Orwell (1949), British novelist and social critic, 1903–1950*

The Geeks Have Inherited the Earth

Zina Moukheiber

Zina Moukheiber's piece is another in this anthology dealing with the broad connection between invention, imagination, and play. For inventors such as Buckminster Fuller (discussed by Lloyd Steven Sieden in Chapter 3), play seems to be a vital source of their technological thinking. In this piece Moukheiber looks specifically at the role of science fiction literature and thus adds a piece to Kathe Davis's analysis of women and sci-fi (in Chapter 15).

At the *Visions* Web Site Links to much more about science fiction.

READING PROMPT

How does Moukheiber describe the connection between technology and the literary imagination? Does his analysis make you feel more hopeful or less hopeful about the future?

With a felt-tip pen, William Gibson crossed out "infospace," then "dataspace," on his yellow legal pad. They were words he'd made up, but they just weren't catchy enough. Then it clicked: "cyberspace," a term with the feel of a buzzword and a hint of poetry. Gibson was writing a short story, "Burning Chrome," and trying to label an electronic network. It was pure science fiction, and the year was 1981—years before the Internet started to make its mark on the global consciousness.

Science fiction is more than inelegant prose for pencil-necked teenagers. Businesspeople and investors looking to discover where technology is taking us would do well to pay attention to it. Through the decades the idea factory of SF, as it's called today, has sparked the imagination of generations of inventors who have tried to nudge the world a little closer to their collectively imagined future.

"It's not just literature," says Robert Stearns, chief technologist at Compaq Computer. "It embeds in our mind certain key expectations about technologies and how we're going to use them."

Today's science fiction goes even further, foreseeing fully realized futures whose outlines are visible within the present—as if the today is just the beta version of the tomorrow.

Spinning scenarios for us to choose or avoid, writers like Gibson have their own impact. Writing in *The Economist,* Professor Paul Krugman of Stanford University recently dismissed a small shelf of economics—heavy books by such notables as historian Paul Kennedy, management consultant Stephane Garelli, military expert Edward Luttwak and businessman Sir James Goldsmith: "If you want to read intelligent, well-written, grimly imaginative speculations about what might happen to our societies if present trends continue you would do far better to read Neal Stephenson or other clever 'cyberpunk' writers like William Gibson and Bruce Sterling." The geeks, it seems, have inherited the Earth.

> "Something made you, Cutie," pointed out Powell. "You admit yourself that your memory seems to spring full-grown from an absolute blankness of a week ago. I'm giving you the explanation. Donovan and I put you together from the parts shipped us."
>
> Cutie gazed upon his long, supple fingers in an oddly human attitude of mystification. "It strikes me that there should be a more satisfactory explanation than that. For you to make me seems quite improbable." The Earthman laughed quite suddenly, "In Earth's name, why?" "Call it intuition. . . ."
>
> —*Isaac Asimov:* I, Robot (1950)

None of this futuristic business is new. In 1865 Jules Verne wrote *From the Earth to the Moon,* describing a huge cannon blasting astronauts into space from a Florida launching pad. In tribute to his ingenuity, a full-scale model of his spacecraft (featuring red velvet seats and a whiskey-stocked cupboard) is on display at the National Air and Space Museum. H. G. Wells, a Verne near-contemporary, wrote about tank warfare in *The Land Ironclads* (1903) and aerial bombing in *The War in the Air* (1908).

Scratch a techie and you're likely to find an abiding love for science fiction. Space engineers at the National Aeronautics and Space Administration readily admit they were drawn into rocketry thanks to the writings of Robert Heinlein, Arthur C. Clarke and Isaac Asimov. NASA's Jet Propulsion Laboratory routinely invited Clarke, Ray Bradbury and other writers to discuss interplanetary encounters, such as the landing of the Viking spaceship on Mars.

"Science literature is cut and dry," says Yoji Kondo, an astrophysicist at NASA and a writer. "Science fiction makes it exciting and allows scientists to dream."

Which is why we say businesspeople and investors ought to pay attention. From these dreams come the wonders of our age. Verne's *Robur the Conqueror* (1886) inspired Igor Sikorsky to build the first helicopter in 1939. His company, Sikorsky Helicopters, is now a division of United Technologies. Nobel Prize–winning physicist Richard Smalley got very big dreams reading Arthur C. Clarke's *Fountains of Paradise* (1978), which includes a description of space elevators. "One of my dearest hopes in life is to build the cable for the space elevator," he says. Now, advances he has brought to nanotechnology—super-tiny manufacturing—have taken lightweight, superstrong materials out of the realm of pure fantasy.

Daniel Hillis pioneered parallel supercomputing and founded Thinking Machines in 1983; he credits Isaac Asimov's landmark *I, Robot* and Robert Heinlein's *The Moon Is a Harsh Mistress* (1966) for his early decision to study artificial intelligence. Hillis was fascinated by Heinlein's concept of an increasingly complex network of computers that achieves true intelligence. "It led to my work in parallel processing," says Hillis, who is now doing R&D at Disney. Science fiction's influence on Hillis' decisions went back even further. He read Heinlein's *Have Space Suit—Will Travel* (1958) when he was in junior high, and decided to follow the college choice of the hero of the tale by going to MIT.

Vital tools that businesses and consumers rely on were inspired by sci-fi scribes. Arthur C. Clarke, who holds a bachelor's degree in physics and math from London's King's College, described the concept of communications satellites in a 1945 article, "How the World Was One." Hughes Space and Communications Co. built the first such satellite, Early Bird, and Comsat Corp. launched it in 1965. Comsat invited Clarke to the launch. "He is the godfather of communications satellites," says Harold Rosen, who built Early Bird at Hughes.

Robert Heinlein foresaw robotic hands controlled remotely by gloves and called them "Waldos," a name that, improbably, stuck. Oh, yes, Heinlein got the literary jump on waterbeds, as well. And Dick Tracy's wrist radio is the Holy Grail of the wireless industry, says James Caile, a marketing vice president for Motorola. Caile hung a framed original 1946 comic strip showing cartoonist Chester Gould's most enticing invention in his office. "We'll get there at some point," he vows.

> Cyberspace. A consensual hallucination experienced daily by billions of legitimate operators, in every nation, by children being taught mathematical concepts. . . . A graphic representation of data abstracted from the banks of every computer in the human system. Unthinkable complexity. Lines of light ranged in the nonspace of the mind, clusters and constellations of data. Like city lights, receding.
>
> —*William Gibson,* Neuromancer (1984)

Many of the people trying to design our high-tech world say Gibson's ideas have guided them toward *Neuromancer*'s as-yet-unrealized marriage of the Internet and virtual reality. The book "put us in a different direction," says Larry Smarr, director of the National Center for Supercomputing Applications at the University of Illinois at Urbana–Champaign. "Instead of just designing supercomputing, we were going to design a cyberspacelike world."

When a college kid named Marc Andreessen (Netscape's cofounder) helped to develop *Mosaic,* software that made it easy to browse the fledgling

World Wide Web, Smarr recalls thinking that he had created a window into cyberspace similar to the Ono-Sendai deck, Gibson's imagined brand name for the machines that let users jack into the Net.

Entrepreneurs have taken their inspiration from Gibson as well. Mark Pesce quit a job designing software to form a company he called Ono-Sendai in 1991. His initial efforts to create a three-dimensional environment on the Internet would eventually collapse. To use a phrase from Paul Saffo, a director of the Menlo Park, California–based Institute for the Future, Pesce confused a clear line of sight with a short distance; the task was too vast for his small company. Pesce recalls the failure with frustration: "Science fiction . . . inspires you, and it can make you nuts. People get impatient."

Pesce eventually teamed up with fellow science fiction buff Anthony Parisi to create a language allowing developers to build their own 3-D applications. Now 35 computer companies—including Microsoft, Netscape, Oracle and Intel—have rallied behind the team's Virtual Reality Modeling Language.

The place Gibson named doesn't really look much like what we call cyberspace today: He envisioned a largely corporate and educational network whose vast depositories of data were displayed visually. What we ended up with is more like a fully realized city, with virtual homes, businesses, libraries, nightclubs—even a red light district. It is closer, in other words, to another science fiction vision, the "metaverse" Neal Stephenson envisioned in his 1992 *Snow Crash.* That particular moniker has not caught on, but many of Stephenson's ideas about what the on-line world can look like are driving a new generation of entrepreneurs to try to match it.

In Stephenson's *Snow Crash,* hackers who have the technical know-how to write their own code sport sophisticated, graphically complex on-line avatars, or three-dimensional personae, who interact in the virtual world. Japanese businessmen, on the other hand, order up minutely detailed, perfect images of themselves in the same dark suits they wear to the office. Newbies coming in through cheap public terminals must make do with grainy, jerky black-and-white models. The techno-elite mingle in an on-line bar called Black Sun.

It's already on its way. Software engineer Franz Buchenberger has raised $8 million in venture money to form Black Sun Interactive, based in San Francisco and Munich, Germany. The company has revenues of almost $1 million, and sells tools to build 3-D environments and avatars. Buchenberger has tried to follow Stephenson's game plan at every turn, from avatars coming together in on-line gatherings to "cybercards" for storing and exchanging personal information. For the Super Bowl in January, Black Sun built an avatar community for SportsLine USA's Web site. Fans chatted in a virtual locker room or in the SuperDome, which was a virtual representation of the New Orleans stadium.

Buchenberger e-mailed Stephenson to get his blessing for use of the Black Sun name and even asked if the author would be interested in promoting the company. Stephenson, who has signed a lucrative *Snow Crash* film deal, demurred. "I think people have figured out that I'm not going to do that," says the goateed 37-year-old.

Worlds Inc., based in San Francisco, is also developing on-line virtual chat environments for customers such as AT&T and Absolut vodka. Founder David Gobel said he had been working on the ideas before *Snow Crash* came along, and when he did read the book, initially he says he was "pissed off, because *Snow Crash* gave my business plan away."

> This is America. People do whatever the f*** they feel like doing, you got a problem with that? Because they have a right to. And because they have guns and no one can f***ing stop them. As a result, this country has one of the worst economies in the world. When it gets down to it—talking trade balances here—once we've brain-drained all our technology into other countries, once things have evened out, they're making cars in Bolivia and microwave ovens in Tadzhikistan and selling them here—once our edge in natural resources has been made irrelevant by giant Hong Kong ships and dirigibles that can ship North Dakota all the way to New Zealand for a nickel—once the Invisible Hand has taken all those historical inequities and smeared them out into a broad global layer of what a Pakistani brickmaker would consider to be prosperity—y'know what? There's only four things we do better than anyone else.
>
> music
>
> movies
>
> microcode (software)
>
> high-speed pizza delivery.
>
> —*Neal Stephenson,* Snow Crash (1992)

Of course, science fiction isn't limited to what are now derisively called "hardware stories," and not every author tries to predict which way we'll go. Sometimes, in fact, they're trying to tell us where not to go. Nearly 50 years ago George Orwell foresaw interactive television in *Nineteen Eighty-four,* but as a tool for eavesdropping and oppression. And would any of us be able to bear the hellish futures of Harlan Ellison, author of harrowing short stories such as "I Have No Mouth and I Must Scream"? These are tales that warn us away from futures that would give our technological servants too much control over our lives—a theme that it doesn't take a Unabomber to appreciate.

Orwell's warning still resonates with Frank Moss, chief executive of Tivoli Systems. Moss had lapped up books by Verne, Isaac Asimov and Arthur C. Clarke on the way to getting his doctorate in aeronautics and astronautics at MIT. In Orwell he saw a world in which "global computer networks were controlling people," Moss says. "So I decided to focus on how we can make computer networking more valuable to people, rather than take away their freedom." In 1984 he put computing power into the hands of individual users with Apollo, the first workstation.

The strength of a vivid idea can inspire scientists even when it's part of a darker vision. Arthur C. Clarke and Stanley Kubrick's film *2001: A Space Odyssey* might have portrayed a computer gone mad, but the HAL 9000's prowess with language turned out to be deeply inspiring to speech recognition researchers. "Knowing that such a thing has been imagined was good enough that at some point in time we can deliver the technology," says David Nahamoo, manager of Human Language Technologies at IBM's Research Division.

> All fiction is metaphor. Science fiction is metaphor. What sets it apart from older forms of fiction seems to be its use of new metaphors. . . . Space travel is one of these metaphors; so is an alternative society, an alternative biology; the future is another. The future, in fiction, is a metaphor.
>
> A metaphor for what?
>
> If I could have said it nonmetaphorically, I would not have written all these words, this novel.
>
> —*Ursula K. Le Guin, introduction to* The Left Hand of Darkness (1969)

The authors don't see themselves as prognosticators or Alvin Toffler–style futurists. Austin, Texas–based science fiction writer Bruce Sterling calls prediction "an occupational hazard" of the trade, but insists "We're not retailing blueprints . . . I'm developing scenarios." Sometimes writers get lucky and look like seers. "When Thomas Edison was popularizing electricity," explains Stephenson, "he anticipated things like the lightbulb, the electric washing machine—but he didn't anticipate the electric guitar." Stephenson says that he and such writers as Sterling and Gibson have been "trying to envision the electric guitars and the rock music."

"For every correct prediction a dozen were wrong, or correct only if the facts are stretched a little," notes Peter Nicholls in *The Encyclopedia of Science Fiction.* Where are all those death rays and humanoid robots, anyway? "It is rather a dubious vindication to point out that laser beams can now be used as weaponry," Nicholls wrote. Who knew that lasers would, instead, be nearly ubiquitous, reading the music off our CDs and relaying our telephone chatter over fiber-optic strands?

So what is the science fiction writer's role? "We are wise fools who can leap, caper, utter prophecies and scratch ourselves in public," says Sterling, who helped form science fiction's gritty "cyberpunk" genre. "We have influence without responsibility. Very few feel obliged to take us seriously, yet our ideas permeate the culture, bubbling along invisibly, like background radiation." Sterling gets red-carpet treatment as a social commentator: He was the featured speaker at a recent event held by the chairman of French telecommunications company Alcatel.

The writers earn their status by yoking their imaginations to disciplined homework. Stephenson's latest book, *The Diamond Age,* deals with the coming wonders of nanotechnology. He enlisted the advice of nanotechnology expert and physicist K. Eric Drexler. Stephenson returned the favor in his book with a passage describing a fresco of scientific giants that includes Drexler. Gibson, too, submits his writings to scientist friends who vet the manuscripts for any erroneous hypotheses. In his *Neuromancer* screenplay, Gibson wrote about "arbitrage engines," software which helps traders spot price discrepancies on world markets. A knowledgeable friend told him to take it out. "He said there won't be arbitrage engines in 10 to 20 years, because arbitrage will disappear. It can't exist in a completely wired world of instantaneous communications," recalls Gibson.

Do they get rich off their prophecies? Even when their predictions end up true, science fiction authors rarely have much to show for their perspicacity beyond the bragging rights. "A patent is a license to be sued," says Clarke. "Besides, if I had patented communications satellites in 1945, it [the patent] would have expired in 1965." That's how long it took Hughes to build Early Bird and for Comsat Corp. to launch it. Clarke received a few token shares of Comsat.

Gibson jokes that real foresight would have led him to establish some kind of ownership rights over the word "cyberspace." But Stephenson suggests that such attempts to restrict the spread of ideas could have a chilling effect: "I don't think 'cyberspace' [the word] would be as widespread right now if Bill Gibson had trademarked it and got around to stopping the people who used it."

Corporations have proved more eager to protect their intellectual property. Viacom has been particularly aggressive. The company owns rights to *Star Trek,* the creation of the late television writer Gene Roddenberry. Predicting that scientists might be inspired by Star Trek creations, he forbade Paramount to charge them licensing fees. Viacom, which owns Paramount, has disregarded Roddenberry's wishes.

That may be why no one at Motorola admits for attribution that its newest folding StarTac cellular phone looks suspiciously like the flip-open communicators from the original *Star Trek.* Viacom has threatened the communications giant with legal action if it does. In 1995 an Ontario-based scientific instrument company named Vital Technologies Corp. brought out a handheld computerized device that measures barometric pressure and other environmental conditions. It called it a "Tricorder," after a similar device on *Star Trek.* Viacom insisted on a royalty in exchange for use of the name. The company paid.

"Wonder if the computer's finished its run. It was due about now."

Chuck didn't reply, so George swung around in his saddle. He could just see Chuck's face, a white oval turned toward the sky.

"Look," whispered Chuck, and George lifted his eyes to heaven. (There is always a last time for everything.)

Overhead, without any fuss, the stars were going out.

—*Arthur C. Clarke,* The Nine Billion Names of God (1953)

Ultimately, science fiction's gift to the world isn't in the inventions that have piled up over the decades. It is in the people who read the works and were inspired to create in their own medium, whether in ones and zeroes or in nuts and bolts. For those of us who are never likely to invent a product that will change the world, the literature of possibility helps to replenish a crucial and all-too-scarce commodity: wonder.

Wonder. That's not a bad thing for investors and businesspeople to indulge in occasionally.

The Coming "Cyberclysm"

Arthur C. Clarke

Arthur C. Clarke is one of the most beloved of science fiction writers, in part because of the enduring popularity of Stanley Kubrick's film version of *2001: A Space Odyssey.*

At the *Visions* Web Site Link to the Arthur C. Clarke Foundation.

READING PROMPT

What is it that Clarke most eagerly anticipates or dreads about the future?

I have seen the future—and it doesn't work. According to a recent estimate, computer games now cost the U.S. some $50 billion a year in lost productivity. Before long, we'll be talking about real money.

Today's primitive interactive toys are only part of a vast spectrum of entertainment and information systems so seductive that they can preempt all other activities. A typical person's day in the 21st century will include two hours of channel surfing, four hours of viewing selected programs, six hours of catching up on prerecorded programs, six hours of exploring the HyperWeb, and four hours of artificial reality.

One piece of good news: The last item, adventuring in artificial reality, will solve today's most pressing problem, the population explosion—because virtual sex will be a lot better than the old-fashioned variety: position ridiculous, pleasure momentary, expense abominable.

The observant reader will have noted that this schedule leaves only two hours for the rest of the day's activities. Much of that time will be spent plugged into the most urgently required invention of the near future: the sleep compressor. (They are still working on the next step—the sleep eliminator.)

Some optimists may argue that we have already experienced one huge media explosion in the past, so we may yet survive another. As Ecclesiastes complained several millennia before Gutenberg, "Of making many books there is no end." Doubtless when the printing press was invented, and the life-long labor of patient scribes could be replicated in minutes, some farsighted monk lamented," I can see the day when there will be hundreds—perhaps even thousands—of books! How could one possibly read them all?"

No one ever did, of course, and it may not be long before reading itself is a lost art. A simpler, and much older, method of communication will allow dumb humans to interact with smart machines. Thousands of icons ("cyberglyphs") will have made literacy an unnecessary skill—except, alas, for lawyers.

The science-fiction writers, performing their traditional role of viewing-with-alarm, have long recognized the siren call of the Dream Machine, especially when it can bypass the body's external input/output devices and be plugged directly into the brain. More than half a century ago, Laurence Manning, one of the most visionary founders of the American Rocket Society, wrote a short story on this theme aptly entitled "The City of the Living Dead."

In *The Joy Makers,* James Gunn developed the concept further, describing a world in which the vast majority of people live "cocooned in life-support systems, experiencing nothing but engineered dreams." In 1954, Evan Hunter, otherwise known as Ed McBain, wrote *Tomorrow's World.* Hunter's work, according to the invaluable *Encyclopedia of Science Fiction,* "is exceptional in defending the supporters of vicarious experience against their puritanically inclined opponents, and one of the few science-fiction stories to assume that the people of the future will sensibly accept the Epicurean dictum that pleasure, despite being the only true end of human experience, ought to be taken in moderation." I wonder if Evan is still so optimistic, more than 40 years later.

All these dubious Utopias assume that someone will run the world while the dreamers enjoy themselves. The dangers of this situation were foreshadowed in H. G. Wells's first masterpiece, *The Time Machine* (1895). In this classic, the subterranean-dwelling Morlocks sustained the garden paradise of the effete Eloi—and exacted a dreadful fee for their stewardship.

There have been many science-fiction stories and, I think, at least one movie about frantic human attempts to unplug disobedient computers. The real future might involve exactly the opposite scenario. The computers may unplug us. And it would serve us right.

Utopia Comes Alive

Elaine Showalter

Elaine Showalter's piece is a report on a series of interviews she conducted for a British radio show with important women writers of the 1990s (the end of the century or, in the French phrase so popular at the end of the nineteenth century, the *fin de siècle*). Showalter begins with a look back at utopian writings of a hundred years earlier, focusing on British women writers of the 1890s noted for popularizing what even then was known as "the New Woman."

At the *Visions* Web Site Link to *New Woman,* a glossy contemporary magazine.

READING PROMPT

Showalter's piece is an extended comparison of women at the end of the last century (the fin de siècle *of the 1890s) and women today. How have things, according to Showalter, changed for the better, gotten worse, or remained the same? Do you agree?*

"It was very heaven to be young when I came to London in the nineties," the journalist Evelyn Sharp recalled in her memoirs. "I arrived on the crest of the wave that was sweeping away the Victorian tradition." While their brothers were gloomy, fearful, decadent, or tormented by PMT (Pre-Millennial Tension), Sharp and other New Women of the 1890s eagerly anticipated a new world of freedom for women and a complete regeneration of the relations between the sexes. New Women writers specialized in the literary genre of the utopia, imagining an ideal future in which they would share their aspirations and desires with New Men. They hoped to see and enjoy that future paradise. "The ninth decade of the last century," the suffragist Emmeline Pethick-Lawrence fondly recalled, "was a time of expansion and vision. . . . We read, discussed, debated and experimented, and felt that all life lay before us to be changed and molded by our vision and desire."

By 1896 observers such as H G Wells thought that the optimism of the New Women had exhausted itself; and the heyday of the feminist novel, with its daring utopian visions, was waning. But even as New Women of the 1890s acknowledged the disappointments and dangers of their vision, they refused to turn back the clock. As Olive Schreiner wrote in her parable *Life's Gifts,* her generation had to choose freedom rather than love, but the day would come when women would have both.

What of 1996? Are women still optimistic? Do they look forward to the millennium with confidence? Have the dreams of *fin-de-siècle* feminism been fulfilled? How do women's lives in the 1990s differ from those of their mothers, and what do they expect for their daughters? These were among the questions I asked some of the Newest Women of the 1990s—novelists, journalists, actors, professors and psychiatrists—including Susie Orbach, Erica Jong, Fiona Shaw, Nicola Barker, Jenni Diski, and Joan Smith, for the BBC series *Femmes de Siècle.* Their answers came as a surprise.

On one hand, women artists of the 1990s have found a genuine freedom to explore fantasy, sexuality, anger and adventure. Those in mid-career exult in the medical and technological advances that have prolonged youth and health, and added, as Margaret Drabble comments, "ten years to every plot." Women's entrance into the medical and psychiatric professions has made psychotherapy a more maternal, nurturing profession, and adventurous women are less vulnerable to the neurotic disorders that plagued early feminists.

On the other hand, depression and breakdown are still significant themes in contemporary women's writing. Although they may be leading healthier lives, with less drudgery and more exercise, not to mention HRT [hormone replacement therapy], they are also more obsessed with eating rituals, dieting and food taboos, and more concerned with what Naomi Wolf has called the PBR—the Professional Beauty Requirement to look and dress well in order to succeed. Passive resistance, in the form of refusal—refusal to marry, to eat, to play the expected feminine roles—is still a dominant negotiating strategy for many, although it may only be a metaphor for self-denial. Women may be feeling sexual and youthful longer, but the sexuality of older women—as the shocked and disapproving reviews of Doris Lessing's novel *Love, Again* suggest—is still regarded as outrageous.

And despite the power they themselves have gained, 1990s *femmes de siècle* are suspicious of women who hold political power. Women of the *fin de siècle* had to work through men in order to have political influence. In Florence Dixie's 1890 utopian novel *Gloriana, or, The Revolution of 1900,* the brilliant Gloriana actually disguises herself as a man, gets elected to parliament and eventually becomes Prime Minister, revealing her true womanhood and bringing prosperity, peace, full employment, and green parks. Only in utopian fiction could a politician be so impossibly beloved and pure, but even in the 1990s many feminists want to believe that a woman can become Prime Minister and appear benevolent and maternal. Having become inured to the venomous attacks leveled at Hillary Rodham Clinton for the past four years, I was still startled by the blanket feminist condemnation of Margaret Thatcher, milksnatcher, betrayer of her sex, ruthless politico.

A century ago New Women wanted sexual freedom, but they also wanted sexual security. Rather than attacking marriage they wanted to regenerate it. In Winifred Colley's *Dream of the 21st Century* (1902), for example, monogamy is not abolished but universal. They thought a lot about utopian forms of child-care, collective housekeeping and inexpensive vegetarian restaurants, but they also assumed that they would have servants to take care of the house and the children while they carried out their profession. Some radical feminists also imagined that women in the future—perhaps on Mars—would have the right to patronize brothels and to keep as many young lovers as they could afford.

To be sure, feminist radicals and dreamers of the 1890s faced personal conflicts as well. Their theory of the free sexual union—a man and a woman living together in mutual love without legal contract as long as they wished to do so—usually led to tragedy when put into practice. Indeed many *fin-de-siècle* feminist intellectuals came to see themselves as a transitional generation, which had to sacrifice love or motherhood in the interests of women's future freedom. With the establishment of socialism, declared Marx's daughter Eleanor and her lover, Edward Aveling, "there will no longer be one law for the woman and one for the man." They were to be the pioneers of the new law. But instead Eleanor discovered that Aveling had betrayed her. She killed herself, aged 40.

The expectations and dilemmas of women in the 1990s are different. "What they want most, as we approach the dawn of a new century," claims the *Sunday Times,* "is not to be Superwoman but simply to be themselves"—meaning to leave ambition to men. A new survey, the *Times* reported on 1 December, shows that British "Millennium Woman" has had enough of juggling family and career, and is looking for ways to simplify her life, such as becoming a full-time homemaker. One sociologist at the London School of Economics believes that in the future, truly dedicated career women will not have children at all, while full-time homemakers will have three or four. Only a sociologist would think this makes your life simpler.

In the US, journalists have been debating the decision of the actress Sherry Stringfield, who plays Susan Lewis in the top TV hit series *ER,* to bow out of her demanding role in order to have more privacy, be with her boyfriend and shop for pretty candles, despite an enormous salary and the adoration of 37 million viewers. Most of the press has been sympathetic to Stringfield, and only the *New York Observer* scolded her for lack of professional seriousness.

The women I spoke with for *Femmes de Siècle* were much less romantic than 1890s women about relationships with men, and more realistic—or cynical—about marriage, single (and double) motherhood, and the rise of the New Man. But these successful women were remarkably and uniformly content with their own lives, and many saw their work, whether running a fashionable restaurant or writing a best-selling novel, as compatible with love and motherhood, and possibly even an extension of an innate feminine style. Of course, one might reasonably contend that practicing psychotherapy, cooking, teaching and writing fiction about relationships are more compatible with traditional femininity than politics or business. I make no claims to be scientific on the basis of this privileged group. But I think that Millennium Women are writing fewer utopian novels and living more utopian lives.

The Future Isn't What It Used to Be

Tim Appelo

One recurring image in Tim Appelo's essay is the Carousel of Progress at Disneyland's Tomorrowland—an actual ride that fascinated park visitors in the 1950s and 1960s as they wondered at scene after scene of the promise of technological progress. In presenting the future that "isn't what it used to be," Appelo surveys the world of popular dystopia films.

Based in Seattle, Appelo writes frequently about literature, film, and pop music.

At the *Visions* Web Site Links to Disney's Tomorrowland; to the films, books, and authors mentioned in the essay.

READING PROMPT

What, according to Appelo, can you learn about the present—about ourselves—from popular film treatments of the future?

Back in the '60s, everyone knew what the future would look like. Everything would get cleaner and sleeker and whiter—the decor of every future-gazing movie since H. G. Wells' *Things to Come* (1936). Folks would ply skyways that laughed at gravity, as in Fritz Lang's 1926 *Metropolis* or 1962's *The Jetsons* or the 1967 update of Disneyland's Tomorrowland, with its elevated PeopleMover.

"When I was a child in the late '50s and early '60s, we got excited about the future," recalls David Gelernter, the author of *1939: The Lost World of the Fair,* a book about the Tomorrowland of its time, the New York World's Fair. "Jets and rocket ships and hovercraft! It made you want to work hard in school and learn science and figure things out and get a grip on the world, and it's sad that we don't have that feeling."

It's ironic that the old optimism has waned, says Gelernter, because the wonders predicted by the World's Fair have mostly come true: nylons, TVs, faxes, synthesizers, fluorescent lights, suburbs.

But today, with technology triumphant and the millennium two years away—or three, for those who insist that it starts Jan. 1, 2001—the word "millennium" has reversed the meaning *Webster's Dictionary* gives it: "A hoped-for period of happiness, peace, prosperity and justice."

The millennium countdown is a big deal lately—it's splashed on the covers of *National Geographic* and a special *Newsweek* issue and in a weekly feature in *Time*. It's the subject of a new book, *Questioning the Millennium*, by Stephen Jay Gould. And it's on so many people's minds that if you haven't made your 1999 New Year's Eve reservations yet, you'd better plan to party at home.

Yet in a deeper sense, America is anything but celebratory about the millennium. The most recent film about New Year's Eve 1999, Kathryn Bigelow's *Strange Days*, was one humongous techno-bummer of a bash. Futuristic films are practically all dyspeptically dystopian, from *The Postman* and *Gattaca* all the way back to *Blade Runner* and *Road Warrior*.

Few kids today are apt to say, as a 21st century moppet in white, pointy-shouldered clothes tells her similarly clad dad in *Things to Come*: "They keep on inventing new things now, don't they, and making life lovelier and lovelier!" New gizmos just seem to make life harder and faster, and even Tomorrowland has turned its back on the future.

Tomorrowland is being bulldozed, and when it reopens this spring, it will be a kind of museum to our past idealism about things to come. In part, this is because science has ruled out some of our rosier futuristic notions. The PeopleMover, for instance, turned out to be the worst possible way to move people. "It was the most energy-squandering exhibit we ever had," says Tony Baxter, a vice president for creative development at Walt Disney Imagineering. "Energy wasn't part of the equation then."

We used to think energy was infinite and that human ingenuity would kick evolution up to a whole new level. Beyond the famous evolutionary procession from ambitious fish crawling onto land to knuckle-dragging ape to upright *Homo sapiens*, we would surely become super-smart future beings with brains so big our skulls would bulge like the 30th century character Brainiac 5 in Superman comics. Smart people no longer think humanity is getting smarter. In fact, our computers are beating our chess masters, and the most advanced scientific ideas—such as superstring theory physics—may turn out to be literally beyond human comprehension.

"I think it's possible that we just aren't infinitely smart, as much of [science fiction] often seems to assume," writes Stephen Baxter, the mathematician-turned-sci-fi-novelist who wrote *The Time Ships*, an acclaimed 1995 sequel to Wells' *The Time Machine*. "We may well come up against fundamental limits of our nature."

The leading sci-fi writer of my youth, *Starship Troopers* author Robert A. Heinlein, would have called this a pantywaist defeatism that played right into the Commies' hands. To Heinlein, as to Wells (the first major fantasy writer to have studied Darwin in college), evolution meant progress, and the future belonged to the strong of will.

Today, scientists like Stephen Jay Gould have drummed it into our heads that we were just being big-headed about the future. Evolution doesn't progress at all, and, as Gould puts it: "Our brain has probably reached the end of its increase in size." If our skulls got any bigger, mothers could not give birth. We're basically exactly as smart as we were while attired in smelly mammoth-skin couture in the Pleistocene Era.

But our governing assumption about the future has sure changed. A Stone Age caveman could dream of a Bronze Age. A '60s kid could cherish the illusion of evolution as progress, especially if he was watching Tomorrowland's all-robot drama the Carousel of Progress, which has enjoyed more performances than any other theater piece in U.S. history. (It's still on view at DisneyWorld.) Now, however, everybody thinks the jig is up for apes like us. Popular entertainments about the future have trained us to dance to a post-apocalypso beat. Disneyland, which is immune to bad news, will get around the downbeat zeitgeist simply by sidestepping it. Instead of updating society's view of the future, the Imagineers will celebrate the retro-futuristic visions of Jules Verne, H. G. Wells and Leonardo da Vinci's notebooks. The old Rocket Jets ride will become the Leonardo-like Astro Orbitor. The Disney version of Wells and Verne shows up in the time-travel attraction the Timekeeper.

"In 1967 we hadn't had the apocalyptic visions of *RoboCop* and *Terminator,*" says Imagineer Tony Baxter. "We're burdened by all we've known since then. So we went back to the dream essence."

"Now Tomorrowland will be a historical treatment of what once were our idealized tomorrows," says historian John Findlay, author of *Magic Lands,* which analyzes Disneyland. "Disney is averse to predicting the future anew, and that makes sense. How can it compete with Bill Gates in this regard?"

"It's kind of creepy," says author Gelernter of Tomorrowland's retro future. But Disney's Baxter says the old Tomorrowland was "a little bit innocent and naive. People were . . . I don't want to say delusionary, but still very optimistic about the future. Doing the future is not always an easy thing."

"There's a very strong barrenness that permeates recent depictions of the future," says Judith Adams, a Jules Verne scholar and author of *The American Amusement Park Industry: A History of Technology and Thrills.*

She's not kidding. The most cynical (and ruthlessly factual) comic strip in American newspapers today is called *This Modern World,* by Tom Tomorrow, whose premise is the ridicule of the very idea of national progress. In Stephen King's new novel *Wizard and Glass,* when the future-era hero finds on some rusty 20th century oil tankers the motto "Cleaner fuel for a better tomorrow," he says bitterly, "Rot! THIS is tomorrow"—which, of course, is a decayed future dystopia.

Visual art used to be obsessed with the idea of tomorrow, but that was yesterday.

"The avant-garde revolution in art has become old news," writes art historian Donald Kuspit, "and nobody expects art to revolutionize—radically change—life. In fact, the idea of revolution itself has gotten a bad name."

This culture-wide reversal has been particularly tough on the American theme park, which has long been entwined with the idea of happy revolutions in our way of life. And the revolutions on exhibit used to pay off big time for society.

"The 1893 World Columbian Exhibition in Chicago made electricity seem safe and exciting," Adams says. In 1904, Parisian doctor Martin Couney came to America to show off the incubator he had invented to reduce premature-infant mortality by about 90%. "He couldn't get the rest of the medical establishment to accept his evidence. So he put it on display at Coney Island. He saved over 6,500 babies of the 8,000 brought to him, largely by the immigrant poor."

Such was the proud background of Tomorrowland's Carousel of Progress, with the refrain that Tony Baxter loves to sing: "There's a great big beautiful tomorrow, shining at the end of every day!" But the shining promise at the end of the Carousel of Progress—a nuclear power plant—has been replaced in pop culture by the nuclear plant where Homer Simpson works.

"Now technology, as shown in the comic strip Dilbert, is a killing thing," Adams says. "The way we use technology in our culture is as a way to destroy your peers, so you can be more successful yourself. You're never caught up with technology, you're never safe."

That's why we look back so longingly at the future that used to be.

"Verne and Wells do represent a lost paradise," Adams says. "I think there's a comfort in the nostalgic view of technology."

Says *Entertainment Weekly* multimedia authority Ty Burr: "In a way, it's nostalgia for the future that never happened. You can see it in Steampunk, a sci-fi subgenre that incorporates the 19th century industrial society of Verne into futuristic settings. Look at William Gibson's *The Difference Engine,* the film *Brazil,* with its fascination with ducts and plumbing and antique technology."

The best-selling books *Time After Time* and *The Alienist* fit into the same cultural mood of retro-tech longing, each inspiring a successful sequel. Burr also notes that in *Riven,* the sequel to the best-selling CD-ROM game *Myst,* which presents a desolate future and invites us to figure out what went wrong, the viewer "rides" in a "cable car out of *Blade Runner* by way of Jules Verne." The dials and levers and such in *Myst* signify pure Steampunk nostalgia.

Even ephemeral teen culture has acquired historical consciousness. When today's kids indulge in '70s retro sounds or fashions or giggle at the "Scream" movies, they do it with acute ironic awareness. The earnestness of youth culture has vanished forever.

We're all like the future dweller Bruce Willis in Terry Gilliam's dystopian film *12 Monkeys,* wailing mournfully, "I love 20th century music!" As critic

Richard Corliss said apropos of that sad film, "Dour sci-fi satire always has this message: I have seen the future, and it sucks."

It's a sign of our times that Gilliam made *12 Monkeys* after working on the Steampunk-ish film *Brazil* with screenwriter Tom Stoppard—and that Stoppard followed *Brazil* with his masterpiece, *Arcadia,* a play about the vanity of human hopes for intellectual certainty and progress in the face of chaos theory. Like *12 Monkeys,* the play concludes with a haunting final scene that offers a vision of the future and the past as a melancholy and eternal cycle.

On his deathbed, Walt Disney reportedly pointed to the ceiling, sketching plans for his Experimental Prototype Community of Tomorrow, then later succumbed to delirium, exclaiming, "Don't be late for the plane!" But the plane to the perfectible future remains stalled on the runway. Disney's linear, technological view of progress as a straight line pointed at a star fit to wish on has succumbed to a cyclical view of time—a tragic awareness of where our little dreams and techno-triumphs fit in the scheme of things.

The Carousel of Progress in its final incarnation in Florida features a significant change in its last act. Instead of heralding the great, big, beautiful nuclear-plant future, it concludes with a comic scene in which a robot grandma gets so excited while playing her grandson's computer game that she shouts out her rising score, inadvertently causing the voice-activated oven to burn their dinner.

This is a brilliant adaptation to modern consciousness, and it's typical of the most forward-looking Disneyland attractions. Instead of the omnipotent, ICBM-like original Rocket to the Moon, the new Tomorrowland will feature George Lucas' Star Tours flight simulator, the Timekeeper show and Honey, I Shrunk the Audience, each of which reassures us by showing technology that behaves like the Keystone Kops—it always screws up, but it's just funny. It can't hurt anybody. The only one upset about the Star Tours ride's near-collision with a toxic waste truck is the robot C3PO spluttering at R2D2.

"It helps people relax to see technology as clunky," says Adams, the Jules Verne scholar. "If technology isn't perfect, perhaps it won't take over their lives."

The Imagineers know we're scared of the future, and they've booted the scary, old-fashioned Tomorrowland machines from their garden. The original concrete and bright, pointy steel buildings are being rounded off, retrofitted in olden-day golden hues, with lots of softening vegetation. "The point being that the future is for humans," Baxter says. "It's almost like returning to Eden."

But the old Disney dream of erecting a futuristic techno-paradise is dead. In each scene of the old Carousel of Progress, ingenious Yankee technology got more powerful, until it formed a dome insulating us from the problem-ravaged Earth around it. Progress would snap us free of the chains of history. Though Walt Disney never had himself cryogenically frozen, the urban legend that he did so contained a germ of truth: Walt, dying of cancer from cigarettes, dreamed of defying time and making society immortal.

The folksy, fatherly robot on the Carousel of Progress said, "Now, most carousels just go 'round and 'round, without getting anywhere. But on this one, at every turn, we'll be making progress." But now, instead of singing about the great, big, beautiful tomorrow, Disney's audio-animatrons ought to be singing the Joni Mitchell lyric [about the carousel of time] that captures the folk wisdom of America today . . . .

The Earthly Paradise

Christopher Columbus

This account of the New World is taken from Christopher Columbus's third voyage, begun in May 1498, the first on which he actually sighted the continent—in this case the coast of what is now Venezuela. Although Asians had moved into practically all areas of North and South America thousands of years before and a few brave northern Europeans had briefly come ashore in Newfoundland, it was Columbus's great accomplishment to be the first to narrate the story of the New World, to try to situate this vast mass of largely unexplored fertile land in European consciousness. This meeting of Europe and Asia on the two American continents is arguably the great story of this passing millennium, although obviously the continuing controversy about the shape of the story (especially who are its heroes, its villains and victims) speaks volumes about ourselves.

At the *Visions* Web Site Links to Christopher Columbus sites.

READING PROMPT

What can we learn about Columbus and his world from how he imagines the earthly Paradise?

The Holy Scriptures record that Our Lord made the earthly Paradise and planted in it the Tree of Life, and then sprang a fountain from which the four principal rivers in the world take their sources: namely, the Ganges in India; the Tigris and the Euphrates in Turkey, rivers which divide a chain of mountains and, in forming Mesopotamia, flow thence into Persia; and the Nile, which rises in Ethiopia and falls into the sea at Alexandria.

I do not find, nor have I ever found, any account by the Romans or Greeks that fixes in a positive manner the site of the terrestrial Paradise, nor have I seen it given in any map of the world, laid down from authentic sources. Some placed it in Ethiopia, at the sources of the Nile, but others, traversing all these countries, found neither the temperature nor the altitude of the sun to correspond with their ideas respecting it; nor did it appear that the overwhelming waters of the deluge had been there. Some pagans pretended to adduce arguments to establish that it was in the Fortunate Islands, now called the Canaries; etc.

St. Isidore, Bede, Strabo, and the Master of scholastic history, with St. Ambrose and Scotus, and all the learned theologians agree that the earthly Paradise is in the east, etc.

I have already described my ideas concerning this hemisphere and its form, and I have no doubt that, if I could pass below the Equator, after reaching the highest point of which I have spoken, I should find a much milder temperature, and a variation in the stars and in the water. Not that I suppose that point to be navigable or even that there is water there. Indeed, I believe it possible to ascend there, because I am convinced that it is the spot of the earthly Paradise, whither no one can go but by God's permission. Yet this land that your Highnesses have sent me to explore is very extensive, and I think there are many other countries in the south of which the world has never had any knowledge.

I do not suppose that the earthly Paradise is in the form of a rugged mountain, as the descriptions of it have made it appear, but that it is on the summit of the spot I have described as being in the form of the stem end of a pear. The approach to it from a distance must be by a constant and gradual ascent, but I believe, as I have already said, that no one could ever reach the top. I also think that the water I have described may proceed from it, though it be far, and that, stopping at the place which I have just left, it forms this lake. There are great indications of this being the terrestrial Paradise, for its site coincides with the opinions of the holy and wise theologians whom I have mentioned. Moreover, the other evidence agrees with the supposition, for I have neither read nor heard of fresh water coming in so large a quantity, in close conjunction with the water of the sea, and the idea is also corroborated by the blandness of the temperature. If the water of which I speak does not proceed from the earthly Paradise, it seems to be a great wonder still, for I do not believe that there is any river in the world so large or so deep.

chapter 17 Explorations

17a Intentional Communities

Collaboration. Share an account of a group you have belonged to (formal or informal) that at one time or another discussed its rules for governing itself.

Web work. There is an Intentional Communities Web site, which defines its key term as "an inclusive term for ecovillages, co-housing, residential land trusts, communes, student co-ops, urban housing cooperatives and other related projects and dreams"—although, if poorly administered, some of these "dreams" may seem more like nightmares to their inhabitants. Yahoo! keeps a list of links to intentional communities at Society and Culture > Cultures and Groups > Intentional Communities; for ecovillages see Society and Culture > Cultures and Groups > Intentional Communities > Eco Villages.

The Better World focuses on communities organized strictly around the principle of sustainability, as does Global Eco-Village Network.

One writing topic. Write a critique of intentional communities, considering their strengths and weaknesses, what positive role they have to play in society, and their limitations.

17b Cults

Collaboration. Share accounts of your experience with or knowledge of cults, either firsthand experience or, as is more likely for most, experience through films or other media.

Web work. Trying to study cults on the Web can be unnerving and at times even a little terrifying. Organizations some would regard as cults themselves are free to present themselves via their own Web pages as anticult organizations; and who is to say that one person's cult is not another's righteous crusade. Scientology has been a most controversial organization in this light, with its sponsorship of *Freedom* magazine and its claim of "investigative reporting in the public interest" ("investigative" here not to be confused with "independent"). Members of the Church of Scientology were involved in efforts to take over the Cult Awareness Network (CAN), a group started in response to the gruesome suicide and killings of hundreds of people in Jonestown, Guyana, in 1978. Because of lawsuits initiated by people opposed to its work, CAN did file for bankruptcy in 1997, and its assets were purchased by members of the Church of Scientology.

FACTNet International, the Fight Against Coercive Tactics Network, provides a wealth of information and links, as does the American Family Foundation (AFF), with its magazine, *Cultic Studies Journal.* Meanwhile organizations like the Peregrine Foundation are in business to help family members retrieve people caught up in overly controlling communities or cults. Given what happened to the Cult Awareness Network in its battle with the Church of Scientology, do use caution in working with any of these sites.

The British newspaper the *Observer* offers a comprehensive online guide to cults.

One writing topic. Write an essay considering the strange possibilities that flow when a strong advocate for one side in an argument is successful in presenting itself to the public as a good source of information for solving the very problem most people think it is causing. What if the tobacco industry sponsored an antismoking Web site, the cattle industry, a vegetarian Web site, the automotive industry a Web site on mass transit? The possibilities are limitless and possibly a little frightening.

17c Childhood

Collaboration. From where do our pictures of a better world come? Some suggest that childhood (either our own happy childhood or the happy childhood we were denied but feel we were entitled to) provides some of the strongest possible utopian images. Share one such picture, drawn in words, from something (a perfect place, time, person, or event) associated with your own childhood.

Web work. No one has had a greater impact on creating an imaginary safe haven of childhood and its rich associations of small-town and family life than Norman Rockwell. Pineapple Publishing has a site featuring famous *Saturday Evening Post* covers painted by Rockwell. Doug Jarrett offers more online images.

Scouting is another force behind the image of ideal childhood. Check out these sites: The Boy Scouts of America, the Girl Scouts, and Campfire Girls and Boys. The two forces come together in Norman Rockwell's scouting pictures.

Meanwhile, child abuse remains a disturbing fact of life, and our awareness of abuse warns us against idealization. The Safe Child homepage, sponsored by the Coalition for Children, and TASK (Take a Stand for Kids) are just two places to look for more information on this subject.

One writing topic. Write a critique of the proposition that our images of future states of happiness, both for ourselves and for people generally, are often based on images derived from our own childhood (real or imaginary).

17d Marriage

Collaboration. Share with others your own feelings about marriage or about its future as a major institution in social life.

Web work. The Marriage Toolbox offers weekly advice about making marriages work, while the Couples Place bills itself as "supporting the adventure of relationship." Met Life offers advice, financial and otherwise, on getting married.

Carnegie Mellon University hosts a site on domestic partnership and same-sex marriages. The Human Rights Campaign is an activist organization supporting equal rights for gays and lesbians.

Meanwhile, not surprisingly, there are almost as many sites on divorce as on marriage. Here are just two: Divorce Central and DivorceNet. Finally, the Louisiana legislature has come up with a new form of the marriage contract, called *a covenant marriage,* that supposedly takes literally the words "till death us do part"; for more see the Web pages hosted by Louisiana divorce attorney Bob Walker.

One writing topic. Take a position on whether or not there should or can be more than one way or more than one set of conditions under which people are legally married.

17e Science Fiction

Collaboration. Share an account of a favorite sci-fi movie.

Web work. The IMDb provides a list of the top 250 films of all time (at least according to Web users); go there to see how many of them are sci-fi films. A good place to gather information on *Blade Runner* is with Murray Chapman's FAQ; other sites include the IMDb page and one by William Kolb. Meanwhile, there is also a FAQ for Philip Dick, the author of the original novel on which the film is based.

Time magazine offers a wonderful seventy-year retrospective of sci-fi films. Yahoo! Internet Life offers Science Fiction Online, while Science Fiction Weekly offers regular news and feature updates.

For a different take on things, check out the Feminist Science Fiction page.

One writing topic. Write a paper, pro or con, on the premise that science fiction films or books deal as much with the present as the future.

17f The Millennium

Collaboration. Share thoughts on what you did or what you had rather been doing at the start of the new millennium.

Web work. The Millennium Institute is dedicated to fostering a sustainable future and is hosted by the progressive Institute for Global Communications. The Millennium Foundation of Canada bills itself as "the first organization in the world dedicated solely to the creation of legacies to mark the year 2000." Other sites to visit include the Third Millennium; the Greenwich 2000 site for information on just when the new millennium started or starts; and Party2000, a site dedicated to planning the "biggest party on earth."

For a backward look at the new millennium, check out the writing of Edward Bellamy, a late-nineteenth-century author who, in his best-selling work *Looking Backward,* tried to imagine life in the year 2000. Andrew Woods has a page devoted to Bellamy. The complete text of *Looking Backward* itself, edited by Geoffrey Saucer, is available online.

One writing topic. One way to celebrate the future is to honor something in the past. Write an essay on the person, place, or institution that you would like to see honored at the start of the third millennium (whenever we agree that is).

chapter 18

Global Concerns

READING QUESTIONS

- How can technology assist or get in the way of simplifying your life?
- How can technology assist or get in the way of fulfilling your life?
- If it is true that without technology we all die, is it also true that with it we all live?

QUOTATIONS

And while the law [of competition] may be sometimes hard for the individual, it is best for the race, because it ensures the survival of the fittest in every department.

— *Andrew Carnegie, American industrialist and philanthropist, 1835–1919*

Progress is the mother of problems.

— *G. K. Chesterton, British author and conservative, 1874–1936*

The empires of the future are the empires of the mind.

— *Winston Churchill (1943), British statesman, 1874–1965*

Civilization and profits go hand in hand.

— *Calvin Coolidge (1920), American president, 1872–1933*

Use what is dominant in a culture to change it quickly.

— *Jenny Holzer (1979), installation artist*

One World, Ready or Not

David Ignatius

This piece by David Ignatius is a review of William Greider's 1996 book *One World, Ready or Not: The Manic Logic of Global Capitalism.* Greider is an unusual author, as Ignatius suggests, in being willing to ask big questions. In this case, the question is truly big indeed, maybe the biggest question facing us all, although for that reason not so easy to grasp: namely, just where is all this movement toward global trade, world markets, and international corporations taking us? Clearly there are many winners in this trend—for now, most notably the white-collar workers who handle the finances and legal arrangements involved with these new markets and, less dramatically, consumers who get a wider assortment of goods at a lower price. But what about the possible losers—notably the workers and their communities that are left behind as corporations shift production from one location to another?

At the *Visions* Web Site Links to other reviews of Greider's book.

READING PROMPT

Have you, or has anyone you know, been affected personally by the issues raised in Greider's book? What parts of Greider's arguments does Ignatius find most convincing? Least convincing?

The issue has been bubbling through American politics for the past decade, surfacing in what often seem unlikely places: Pat Buchanan's attacks on multinational corporations in the 1996 presidential campaign; Ross Perot's assault on NAFTA; Ralph Nader's campaign against the World Trade Organization as a threat to American workers and consumers; the home-grown radicals, left and right, whose literature attacks organizations like the Trilateral Commission and the Council on Foreign Relations for supporting free-trade policies.

It has become an underground river, this populist anger about the effects of economic "globalization." Tens of millions of working Americans believe that their jobs are being exported overseas to countries where labor is cheap—and that journalists and other members of the meritocratic elite, who don't share their plight, don't give a damn about it. This anxiety is everywhere in working America, as more companies "downsize" to cut costs and remain competitive in the global economy. But until recently, it has lacked an articulate spokesman. Now, in William Greider, it has one.

Greider dares to think big in his new book, *One World, Ready or Not*—on the same world-historical scale as the social theorists who struggled a century ago to understand the first capitalist explosion. His topic is nothing less than the fate of working people as global capitalism, "strong and supple, a machine that reaps as it destroys," accelerates into the 21st century. Greider is not optimistic, to put it mildly. "If my analysis is right," he declares, "the global system of finance and commerce is in a reckless footrace with history, plunging toward some sort of dreadful reckoning with its own contradictions, pulling everyone else along with it."

Greider even pays a brief but unembarrassed homage to the most potent of the 19th century theorists, Karl Marx. Although "the communist experiment failed utterly," he notes at one point, "the contradictions of capitalism that originally inspired Karl Marx's critique are enduring, largely unchallenged and uncorrected."

This is heresy, especially among economists, who tend to view Marx's economic ideas much the same way that astronomers regard astrologers. Indeed, the current high priest of economics, MIT professor Paul Krugman, denounced Greider's book in *The Washington Post* as "not only reckless but simplistic, and remarkably ill-informed." To Krugman, Greider is worse than illiterate—he's innumerate. Krugman is right that the book contains some dubious economic assumptions and technical mistakes, but these criticisms miss the point. This isn't an economics text. It's a work of grand social theory, with a startlingly broad intellectual ambition.

Big, woolly questions like the ones Greider is analyzing get relatively little attention from public intellectuals these days. We have become a nation good at snipping and sniping, but instinctively mistrustful of grand abstractions. The fate of the world; the global contradictions of capitalism; the unresolved legacy of Karl Marx—try talking about topics like those on CNN! The range of respectable opinion these days is, as Greider says, amazingly narrow. The desirability of global trade is a settled issue for most intellectuals, and not just economists. The Cold War is over; the forces of the free market have triumphed.

But Greider is insistent—earnest as an undergraduate, unafraid of being labeled a Lefty, determined to awake us to the cataclysm he sees ahead. Greider's book is "thinking outside the box," as management consultants are always recommending. It's big, it's brave, it's provocative. It gives voice to a powerful current in American politics.

GLOBAL WARNING

But is it right? That's a much trickier question, and my guess is that Greider's account is wrong. Not because he makes mistakes in his economics, but because he makes mistakes in understanding the ingenuity and adaptability of human nature. Like the grand theorists of the 19th century, Greider posits implacable forces shaping human history. But history isn't a straitjacket. It doesn't move in one direction. Things that look ruinous in the abstract often turn out not to be, in the practical ebb and flow of the real world. The great convulsions of human affairs tend to come from places where the theorists haven't been looking: the rise of anti-Semitism in Germany, the splitting of the atom, the invention of the microchip.

Greider's argument bears careful study, for it is radically pessimistic. "The global system is astride a great fault line," he writes. "The wondrous new technologies and globalizing strategies are able to produce an abundance of goods, but fail to generate the consumer incomes sufficient to buy them."

That's the central contradiction in Greider-economics. The specter haunting the capitalist world in 1997 is overproduction. As globalization spreads capitalism to every nook and cranny of the world, there are too many factories producing too many goods; and because workers around the world aren't sharing adequately in the new prosperity, they don't have enough money to buy all the products being churned out from Detroit to Kuala Lumpur. As a result, manufacturers face a permanent shakeout. It's downsize or die.

In this relentlessly competitive world, companies don't have the luxury to be generous. They can produce products anywhere, so they go where labor is cheap. Greider calls this "labor arbitrage," in which managers move whole factories to take advantage of differences in wage rates. This is Ross Perot's "giant sucking sound" amplified a thousand times. Workers suffer—both in the countries from which capital is fleeing and in the countries to which it temporarily bestows low-wage jobs. Greider's thesis echoes Marx's famous notion of a "reserve army of the unemployed" that would relentlessly force wages down and produce a final, devastating crisis for capitalism. (Marx was wrong about that, for some of the same reasons that Greider is wrong, as I'll explain later.)

Greider argues that what keeps this system going is the benevolent (and misguided) tolerance of the United States. In our devotion to free markets and free trade (an "enthralling religion," he calls it) we have become the world's leading dupes. Following the analytical path marked by James Fallows, he contends that the great economic success stories of the late 20th century—Japan, China, Korea—owe their success not to free trade, but to protectionist policies that allowed their domestic industries to develop and export. In Greider's view, the world economy is a zero-sum game, in which Asia's new prosperity (or at least, that of its capitalists) has come at America's expense. America has kept the unwieldy system going by acting as the "buyer of last resort," consuming a huge chunk of the world's overproduction. But we can't afford to continue our generosity.

The only beneficiaries of the new order are the financiers—the capitalists, if you will. The returns to financial capital have increased in recent years, says Greider, and so has its political power. So potent have the financiers become in the United States, he contends, that they have overwhelmed the business interests that actually make and sell things—and who would benefit from economic policies that fostered more rapid economic growth.

To bolster his case, Greider quotes various Clinton administration officials bragging about their success in keeping wage increases low. It is one of the virtues of this book that it helps you see such comments, which normally seem as ordinary as the nightly summary of the Dow Jones index, in their true outrageousness.

Here is where Greider is powerfully right. As our economy has shifted gears over the last 20 years, the richest Americans have been capturing an ever-larger share of the national income—more than at any time since the 1890s. At the same time, our market economy has developed a sharper edge. Companies that were once run comfortably for their workers and managers are now run for the benefit of the owners, the shareholders—which means the Wall Street money managers who invest our pension funds and 401k plans. This process has pushed U.S. companies to become better managed, more aggressive, and more competitive in international markets. But it has also thrown a lot of people out of work. The resulting insecurity may be good for our economy in the long run, but it's still unpleasant if you're on the receiving end. Moreover, you don't have to be a socialist to think that if the distribution of income in the United States continues to be skewed toward the wealthy we're going to have growing social and political problems.

THE GLASS HALF EMPTY

So what then is wrong with Greider's argument? The fundamental weakness, I think, is that in focusing on the "reckless footrace with history," it fails to appreciate the global economy as a dynamic, complex system that is constantly adapting to changing circumstances. That limited vision comes through in Greider's descriptions of what he found as he toured the world doing research for his book. He goes to a booming Malaysia and finds exploited village girls working in electronics factories; he goes to Germany and sees heartless financial traders who want to lower the wages of the people sweeping the floors. He goes to Poland and finds workers who have been the victims of capitalist "shock therapy." This is a man who is capable of seeing a cloud in every silver lining.

Greider's descriptions don't match my own observations of economic development in the Third World. The state-run socialist regimes that are slowly giving way to the market in places like Egypt, India, Indonesia, and Argentina were corrupt oligarchies. Their class systems, based on nepotism and political cronyism, were as rigid as those of feudal kingdoms. They offered a measure of security, but at a level that provided little more than subsistence wages for most people. Anyone who visited the oppressive, somnolent China of fifteen years ago and now sees the outpouring of energy, creativity, and wealth of the late 1990s cannot easily disparage the impact of global capitalism on the lives of ordinary people.

For all his reporting on the ground, Greider is intellectually most comfortable in the upper atmosphere, where those supra-human forces of history operate, pushing us toward ever-deepening contradictions and crises. That was Marx's mistake. He failed to understand the adaptability of modern capitalism, in part because he didn't reckon on technological ingenuity. Marx wrote before Thomas Edison, Alexander Graham Bell, the Wright brothers, Henry Ford, Thomas Watson, Bill Gates. Each of their inventions amplified productivity, lowered costs, created jobs and wealth on an astounding scale. Indeed, one reason for the relentless innovation of the modern economy is precisely that companies want to escape Greider's hyper-competitive, zero-sum world. So they innovate, and the world gets richer.

Another weakness of Greider's book is that he doesn't pay enough attention to the essential adaptive mechanism of a capitalist economy—which is the price system. In his repeated discussion of "overproduction," he glosses over the fact that prices are what bring supply and demand in line. People don't build factories unless they think they can make money. If there are too many factories producing too many goods, prices and profits will fall—and people will stop building factories until supply and demand are in sync again.

In his emphasis on static units of account—those stacks and stacks of products being produced by all the factories—Greider makes a mistake that's reminiscent of the environmentalists who warned a few years ago that the world was running out of natural resources. These analysts forgot that as resources become more scarce, their prices rise—stimulating more exploration and production (and development of substitutes), with the new supplies eventually lowering prices again. That isn't a hypothetical textbook example. It actually happened in the 1970s and '80s, with oil.

The adjustments required by the price system are sometimes painful—and Greider is right in worrying that the real wages of low-skilled American workers are likely to fall toward the levels of their global competitors in Mexico or China. That's where the Greiders and Perots and Buchanans have it right. Low-skilled American workers are getting screwed in the global economy, and the country needs to help them adapt. Clinton's emphasis on education and job training is just the beginning of what's needed. But the seriousness of this problem doesn't mean we should junk our economic system, and its built-in mechanisms of adaptation.

REGULATION, REGULATION, REGULATION

Because Greider doesn't like the human costs that the market imposes in times of transition, he argues for rigidities. In particular, he wants to see the U.S. government intervene to protect workers—with capital controls, tariffs to limit competition, taxes on wealth, debt forgiveness.

These are bad ideas, for the simple reason that they're likely to make things worse. The evidence against such government controls is persuasive: The Great Depression was created in part by precisely the kind of tariffs that Greider is advocating; the stagflation of the 1970s was created in part by wage and price controls; the capital flight that, until recently, was endemic in countries like Brazil and Argentina was caused in part by those governments' attempts to control capital.

Greider takes the Mexican peso crisis as evidence that economic globalization can have disastrous consequences. He speaks disparagingly of the Harvard-educated economists who tried to instill this thinking in Mexico—as if they were doing something wrong in trying to open a corrupt state-socialist regime to competition. Yet Mexico's real problem under President Salinas was that the free market never penetrated far enough—not that it went too far, as Greider maintains. The economic reformers were skating on top of a corrupt, one-party state. Even the President's brother was allegedly involved in drug running and money laundering.

FLEXIBLE STEEL

The only industry I can discuss with any confidence is the steel industry, which I covered as a reporter for *The Wall Street Journal* in the late 1970s. If ever there were an example that, at first glance, might seem to support Greider's thesis, it's steel. But Big Steel actually offers the clearest evidence that Greider's doom-saying is wrong

The American steel industry was an early "victim" of globalization. The United States had built up a vast steelmaking infrastructure, with enormous overcapacity. The men who ran the steel companies never liked to close a plant; they always wanted to keep a margin of extra capacity to take care of their best customers at the top of the economic cycle—even if that meant keeping open ancient open-hearth furnaces that should have been banked twenty years before. Like Greider, the steelmakers tended to think in terms of absolute units of output, rather than profitability. If the American steel industry was pouring 100 million tons of raw steel a year, that was inherently good—regardless of the profitability per ton. The leader of the industry, U.S. Steel Corp., was afraid of being too innovative, lest it increase its market share and draw an antitrust suit from the Justice Department.

Workers in this cosseted industry prospered. Represented by a strong union (which probably exerted greater control over the industry's operations than did management), steel labor was able to bargain for unusually high wages and benefits, totaling an average of nearly $20 an hour by 1980. Arguably, these high wages represented a tax on American consumers, who paid the bill. But steelworkers were blue-collar royalty. The only force that could upset this neat little world was foreign competition. To protect against precisely that threat, union and management bound themselves together in a no-strike agreement. And for nearly a decade, they used their joint political muscle to keep most foreign steel out.

When Japanese steelmakers first began to make inroads in the U.S. market in the mid-1970s, the U.S. industry argued that this competition was "unfair."

I spent much of my journalistic energy in those days demonstrating that this wasn't so. The Japanese were gaining market share because they were producing better, cheaper steel. They weren't clinging to old open-hearth furnaces. They were operating from new, "greenfield" plants that were the most efficient in the world. It may have been "unfair" that we bombed their old steel plants to rubble during World War II, forcing them to build new ones, but that's life.

Inevitably, the U.S. steel industry began to shrink. Companies laid off thousands of workers as they began to retrench to their best, most-efficient plants. The intricate web of steelmaking that had lined the banks of the Monongahela River since the days of Andrew Carnegie—the coal barges, coke ovens, blast furnaces, and open hearth furnaces, and the rolling mills, plate mills, wire mills—were all shuttered and closed.

It looked like an American tragedy in the making—a microcosm of the zero-sum world Greider describes. Japan had won; America had lost. And for many steelworkers and their families, the transition from that old world into a new one was wrenching. But for Pittsburgh as a whole (not to mention the United States) it proved a blessing. Many steelworkers (and more importantly, their sons and daughters) retrained and found jobs in new, more productive industries. As the workforce developed new skills, the region attracted new companies and whole new industries. The new Pittsburgh was less of a blue-collar, union town, but it was no less prosperous than before. Indeed, by every measure, real income grew. Even the steel industry was arguably better off for the changes. The most productive steel plants in the world are now found in the United States, and though there are far fewer steelworkers, the real take-home pay of those that remain is higher than ever.

Take a trip up the Monongahela River now and, true, you'll see an economic gravesite. But if you see only that, as Greider does, you've missed the point.

Toward an Opportunity Society

Newt Gingrich

Newt Gingrich has filled an extraordinary and yet somewhat typical place in American public life in the 1990s. Using his assertive personality, adroit political skills, and upbeat message about solving our national problems without government, Gingrich burst into the national attention, leading the Republican party to a historic 1994 victory that gave them a rare majority in the U.S. House of Representatives and made him the first Republican Speaker of the House of Representatives in forty years (and next in the line of succession to the presidency, after the vice president) and in the process becoming *Time* magazine's Man of the Year for 1995. Then, in 1997, he was reprimanded by the House for ethics violations and agreed to pay a $300,000 fine (covered in depth at CNN's AllPolitics). A year later, following the unexpectedly poor showing of the Republican Party in the midterm congressional elections of 1998, Gingrich stepped down as Speaker and gave up his House seat.

At the *Visions* Web Site Links to Gingrich sites, both friendly and hostile.

READING PROMPT

What is "American Civilization" in Gingrich's eyes, and just where does he locate it in its true form—in the past, the present, or the future?

In December 1992, I was working on long-range planning with Owen Roberts in Tampa, Florida. Owen is a brilliant student of free markets and the free enterprise system. He believes deeply in American Civilization and in the extraordinary range of opportunities that America has given so many of its citizens.

Yet that December day, Owen Roberts was in a rage of despair and frustration about the current limitations on the American ability to help people. The United States was going into Somalia to help save innocent Somalians from the starvation and death that their own warlords were inflicting upon them.

Roberts was not opposed to humanitarian missions and to American intervention to help innocent victims. Far from it. His frustration lay in the fact that our intervention would be temporary, our help would last only a little while, and we would lack the knowledge and commitment to truly transform Somalian society.

The things we could not teach Somalia were obvious, yet they are at the heart of the successes we Americans have had as a people. Free enterprise in a free market, entrepreneurship, productivity, incentives and the work ethic, citizenship, the rule of law, the right to free speech, and free elections were the kind of core values, principles, habits, and institutions that were necessary if Somalia was to develop a decent future for its citizens.

This collection of values, principles, habits, and institutions could be grouped together as "American Civilization." They were precisely the patterns de Tocqueville described in *Democracy in America* after his visit to America in 1831 and 1832.

I asked Roberts why we would think we could teach American Civilization to Somalians when we no longer teach it to young Americans. He agreed with me that we did not have a commitment to American Civilization in our schools. He then asserted that, if we could not teach Somalians because we no longer taught ourselves, maybe the answer was to begin by reasserting American Civilization here at home—to once again insist that America was a civilization worth studying and that renewing American Civilization was possibly the most important challenge our generation had to meet.

The more I watched the evening news of violence, brutality, drug addiction, and child abuse, the clearer it became that American Civilization has been decaying.

I talked with people about the challenge of helping the poor leave poverty and the vital goal of giving every American an opportunity to pursue happiness, live in safety, get a good education, and find a good job. It became clear that the welfare state had failed, and we had to return to the values and principles of American Civilization if we were going to truly help the poor leave poverty.

Furthermore, as we wrestled with balancing the budget in Washington, it became obvious that only a transformation of government functions and duties would allow us to eliminate the deficit. Yet such a transformation would require a deep shift in popular thinking about the role of the federal government compared with the roles of the citizen, voluntary associations (including religious institutions), private business, and local and state government. In effect, balancing the federal government's budget would require a return to the pre–New Deal (pre-1933) model of America, a model that had worked for 144 years as the most decentralized and least governmental society in the world. That transformation would require a lot of thought and a lot of dialogue with the American people. No change so profound and so far-reaching could be developed by so-called experts or imposed by elected officials on a free people. Only a thorough dialogue involving a lot of citizens could create an acceptance for a transformation on this scale.

A FOCUS ON THE FUTURE

It was also clear that our effort to revitalize American Civilization could not simply be a look backwards and an effort to return to some kind of "Golden Age" of a past America.

The power of the Information Revolution by itself would require us to rethink the principles of American Civilization within a new framework. Yet the Information Age is facilitating a second breakthrough that will prove to be equally decisive in transforming our lives. We are impacted both by a knowledge revolution and by the rise of a genuine world market.

The growth of the world market is a phenomenon with a long history. From fairly early times people have traded with each other. In the ancient world the Asian camel caravans connected Europe and China. Seagoing trade has been a constant for several thousand years. Yet there has been a long, steady rise in the importance of international trade that has begun to reach critical mass only in our lifetime.

As late as the 1950s, economics could still be described as a largely national phenomenon. Economists described the gross national product. Theories of inflation and economic cycles were largely national theories with a national focus. The Federal Reserve Board in America, the Bank of England, and the Bundesbank in Germany were each focused largely on national problems indicated by national data and solved with national tools.

The economic world has been transformed by the rise of a worldwide real-time information system and the growth of faster and less expensive means of communication. As people know more and more about opportunities across the planet, they shift their money to pursue marginal advantages without regard to national boundaries. In recent years the private flow of money has dwarfed anything the various national banks could do.

The economic change is about far more than merely money. As the cost of communication and transportation has declined, it has become possible to order goods and services from across the planet and have them delivered to your front door as easily as ordering things within the United States. The result is a growing connection between goods and services across the planet. In this new global market we will create American jobs through world sales.

It is a simple fact that future standards of living will be directly affected by what happens in the world market. Young Georgians are not being educated to compete with young Alabamians and young Floridians. Our young people are being educated to compete with the German, Japanese, Chinese, and Indian workers of the next generation. The challenge of competing in the world market will require us to rethink our approach to education, but that will not be nearly enough. The pressures of competition will force us to rigorously reexamine much more. Taxation, regulation, litigation, the welfare system, the cost of health care, and the very structure and inefficiencies of the current bureaucratic government will all have to be rethought if Americans are to compete in the world market of the twenty-first century.

This combination of the Information Age and the rise of a true world market requires that any effort to renew American Civilization must be forward looking rather than merely an effort to return to the past. It must represent a "conservative futurist" approach that might, at first glance, appear to be a contradiction in terms. However, combining an interest in what lies ahead with an effort to understand the lessons of what has happened in the past might prove to be the most powerful combination of all.

After all, our fascination with the potential of the future must not blind us to the value of the past. The American Civilization that we inherited has provided more opportunity to more people of more ethnic backgrounds than any civilization in history. People from across the planet come to America to learn to be American.

As one first-generation immigrant explained it to me recently: "Americans are optimistic and believe the future can be better. They tolerate each other within a remarkable diversity of personalities and habits. Americans believe everyone has the right to dream big dreams and to pursue happiness, and everyone has the right to define those dreams and that happiness on their own terms. In America, the government is subordinated to the people, and both politicians and bureaucrats are supposed to serve the citizens rather than vice versa."

Yet despite his positive, deeply emotional feelings about America, this man told me that, in his 30 years in America, he thought the unique characteristics that had so attracted him had grown weaker. "People are now less optimistic. They are less convinced that their children can lead full lives and actively pursue happiness. They are more frustrated and intimidated by government and have less of that independent spirit and assertiveness that have been the hallmark of Americans for nearly 400 years." This particular observer of the American scene (and a man who deeply loved America and thought it the best hope of mankind) was very worried that the key characteristics of America were beginning to fade and we were beginning to be "normal" in the historic sense of traditional societies with less hope, less freedom, and less opportunity for the average citizen.

In those observations my immigrant friend has captured the challenge we face: How can we both take into account the extraordinary opportunities of the information society and the emerging world market and combine them with the extraordinary strengths of American Civilization as it has existed over the past 400 years?

By combining the potential of the future with the lessons of the past to enrich the present, I think we can develop a tremendous increase in the opportunities for virtually every American.

Yet I believe the only way to get the scale of change we really need is to focus on educating citizens so they have the principles, the framework, and the tools necessary to effect change. I deeply believe that government cannot solve the problems by itself.

Indeed, the track record of the last three decades (beginning with Lyndon Johnson's Great Society) proves decisively that if government were the answer, and more taxes and more bureaucracy would solve things, then we would already have found a path to a more prosperous, safe America. Yet three decades of investing more and more resources in government answers has been a disaster for precisely the human beings we have been trying to help.

It is impossible to maintain civilization with 12-year-olds having babies, 15-year-olds killing each other, 17-year-olds dying of AIDS, and 18-year-olds getting diplomas they can't even read. Yet that is precisely where three generations of Washington-dominated, centralized government, welfare-state policies have carried us.

A RETURN TO CIVIC RESPONSIBILITY

Now we must find a new path to replace the welfare state with an opportunity society. We must replace our centralized government approach with a dramatically, even radically, decentralized approach—one that relies on each citizen and each community to provide leadership and creativity. This requires a degree of devolving power out of Washington that virtually no one has thought through at this point. Furthermore, this devolution of power cannot just be to shift responsibility and resources between Washington and the state capitals. It is not enough just to return power to state and local governments. We must think through the process of returning power to local citizens, local voluntary associations, private businesses, and only then to the local, state, and finally federal governments.

America is now facing a series of changes so profound that only an aroused, informed citizenry will be able to think through, decide, and implement enough changes on a large enough scale and with enough understanding of local realities.

It is impossible for a small group of politicians or bureaucrats to understand a country so vast and a people so diverse and numerous. Only a boldly decentralized system will be able to work through the many changes that the Information Age and the world market will make necessary in the next few years.

Yet a decentralized system that relies on citizen leadership and on voluntary activities actually requires a more thorough approach to developing principles, tools, habits, and values. A decentralized system has to have some core beliefs and core principles that are widely understood and agreed to if it is to be effective.

Decentralization only works if people have a clear sense of their general direction and the principles, values, framework, and habits that will mark that direction. Defining the rules of the game, outlining what it means to be an American, creating a framework of expectations so people will know what their civilization expects of them and what rewards they will get if they meet the expectations: Each of these core concepts has to be outlined and clearly understood if America is to work as a society.

Precisely because we want our central government to be limited in its powers and authority, it is vital that we bind ourselves by intellectual principles rather than governmental rules.

In a course I give on renewing American Civilization, I make the following case:

1. There is an American Civilization.
2. American Civilization has created the greatest range of opportunities for people of all backgrounds, from all ethnic groups, in the history of the human race.
3. There are principles of American Civilization that are worth studying because they are the keys to sustaining and maintaining the freedom, the prosperity, the safety, and the opportunities that were historically the birthrights of being American.

4. Every generation has two waves of immigrants who have to study American Civilization if it is to survive. One wave is geographic and are called foreigners. The other wave is temporal and are called children. Both waves have to learn the principles and framework of American Civilization if they are to function within it and to be successful in maintaining it for another generation.
5. The pressure to renew American Civilization is compounded by the pressure of the transformation into an Information Age and by the competitive pressures of the emerging global market.
6. Creating the optimal American society for the twenty-first century with the greatest opportunities for every American will require a determined effort by millions of citizens, all acting on their own as they define their duties and obligations. America will succeed by the renewal of civic responsibility, not by the centralized effort of government bureaucracies.
7. Therefore, the development of an empowering set of ideas that provide the principles, tools, habits, and structures of thought is the foundation on which civic responsibility can be built. Without this intellectual framework it will be impossible for individual citizens, civic leaders, or government officials to have a common framework and vision for their independent, decentralized efforts.

The rapid pace of change makes it vital to develop a common, general direction and a set of shared principles. The challenge of simultaneously renewing American Civilization, entering the Information Age, and becoming competitive in the world market is so great that it could not possibly be achieved by a centralized, bureaucratic approach.

The necessary changes are so complex and involve so much work that they could not possibly be done in Washington by a government-dominated process. If we are to create an opportunity society—an Information Age, world-market-oriented America with a renewed civilization—millions of Americans will have to be involved.

This approach is based on the fact that the experiment with professional politicians and professional government has failed. It is impossible to hire someone to make decisions for your society and then ignore the process and have an acceptable product. It is impossible to hire someone to educate your child and then ignore the process and have a satisfactory result. It is impossible to hire someone to ensure your safety and then ignore the process and be safe. We simply have to reassert civic responsibility and a full sense of each individual and each family and neighborhood having real power and real responsibility to contribute to their own future and to contribute to solving their country's problems within their own community.

This model for renewing American Civilization is very different from the big government, big bureaucracy model that focused on politicians and their ideas for solving problems. It is an active citizen model that returns power to the people to whom it belongs in the first place as "unalienable rights." It then calls on them to make a prosperous, safe America in which every American has a sense of personal strength and unlimited opportunity and in which the pursuit of happiness leads the American people to create a nation economically, militarily, and morally able to help the entire human race achieve self-government, safety, and prosperity.

Ecofeminism

Vandana Shiva

Vandana Shiva, a physicist, ecologist, and activist, is one of the most articulate voices for both women's rights and environmental issues in the Third World. Her numerous books include *Staying Alive, Ecology and Politics of Survival, The Violence of the Green Revolution,* and *Monocultures of the Mind.*

At the *Visions* Web Site Links to other Shiva essays and interviews, including one from Australia discussing the global impact of McDonalds.

READING PROMPT

Why does Shiva so oppose globalization? To what extent do you agree or disagree with her?

Q: How did you become an activist?

Vandana Shiva: My personal background is actually very unusual for the kind of career I chose. I didn't meet anyone who had ever done physics in my life. I grew up in the Himalayan forests. My father was a forest conservator, which meant that if I wasn't in school I was in the forests with him. That has been very largely responsible for my ecological inclinations.

One particular spark was when I went back to my favorite spot in the mountains where my father always used to take us before my graduate studies in Canada and finding that the stream I had gone swimming in wasn't there. The forest had been converted into an apple orchard with World Bank financing. The entire place, literally, had changed.

A second trigger for me was when I did a study on social forestry. It turned out that the World Bank was basically financing the conversion of food-growing land to timber-growing, pulp-growing land with huge subsidies. That study created a whole movement. The peasants and farmers reacted. They started to uproot eucalyptus. It created a major discussion of industrial forestry, and it was the first major challenge to a World Bank project in India. This was way back in 1981.

The director of the institute where I was working apologized about these young, enthusiastic researchers when the World Bank visited because he was afraid the institute would lose the World Bank consultancies. That's the day I decided that I had to follow my mind and heart. I couldn't be working for the bosses who were apologizing for the fact that I was following my conscience.

I went back home and started the Research Foundation for Science, Technology, and Ecology—an extremely elaborate name for the tiny institute that I started in my mother's cow shed. My parents handed over family resources and said, "Put them to public purpose." That's how I survived.

Q: How did you get involved with the Chipko movement?

Shiva: The Chipko activists have always been close to my parents, since my father was among the few forestry officials who supported them within the bureaucracy. And I was involved with the Chipko movement in my student days. The Chipko movement started up in Alakananda Valley, which is in the Himalayan foothills. The women were protesting against logging, which was destroying their fuel and fodder base, and making their springs disappear. They were walking longer distances to collect water. It was creating a very direct threat to their lives because landslides were occurring and floods were increasing. In the 1970s these women would go out in the hundreds and thousands and say, "You will have to chop us up before you chop this tree." These actions spread village to village.

In 1978 we had a huge flood. An entire mountain slipped into the Ganges River, formed a four-mile-long lake, and when that lake burst, we had a flood in the Gangetic basin all the way down to Calcutta, where homes were under one or two feet of water. It became quite clear that these were not illiterate village women just acting out of stupidity—which is what it was made to look like in the early days. Delhi, the central government, realized what the women were saying had something to it. We got a logging ban in the mountains as a result of that.

Q: You write that the Chipko movement flew in the face of the traditional paradigm of a charismatic leader.

Shiva: Absolutely. It was ordinary women who created this ingenious strategy and said, "We'll block the logging by embracing the trees." And the message spread from village to village, literally by word of mouth. There was no organized external leadership doing this. People like me came in to support the movement long after it had been given its articulation by the women.

A particular incident stands out very clearly in my mind. The government realized that the women were getting too strong. Their slogan was, "Why all these profits to these contractors and timber merchants?" The officials thought what they would do was put in place logging cooperatives made of local men. Then the government would get the revenue. They said, "We'll cooperatize this sector. We'll make it into a state sector. We won't have private contractors." They sent out the logging teams. In a particular village the logging unit was being headed by the chief of the village, and the protest against logging was headed by his wife, Bachni Devi. It was a tremendous conflict. The women were saying, "To us, it's the destruction of our forests. It doesn't matter who holds the axe. We want these trees to live."

Q: You wrote a book in the late 1980s with the German sociologist Maria Mies called *Ecofeminism.* Do you still use that term?

Shiva: Ecofeminism is a good term for distinguishing a feminism that is ecological from the kind of feminisms that have become extremely technocratic. I would even call them very patriarchal.

I saw some women had written that the cloning of Dolly was wonderful since it showed that women could have children without men. They didn't even understand that this was the ultimate ownership of women—of embryos, of eggs, of bodies—by a few men with capital and control techniques, that it wasn't freedom from men but total control by men.

Q: One of the things you talk about is *stri shakti,* women's power. What's that about?

Shiva: It's about the power that women of Chipko have. It's the power the women of the Narmada Valley have when they stand there and say, "Narmada mata is our mother. We will not let you either dam her or displace us." It's the *shakti* in the women who are blocking the industrial shrimp farms on the coasts of India. That amazing power of being able to stand with total courage in the face of total power and not be afraid. That is *stri shakti.*

Q: Tell me about your mother.

Shiva: My mother was a tremendous woman. I was just cleaning up old trunks and I found a book with her notes written during the war years, in the 1940s. She was studying in Lahore, which became Pakistan. She was writing about how women alone could bring peace to the world, that the men with all their greed and egos were creating all these tensions and violence. I always knew she was a feminist, ahead of her time. She brought us up that way, to the extent that we never felt that we had to hold ourselves back because we were girls. She didn't put pressure on us because she had spent her life removing that pressure from herself. In her day she was highly educated. She was among the few women in her community who became a graduate. She was an inspector of schools in the education department.

When partition took place in 1977, she came to India and decided to leave a highly privileged career and become a farmer. She spent some time doing politics on the side to build a new India. By the 1960s she had reached saturation with politics and did a lot of writing on spirituality in nature. She has always been a major influence for me in never feeling second rate because you're born a woman, never feeling afraid of any circumstance in life. I never saw her afraid, and yet, with all that, she was so deeply compassionate. She taught that if anyone needs you, you should be available to them.

Q: You often invoke Mahatma Gandhi. Why is he so important to you?

Shiva: I have a deep connection with him, partly because my mother was a very, very staunch Gandhian and brought us up that way. When I was six years old, and all the girls were getting nylon dresses, I was very keen to get a nylon frock for my birthday. My mother said, "I can get it for you, but would you rather—through how you live and what you wear and what you eat—ensure that food goes into the hands of the weaver or ensure that profits go into the bank of an industrialist?" That became such a checkstone for everything in life. We used to always wear *khadi* [handspun or handwoven cloth] as children. The fact that I still find so much beauty in a handicraft is because my mother taught us to see not just the craft as a product but the craft as an embodiment of human creativity and human labor.

My links with Gandhi now are very political links because I do not believe there is any other politics available to us in the late twentieth century, a period of a totalitarianism linked with the market. There is really no other way you can do politics and create freedom for people without the kinds of instruments he revived. Civil disobedience is a way to create permanent democracy, perennial democracy, a direct democracy.

And Gandhi's idea of *swadeshi*—that local societies should put their own resources and capacities to use to meet their needs as a basic element of freedom—is becoming increasingly relevant. We cannot afford to forget that we need self-rule, especially in this world of globalization.

Q: You write a lot about biodiversity. What do you mean by that?

Shiva: It's basically the diversity of all life forms around us: the plants, the animals, the microorganisms, both the cultivated and the wild. We have a very old conservation movement, particularly in the United States, which has focused on campaigns to protect endangered species: the spotted owl, the old-growth forest. But usually it stops there. To me, biodiversity is the full spectrum. Species conservation is not only about wilderness conservation. It's also about protecting the livelihood of people even while changing the dominant relationship that humans have had with other species. In India, it's an economic issue, not just an ecological one.

Q: How is globalization affecting India?

Shiva: American firms are beginning to reproduce nonsustainable systems, to force the elite of India to become energy consumers of the kind that the U.S. has become. That's what globalization is about: Find markets where you can. If China has markets, rush there. If India is an emerging economy with millions of new consumers, sell them the Volvo. Sell them the Cielo car. Sell them whatever you can, hamburgers and KFCs. It's the middle classes who have moved into being able to own a car, a refrigerator. For them there is this mantra that the General Electric refrigerator is better than some other model, that the Cielo car is fancier than the Ambassador.

Because of these new car models there is suddenly on the streets of Delhi a new intolerance by the motorists for both the cows and the cyclists. So for the first time the sacred cow in India, which used to be such a wonderful speed-breaker, is now seen as a nuisance. For the first time, I've seen cows being hit and hurt. These guys just go right past, and if the cow is sitting on the road, they don't care. We can't afford to have a sacred car rather than a sacred cow.

The other thing they're working very hard at doing is to try and make cyclists—including all the people who do servicing and sell vegetables on every street—declared illegal because they're getting in the way of the fast cars. It means robbing the livelihood of millions of people who are more ecological, who are helping save the climate for all of us. I hope that in the next two months I'll be able to work with some of these cyclists and vendors who are being made illegal on the streets of Delhi.

Q: The United States, with 3 to 4 percent of the population, consumes upwards of 40 percent of the world's resources. Why is no one talking about restructuring the economic system, the patterns of consumption?

Shiva: Amory Lovins has said that the only reason Americans look efficient is that each has 300 energy slaves. Those 300 energy slaves will now be reproduced among the elite of India.

The poorest of families, the poorest of children, are subsidizing the growth of the largest agribusinesses in the world. I think it's time we recognized that in free trade the poor farmer, the small farmer, is ending up having to pay royalties to the Monsantos of the world. It's not that Monsanto is making money out of the blue. It's making money by coercing and literally forcing people to pay for what was free. Take water, for instance. Water has always been free. We've never paid for drinking water. The World Bank says the reason water has been misused is because it was never commercially priced. But the reason it's been misused is because it was wasted by the big users—industry, which polluted it.

Today you have a situation where now the prescription is: People who don't have enough money to buy food should end up paying for their drinking water. That is going to be the kind of situation in which you will get more child labor. You will get more exploitation of women. You're going to get an absolutely exploitative economy as the very basis of living becomes a source of capital accumulation and corporate growth. In fact, the chief of Coca-Cola in India said: "Our biggest market in India comes from the fact that there is no drinking water left. People will have to buy Coca-Cola." Something is very, very wrong when people don't have access to drinking water, and Coke creates its market out of that scarcity.

Everything has been privatized: seeds, medicinal plants, water, land. All the land reforms of India are being undone by the trade liberalization. I call this the "anti-reform reform."

Q: You mentioned Monsanto, a major U.S.-based multinational corporation. You write that "the soya bean and cotton are now Monsanto monopolies." How did that happen?

Shiva: If you read Wall Street's reports, they don't talk of soya bean as originating in China. They don't talk of soya bean as soya bean. They talk of Monsanto soya. Monsanto soya is protected by a patent. It has a patent number. It is therefore treated as a creation of Monsanto, a product of Monsanto's intelligence and innovation.

Monsanto makes farmers sign a contract for Roundup Ready soya because the soya bean has been genetically engineered to tolerate high doses of herbicide, which means that it will allow increased use of Roundup by farmers. It's projected to reduce chemical use, but it will increase Roundup use. The reason Monsanto's done this is because the patent on Roundup runs out in a few years, and it's their biggest selling commodity. They sell $1 billion a year of Roundup, the herbicide. The contract with farmers forces farmers to use only Roundup. They cannot use any other chemical. Monsanto can come and investigate the farms three years after planting to see if farmers have saved the seed. Saving the seed—having even one seed in your home—is treated as a crime in which you are infringing on Monsanto's property.

The kind of capitalism we are seeing today under this expansion of property into living resources is a whole, new, different phase of capitalism. It is totally inconsistent with democracy as well as with sustainability. What we have is capital working on a global scale, totally uprooted, with accountability nowhere, with responsibility nowhere, and with rights everywhere. This new capital, with absolute freedom and no accountability, is structurally anti-life, anti-freedom.

Q: A majority in the United States opposed NAFTA [the North American Free Trade Agreement] and GATT [the General Agreement on Tariffs and Trade]. But this didn't stop those agreements. Why do you think that is?

Shiva: That's the most significant crisis in the world today. We have reached a stage where governments and political processes have been hijacked by the corporate. world. Corporations can within five hours influence the vote in the U.S. Congress. They can influence the entire voting patterns of the Indian Parliament. Ordinary people who put governments in power might want to go in a different direction. I call this the phenomenon of the inverted state, where the state is no longer accountable to the people. The state only serves the interests of corporations.

Governments have a favorite phrase: "lean and mean." But they've been made very, very fat for corporate interests. Look at the way your Patent Office is increasing in size. It's an arm of government. It's not getting thin. But to create the protection for the corporations, government is actually growing bigger than ever before, in every part of the world. Yet it's growing extremely thin as a protector of people.

GATT today is in my view the counterpart of the Papal Bull [the 1493 edict that legitimized European conquest of the world]. Renato Ruggiero, who is the director general of the World Trade Organization that came out of GATT, basically said that GATT is the world constitution. It's interesting that the people were not involved in writing this constitution. Do you want a world constitution whose only yardstick and measure is freeing up capital and commerce from any limits, whether it is social responsibility, the rights of workers, or restrictions on either exploitation of resources or dumping of toxins? And the free-trade treaty that we have is a treaty for the annihilation of life on this planet if we don't very, very quickly change the terms of politics and economics in the world.

Q: The slogan, "Think globally, act locally," isn't enough for you. Why?

Shiva: For me, both thinking and acting have to be local and national and global all the time. That's why I travel across oceans, miss flights, and sit at airports—for the simple reason that I believe the only way globalization can be tamed is by a new internationalism which recovers the local and recovers the national.

Q: It looks like you have a lot of fun, even though you carry on an exhausting schedule.

Shiva: I do have fun. Even when I'm fighting I'm enjoying it, for two reasons: I think there's nothing as exhilarating as protecting that which you find precious. To me, fighting for people's rights, protecting nature, protecting diversity, is a constant reminder of that which is so valuable in life. That is recharging. But frankly, I also absolutely get thrills from taking on these big guys and recognizing how, behind all their power, they are so empty. I just keep going at it. Each of these balloons does deflate. I've seen a lot of balloons get deflated in my life.

The Future of Marketing

Bob Wehling

The Cincinnati-based Procter & Gamble is a huge multinational corporation with more than 100,000 employees (now defined on its homepage as a "global community"), best known through its many brand-name household products, such as Tide, Charmin, Bounty, Pampers, Cover Girl, Head & Shoulders, and so forth (there are over 300 brands sold in 140 countries). With a Web page devoted to the company's purpose, principles, and core values, P&G's current motto is "We make every day better . . . in every way we can." P&G also hosts a Community Center Web page with information on volunteering. Bob Wehling was a senior vice president for advertising at Procter & Gamble when he gave this speech about the Web and the personal products industry to a global conference of advertisers held in Sydney, Australia, on October 31, 1995.

At the *Visions* Web Site Links to Procter & Gamble; to its major competitor, Colgate-Palmolive.

READING PROMPT

Calvin Coolidge is famous for having said that what was good for General Motors was good the country. To what extent do you feel this sentiment also applies to Procter & Gamble?

Over the next half hour or so, we're going to have some fun. We're going to take a test drive—at warp speed—down the information superhighway.

Our destination is the future: the future of marketing, actually, a future where smart marketers will be surfing across the internet into the homes of billions of consumers in hundreds of countries, at the speed of light.

We'll be driving by the virtual storefronts of companies that are already creating this future. Pay attention. Because when our trip is finished, you'll be ready for more than a test drive. You'll be ready to get wired into this exciting future yourself. I know, because that's exactly what happened to me—and to a lot of us at Procter & Gamble—after we took this trip.

PREPARING FOR THE NEW MEDIA FUTURE

Before we get started, let me give you just a little background.

There's been a lot of hype about the information superhighway for several years now. P&G got into the fray a year and a half ago when our chairman, Ed Artzt, gave a watershed speech to the American Association of Advertising Agencies.

Here's a brief clip that captures the essence of what he had to say.

"From where we stand today, we can't be sure that ad-supported TV programming will have a future in the world being created—a world of video-on-demand, pay-per-view, and subscription television.

"If that happens, if advertising is no longer needed to pay most of the cost of home entertainment, then advertisers like us will have a hard time achieving the reach and frequency we need to support our brands."

That talk created quite a stir. Dick Hopple at DMB&B [ad agency D'Arcy Masius Benton & Bowles] called it "a seminal speech for the industry." *Advertising Age,* the *New York Times,* the *Wall Street Journal,* and the *Financial Times* all covered the talk extensively. It was truly a wake-up call that got the industry moving.

The first thing that happened was the creation of CASIE [the Coalition for Advertising Supported Information & Entertainment], an industry coalition led by senior advertiser and agency executives from the Four A's [the American Association of Advertising Agencies] and the Association of National Advertisers.

They got to work in several areas. They kicked off a legislative and regulatory game plan both to protect ad-supported broad-reach television and to ensure advertiser access to new media.

They started work to establish common technical standards for new media. The lack of standards is one of the biggest barriers to advertiser participation in new media. Currently, applications developed for one new media provider cannot be used on another, without the advertiser incurring significant re-authoring costs.

But, perhaps most important, they did an inventory of new media research, looking at which new media technologies were closest to commercialization and how their emergence would likely affect consumer behavior.

The Future of Marketing

The most valuable learning that has come from this research is that the technologies that had seemed most imminent when Ed gave his speech—interactive TV, pay-per-view movies, video game channels—are being eclipsed by computer-based media.

This has happened principally because the economics of building and operating interactive TV systems are out of line with what consumers will pay. And furthermore, while only a few thousand households currently have access to interactive TV, more than 20 million people are on the internet today—and that number is growing fast.

As a result, the internet is now positioned to influence the future of marketing as much as—and maybe even more than—any other medium we've ever known.

The internet originated in the early '70s as a network to connect university, military and defense contractors. It was an information sharing tool that was never intended for broad public use. But throughout the '70s, as its capabilities were expanded, the network grew.

Then, in 1993, two things happened that dramatically accelerated the net's growth. First, the World Wide Web was created. The Web is the graphical portion of the internet. And, at about the same time, the first, easy-to-use navigation tools were developed. Together, these two breakthroughs turned the internet—or, more specifically, the Web—into a medium that consumers were attracted to and could find their way around.

As a result, the Web has grown—and continues to grow—at extraordinary rates. The number of Web sites is doubling every 53 days—that's growth of about 50% per month. Today, millions of consumers are "surfing" the net through online commercial services, like America Online and Prodigy, and through direct connections with navigators such as Netscape.

This is unprecedented. As the *Economist* pointed out recently, "no communications medium or consumer electronics technology has ever grown as quickly—not the fax machine, not even the PC. At this rate, within two years, the citizens of cyberspace will outnumber all but the largest nations."

What does all this frenetic activity mean for us, as marketers? It means that an entirely new form of marketing—interactive, online marketing—is emerging as a breakthrough way to sell products and services to consumers.

UNPRECEDENTED OPPORTUNITIES FOR MARKETERS

There are different types of marketing opportunities online. The most common, on the Web, are corporate sites, or home pages. A home page is a company's online storefront. It's a place where consumers go for product and corporate information. At a minimum, a company's home page can be an extremely effective public relations tool. At its best, a web site captures the personality of the company and presents such useful information and services that consumers not only visit but return again and again.

Another opportunity is "webvertising"—banner ads placed inside editorial or other content in Web-based media. What makes these banner ads so effective is that, by clicking on them, a reader can jump instantly to the marketer's own site.

This is a technology called "hyperlinking." And it works not only for banner ads, but even for straight editorial. For example, the *Economist* did an article about the internet earlier this year. In the online version of that article, a reader interested in a particular example—say, this reference to Zima beer—could simply click on the word "Zima" and, in an instant, be taken directly to the Zima home page.

No other marketing medium can accomplish this. It elevates the power of right-time marketing to an unparalleled level.

There's one additional marketing niche that's just beginning to evolve on the Web, and that's sponsored programming.

The most common programming right now are celebrity interviews. Oldsmobile, for example, features online discussions with a range of celebrities as a way to attract consumers to its corporate site.

But an even more innovative form of content is the virtual magazine—which, at its best, is far more than an electronic version of a print publication. HotWIRED is a good example. It's a highly-interactive journal of online culture that incorporates banner ads and commercially-supported hyperlinks throughout its editorial. In fact, they reportedly made $2 million in ad revenue last year from these kinds of placements—and you can bet they're just getting started.

As you can see, these are far ranging and very significant new marketing opportunities. Now, to be clear, mass media isn't going away. There will always be a need for the reach and frequency of mass advertising. But online marketing represents the next generation of advertising—and smart marketers are already using it as a powerful new way to create consumer awareness, to stimulate trial, even to sell their products and services.

P&G on the Net

At P&G, we're just getting online. We launched an experimental web site in Germany about five months ago. This site includes information about P&G, it announces employment opportunities and, most important, it contains information about a few of our brands.

For example, consumers can request a handy dispenser for our leading laundry detergent brand, Ariel, and find solutions to common laundry problems. As you can see, this first experiment is pretty basic.

We made our second—and somewhat more sophisticated—venture onto the net just two weeks ago, with our new WordSlam site tied to the U.S. introduction of Hugo, the new Hugo Boss men's fragrance.

This site is an integral part of our product launch and publicity campaign. The selling line for this new fragrance is, "Don't imitate . . . innovate."

The Hugo page builds on this theme with the first online Spoken Word poetry contest. Spoken Word competitions are a hot trend among Generation Xers and this unique site should attract young "surfers" from around the world. While they're visiting, we'll give them a chance to learn about the new Hugo line of products. We'll invite them to join in online forums and chats with celebrity guests. And we'll provide links to other sites, as well.

We're planning to expand our Web presence over the next several months. One prototype we're experimenting with is for our leading laundry brand in the U.S. We call it the Tide Stain Detective. If you spill red wine on your favorite cotton shirt, you can ask the Tide Stain Detective how to get it out. And he'll give you a fast—and proven—solution. This is the kind of value-added information consumers want, which builds loyalty to our brand.

It's premature for me to talk much more about our plans, but—as we take our test drive into the future—I can tell you some of what we've learned over the past few months while we've prepared to expand our presence on the Web.

So let's get going. We see six core benefits of marketing online.

Instant Access to the World

First, the net is global. It reaches consumers in literally every part of the world. It's hard to get a precise number of users, but we know that there are somewhere between five and six million host computers that provide Web access to at least 20 million consumers in over 100 countries.

The distribution is very uneven—90% of the host computers are in North America and Western Europe—but interest and access is growing around the world.

For example, there are now over 300,000 host computers in Asia/Pacific. In fact, Australia is the third most wired country in the world. In terms of the number of computer hosts per 1,000 people, Australia ranks behind only Finland and the U.S.

In Central and South America, there are only 16,000 hosts—but that's more than double the number that existed just a year ago.

One major restriction to the Web's global growth is language. The net is principally an English-language medium today, which makes it inaccessible to non-English-speaking users. But smart, global marketers like IBM are changing that by providing multi-lingual sites.

IBM offers a unique feature called Planetwide. By clicking on Japan, for example, a user can get information about IBM, its products or its operations in Japan—in Japanese. Or in Italian. Or Spanish.

More than any other I've come across, the IBM site demonstrates conclusively that this is truly a global medium. There is no other medium that enables you not just to reach but to interact with consumers in virtually every country on earth—if you do it right.

And its global reach is not just for the IBMs of the world. Because the price of entry onto the Web is relatively low, a corner flower shop, for example, has the potential to become a global flower powerhouse virtually overnight with the right kind of approach.

Unprecedented Depth of Sale

Second benefit: online marketing is self-selective. You know the consumers who visit your site are interested. They want to know more about your products. Not only does this help identify your highest-potential consumers, it permits a depth of sale that no other medium can provide.

Auto makers have been among the first companies to tap this potential. Virtually every major car company is online today: Ford, General Motors, Toyota, Nissan, Honda, Mitsubishi. They provide a world of information, from a lineup of models to lists of options to comprehensive dealer directories.

But the best among them, from what I've seen, is the Chrysler Technology Center. You can get all kinds of information on Chrysler—the company's environmental record, its financial performance, you can even chat with CEO Bob Eaton.

But most interesting, you can look at cars.

Want to see the cars of the future? Go to the Technology section and look at concept cars. For example, the Plymouth Back Pack—a part utility, part pickup and part sporty coupe that's likely to be a real hit with young buyers.

Or you can go to the showroom and look at the latest cars on the market today. If you're in the market for a Jeep, you've got your choice of three models. Click on the Jeep Grand Cherokee and you can look at the vehicle, pull down specs, compare it feature-by-feature to its main competitors—even read reviews from the auto press.

And that's not all: soon you'll be able to custom price the Jeep of your dreams—or the Chrysler or the Dodge or the Eagle. Choose the features you want and the system will automatically calculate the suggested retail price and even tell you what your monthly payment will be.

Together, these features are the perfect right-time marketing tool: they give you the chance to reach consumers who are interested—when they're interested.

Opportunities to Engage

The third benefit of online marketing is that it is interactive. It enables you to engage consumers in a way that no other medium can. The Chrysler site I just mentioned is one good example. Another is American Express ExpressNet on America Online.

You can apply for a card, check your account status, pull down photos of exciting travel locations, plan a trip and even make reservations. If you're travelling to Asia and want to know what's happening in Beijing this week, you can choose from sports to shopping to restaurants and nightlife. You can order a customized Fodor's travel guide that helps ensure you get to see and do exactly what you want to do no matter where you go.

Whether you're planning a trip or charting your expenses after one, ExpressNet puts you in charge. And reminds you that, no matter where you're going or where you've been, you can always count on American Express.

Fully Integrated Marketing

The fourth benefit is that online marketing is fully integrated. It combines the activities of every marketing discipline, from advertising to PR to direct mail.

Sony, for example, uses its web site to promote new products, from CDs to movies to electronics. It's even being used as a sampling device.

Click on Wiretap and check out the CD from Mariah Carey, one of the biggest recording artists in the U.S. You can even sample the video of her #1 hit, Fantasy.

You can find out where she's touring, enter a contest for a "Fantasy" weekend in New York, or win tickets to her live TV special.

This is truly integrated marketing. There is no other medium in which you can integrate so many different marketing tools at once.

Marketing One-to-One

The fifth benefit of being online is the medium's unique one-to-one marketing capability. This is the most important loyalty-building benefit of online marketing, because it gives you the ability to establish enduring relationships with individual consumers.

McDonald's McFamily, on America Online, is a good example. This site reinforces the idea of McDonald's as a parent's best friend. Parents can get information on the latest Happy Meal or nutritional information about McDonald's products.

But even more valuable is the community McDonald's has created. For not only can parents talk directly to McDonald's, they can also talk with experts and with each other on a whole range of parenting issues—from how to keep the family healthy and safe to how they can spend more quality time with their kids.

McDonald's has proven the benefits of relationship marketing through a number of direct mail programs they've used in recent years. And they believe the interactivity of online marketing will make it even easier to build relationships with consumers. In fact, the level of relationship building they're achieving online would have been almost impossible to create even two or three years ago.

The Virtual Store

Finally, there is one last and very important benefit: you can sell.

A great example of this is right here in Australia—the Flag network of hotels, inns, resorts and apartments. Flag is Australia's largest independent accommodation group with properties in Australia, New Zealand, Papua New Guinea, Fiji, Western Samoa as well as the U.S. and the U.K.

They make it easy to stay at a Flag hotel or inn. A traveler can navigate by map to any of the 450 facilities in the Flag network. If you're on your way to Australia, simply click on the region—say New South Wales—and the city—perhaps Sydney—and you're ready to select your hotel.

You can pull down details and ratings of every Flag hotel in the area and, once you've made a decision, enter your reservation.

Flag will confirm within 24 hours, either by phone or e-mail—whichever you request.

There are other benefits of being online, as well. But these six are the most important, because they demonstrate that everything we do to build consumer loyalty is affected by this new medium: from awareness to trial to purchase and repurchase. It is a major step forward in the evolution of marketing.

NEXT STEPS: TAPPING THE NET'S POTENTIAL

The key question is, how do we tap the full potential of this exciting new medium? How do we make it a global bonanza for advertising and a rich source of entertainment, information and community for consumers?

To do this, we have to resolve a number of issues—from the security of online transactions to consumer privacy. But there are a few things, in particular, that I would urge you to focus on most.

Measuring Online Effectiveness

As an industry, we need to establish clear, broadly-accepted standards for measuring the cost effectiveness of the net versus other media.

All the examples I've just shared demonstrate that the Web can be a highly-effective marketing medium. But, for it to become a truly valuable tool for the industry, it must also be highly-efficient. And we need accurate, reliable measures to gauge that efficiency. CASIE has provided a good starting point. Just a month ago, they presented a set of guiding principles for measurement that I strongly endorse. We have copies of the executive summary in the back, but let me just mention the highlights.

First, audience measurement of interactive media should be directly comparable to the measures used for other media. This is fundamental to the reliability of audience research.

Second, audience measurements should be taken by objective third-party research suppliers and not by the medium being measured. What little measurement there is today rarely follows this principle.

And one more: interactive media research standards must be set by a broad representation of the advertising industry, including advertisers, agencies, media, research companies and industry groups.

In all, CASIE has developed 11 guiding principles. We need to use them to establish clear measurement standards as quickly as possible.

I've been told by some of our agencies that it may take three to five years to iron out the measurement issue. That's too long. We need to do it by next year so we can begin building on and justifying our investments in this new medium.

This is a global issue and it needs broad, global leadership from the advertising industry. So today, I'm calling for the WFA [the World Federation of Advertisers] to join forces with CASIE to accelerate the creation and adoption of measurement standards for interactive media.

By this time next year, at the '96 WFA conference, I'd like to see the WFA and CASIE jointly endorse a set of broadly accepted standards that will enable us to measure and compare the effectiveness of interactive media in any part of the world.

Getting Up Close and Personal

In addition to this industry-wide effort, there are a few important things that I encourage you to do as individual advertisers.

First, get up close and personal with the internet.

Any advertising professional who hasn't been online should get online. You simply cannot appreciate the potential of this medium until you've "surfed" it yourself.

In fact, seeing first hand how limited the current online efforts are will convince you that, even if you're not online today, you're not that far behind. No one out there—even the best, like those I've mentioned here today—is very far out in front at this point. Now is the time to get online. Not just to catch up but to jump ahead. It's a big opportunity.

Another thing you can do, when you get back home, is to be sure that your governments are not establishing regulations that will inhibit the growth of this medium or advertiser access to it. If that's happening, it is in your interest to try to change it. This is as strategic a public policy issue as any advertiser will face.

SIX QUESTIONS TO ASK BEFORE YOU GO ONLINE

Now, if you go away from this meeting thinking that Wehling's talk was interesting, but I'm not sure it has anything to do with me, I encourage you to think about it harder. Think about what the *Economist* said recently: "As a new medium with almost no distribution costs, the internet has the potential to reshape the media world, letting new competitors in and forcing established giants to evolve or die." That's a pretty clear call to action, in my book.

If, on the other hand, you go away from here fired up about the potential of this new medium—convinced that you've seen the future and it works—then I have some parting advice for you.

The key to winning online is content. That's true for any marketing medium, and it is especially true for the internet. And what's important to remember about online content is that it's not just a matter of piping a 30-second ad over the internet or converting printed materials into electronic form. It requires a fundamentally different approach.

My suggestion is to start with the six benefits I listed earlier. They should provide a valuable framework to guide your thinking. Take a look at your marketing messages and at how you're delivering them today. Then ask yourself six questions:

1. How can I make my message global?
2. How can I offer such useful and interesting content that consumers will come back to my site again and again?
3. How can I make it interactive?
4. How can I fully integrate the full range of marketing disciplines into the delivery of my message?
5. How can I make my message more compelling by using the net's one-to-one capability? And,
6. How can I use this technology to sell my products?

I think you'll find that the answers to these questions will provide a creative blueprint as you think about how to tap the full potential of the net.

FIND A WAY TO GET ONLINE

The key is: find a way to get online. Experiment. Learn about it. Get prepared. You may ultimately decide it's not right for your business. But make it an informed decision.

In fact, I echo what the *Economist* had to say. I believe that—over the long term—marketers who remain unprepared for the sea-change we're about to experience won't survive. Marketers who understand the implications and get ahead of the curve will not only survive, they'll thrive. They'll emerge more competitive than ever and they'll build relationships with consumers that are deeper and more enduring than any we can create today.

The future of marketing is bright and the smartest among us are going to take tremendous advantage of that potential. I encourage you to be among the leaders.

Thanks for the chance to talk with you today. I look forward to seeing you online.

chapter 18 Explorations

18a Personal-Care Products, or Selling Hygiene

Collaboration. The range of products individuals need to purchase in order to take care of their daily hygiene needs continues to change, in part because of intense marketing by corporations that produce personal-care products. Make and share anonymously a list of products that you have started to purchase only in the last few years or, conversely, of products you no longer buy.

Web work. Yahoo! lists nearly 400 companies offering skin-care products at Business and Economy > Companies > Personal Care > Skin Care and has large numbers of links under hair care, fragrances, cosmetics, and (obviously, with aging baby boomers, a growing category) hair replacement (104 listings in September 1998). The British-based Body Shop has grown into a large multinational corporation with 1,500 stores by offering consumers the right personal-care products.

One writing topic. Take an in-depth look at one particular personal-care product you now use or have used in the past, and see how many different competing products you can find and on what basis, if any, they compete with one another. What would make you change brands or alter your pattern of use? What role does advertising and image manipulation play in your consumption?

18b Barter and Resale

Collaboration. Share a detailed account of a time or times when you have traded for something instead of paying cash.

Web work. Open-Air Market Net offers what it calls a "world wide guide to farmers' markets, street markets, flea markets and street vendors." Also check the Flea Market Guide of US Flea Markets and the Web site for what claims to be the world's largest flea market, that in San Jose, California. Meanwhile, online auctions, led by eBay, have become unexpectedly popular Web sites.

Joseph Soares of Yale University's Sociology Department has a site on advertising in America; also check out the homepage of the PBS special on overconsumption, Affluenza.

One writing topic. Write an essay comparing regular stores in your community with the flea markets, garage sales, and other venues of reselling or barter.

18c The Year 2000: Start of the Asian Century?

Collaboration. Share a detailed account of your strongest impression of or association with any country of Asia.

Web work. Cultural Bridges provides what it calls "opinionated travelogues" describing numerous different spots, many in Asia. AskAsia is a site for students and educators. Sawnet, the South Asian Women's Network, provides much helpful information and links.

Everything about Asia is huge—the population, the potential, the problems—and within Asia everything associated with China takes on special import. Discover China is a good place to start; Inside China is a news service of the European Internet Network.

North Americans, at least as far back as Walt Whitman's great poem "Passage to India," have had a special interest in the Indian subcontinent. Check out the *Times of India* newspaper site.

Asia is regularly looked at by Western industries as the place in the world with the most people but not the most products, and hence as a huge untapped market of consumers—for instance, millions upon millions of potential new customers for automobiles. The Ford Motors page has links to all its worldwide subsidiaries, many in Asia.

One writing topic. For centuries writers in the West have turned to Asia, which they saw as so different from their own countries, as a source of lessons they wanted to teach their fellow Europeans or Americans, often lessons about living more harmoniously with nature. One famous example, twice made into Hollywood films, is James Hilton's 1933 novel *Lost Horizons.* Write an essay in which you describe an imaginary country (Asian or otherwise) that has values you see as lacking in your own community.

18d Sneakers and Free Trade

Collaboration. Make and share lists of arguments commonly made in support or in opposition to the United States' extending free-trade agreements with other countries.

Web work. Anti-Nike forces are led by the Boycott Nike page; an Australian site has the full text of "Sweating for Nike" by Tim Connor and Jeff Atkinson. For now (at least until Nike offers Yahoo! its own shoe contract), Yahoo! has an entire page of anti-Nike links at Business and Economy > Companies > Apparel > Footwear > Athletic Shoes > Brand Names > Nike > Consumer Opinion > Anti-Nike, as well as a pro-Nike page.

Nike's homepage has company information, including a page with labor updates. Sneaker Nation offers a lightweight look at sneaker fashions and trends.

The Victorian Web at Brown University offers both a brief overview of child labor practices in nineteenth-century Britain and extensive excerpts from nineteenth-century eyewitness reports.

The Institute for Global Communications, an Internet clearinghouse for progressive groups, has a site on labor issues, Labor.Net. More information on child labor is available from the World Confederation of Labour, founded in the Hague in 1920 and representing twenty four million workers, mostly in Third World countries. Another good source of information is the Green Culture homepage.

The Calvert Group has led the way in creating socially responsible mutual funds. Green Money Online has the motto "Responsibility from the Supermarket to the Stockmarket."

One writing topic. Write an argument on the role that working and environmental conditions in other countries should play in our decisions about buying products manufactured in those countries.

18e Progress, Global and Local

Collaboration. Make and share a list of specific ways you feel your life and your local community are generally getting better or getting worse.

Web work. There is a strong body of U.S. political thought, now largely associated with the right, that sees progress mostly in global terms, involving continued economic expansion fueled by new technologies, innovative thinking generally, and limited government. Two policy groups that generally support this view are the Progress & Freedom Foundation, formed in 1993 on the belief, as stated in its mission statement, that "Mankind has advanced in the past, is presently advancing, and will continue to advance through the foreseeable future," and the Cato Institute, an openly libertarian association founded in 1977.

Likewise, there is an opposing body of political thought, now largely associated with the left, that sees progress as mainly a matter of improving things locally, specifically in terms of quality-of-life issues that affect everyone in a community, such as curbing air and water pollution and providing as an entitlement local community services in areas such as health, transportation, and recreation. Two policy groups that generally support this view are the Center for Policy Alternatives and the Sustainable Community Network.

Policy.com advertises itself as the site that directs people to policy and other advocacy groups on both sides of the issues.

Starting in 1995 the editors of the *Utne Reader* attempted to identify what they called 100 visionaries, although some of the people on their list are more notable for finding their inspiration in the past.

One writing topic. Consider one important way that the bumper sticker "Think globally, act locally" might be applied to a situation in your home or campus community. To do this you will need to combine your own personal research into local problems with the larger issues raised by organizations in the Web work above and by the readings in this chapter.

18f The Simple Life

Collaboration. Many of us—futurists and Luddites alike—lead similar lives, shaped as our lives are by the same basic technologies: the automobile, television, telephone, personal computer, and so on—just as much of Thoreau's life, notwithstanding the relatively short time he spent alone at his cabin on Walden Pond, was not all that different from the life lived by his fellow citizens of Concord. What made Thoreau's life special, and what makes many people's lives special, is not the major part of their lives that is shaped by the prevailing technology but that one aspect of their life in which they insist on doing things their way. Make and share two lists of activities you enjoy: *(a)* activities that were not available to your grandparents or even your parents and *(b)* activities that are essentially unchanged from what your parents or even your grandparents also enjoyed.

Web work. One issue to investigate here is that of sustainable development: the ideal of maintaining, even increasing, the standard of living in a community while keeping the rate of change such that the future imposes no greater burden on the environment than the present.

EcoVillage in Ithaca, New York is one such effort.

Simple Living magazine, "the journal of voluntary simplicity," is a product of the Simple Living Network, which also hosts Dave Wampler's Web of Simplicity, offering links to other simple-living Web sites.

One writing topic. Write a letter either to your grandparents or to your grandchildren focusing on one of two things: (1) one or more enjoyable traditional activities that have been passed down and that you hope will continue to be passed down between generations, or (2) one or more enjoyable activities that are new to your generation or that you expect will be new when your grandchildren are your current age.

"The Wild" *Henry David Thoreau*

Things seem to have become backward, with many nominal conservatives such as Newt Gingrich (in Chapter 18) extolling the virtues of a largely unknown future (to be created by the unleashing of corporations) and many nominal progressives like Jay Walljasper (in Chapter 14) nostalgic for an earlier, less technologically advanced society—and with still others yearning for the earlier industrial age of roadside diners and others for the agricultural perfection of the Amish. Progressives lost in the past; conservatives pining for the future! Meanwhile, the great Henry David Thoreau—a hard-to-pin-down mixture of radical and conservative—is at his most intensely poetic in the following piece, writing about the need to abandon all technology in order to allow himself (and perhaps to force all the rest of us) to confront the natural world as it truly is, devoid of all human interference: what he calls "the wild."

There is a story about an ancient people who, during times of great trial, sought spiritual renewal by following a time-honored pattern of going to a secret spot deep in the woods and there performing a sacred dance while reciting a mystic chant. Over years the chant was forgotten; then the people went to the sacred spot in the woods and did only the dance. Then over the years the dance was forgotten, as eventually was the location in the woods—and yet the people were not lost, for they still had this story they could repeat of an ancient time and an ancient people who found solace deep in the woods. And so it was with Thoreau, as he attempted to renew the spirits of the most technologically advanced generation the world had ever known by retelling his own account of this primal encounter with the world.

In the twentieth century the word *primitive* has regularly been used to express a longing for emotional identification with a world with only the simplest forms of technology. The French painter Paul Gauguin popularized interest in the primitive when he moved to and started painting pictures of Tahiti.

A few random *primitive* sites, just to show how far we have come from Thoreau's vision a century and a half ago: Urban Primitive, "a body design studio specializing in original, high quality, body enhancing tattooing, scarification, branding, body piercing, mehendi, permanent makeup and cosmetic retouching of scars"; Future-Primitive, "Professional graphic artists specializing in custom 3D artwork. We create realtime 3D environments, characters, animations, 2D and 3D graphics for games and the Web, VR simulations, VRML worlds, user interface graphics, and 3D architectural design, modeling and walkthroughs"; the alternative band Primitive Puppet; and the Baybarch Primitive Baptist Church in Claxton, Georgia.

Nicholas Roeg's 1971 film *Walkabout,* made available on video in 1996, tells the story of two normal (civilized) Australian youths (a fourteen-year-old girl and her six-year-old brother) forced to confront life in the wild with the aid of an adolescent aborigine. The title refers to the aboriginal ritual of sending youths out in the wilderness for six months to test and develop their survival skills.

At the *Visions* Web Site Links to all things wild.

READING PROMPT

What is it about the wild that so fascinates Thoreau?

The West of which I speak is but another name for the Wild; and what I have been preparing to say is, that in Wildness is the preservation of the World. Every tree sends its fibers forth in search of the Wild. The cities import it at any price. Men plow and sail for it. From the forest and wilderness come the tonics and barks which brace mankind. Our ancestors were savages. The story of Romulus and Remus being suckled by a wolf is not a meaningless fable. The founders of every state which has risen to eminence have drawn their nourishment and vigor from a similar wild source. It was because the children of the Empire were not suckled by the wolf that they were conquered and displaced by the children of the northern forests who were.

I believe in the forest, and in the meadow, and in the night in which the corn grows. We require an infusion of hemlock, spruce or arbor vitae in our tea. There is a difference between eating and drinking for strength and from mere gluttony. The Hottentots eagerly devour the marrow of the koodoo and other antelopes raw, as a matter of course. Some of our northern Indians eat raw the marrow of the Arctic reindeer, as well as various other parts, including the summits of the antlers, as long as they are soft. And herein, perchance, they have stolen a march on the cooks of Paris. They get what usually goes to feed the fire. This is probably better than stall-fed beef and slaughterhouse pork to make a man of. Give me a wildness whose glance no civilization can endure—as if we lived on the marrow of koodoos devoured raw.

There are some intervals which border the strain of the wood thrush, to which I would migrate—wild lands where no settler has squatted; to which, methinks, I am already acclimated.

The African hunter Cumming tells us that the skin of the eland, as well as that of most other antelopes just killed, emits the most delicious perfume of trees and grass. I would have every man so much like a wild antelope, so much a part and parcel of nature, that his very person should thus sweetly advertise our senses of his presence, and remind us of those parts of nature which he most haunts. I feel no disposition to be satirical, when the trapper's coat emits the odor of musquash even; it is a sweeter scent to me than that which commonly exhales from the merchant's or the scholar's garments. When I go into their wardrobes and handle their vestments, I am reminded of no grassy plains and flowery meads which they have frequented, but of dusty merchants' exchanges and libraries rather.

A tanned skin is something more than respectable, and perhaps olive is a fitter color than white for a man—a denizen of the woods. "The pale white man!" I do not wonder that the African pitied him. Darwin the naturalist says, "A white man bathing by the side of a Tahitian was like a plant bleached by the gardener's art, compared with a fine, dark green one, growing vigorously in the open fields." Ben Jonson exclaims, "How near to good is what is fair!"

So I would say, "How near to good is what is wild!"

Life consists with wildness. The most alive is the wildest. Not yet subdued to man, its presence refreshes him. One who pressed forward incessantly and never rested from his labors, who grew fast and made infinite demands on life, would always find himself in a new country or wilderness, and surrounded by the raw material of life. He would be climbing over the prostrate stems of primitive forest-trees.

Hope and the future for me are not in lawns and cultivated fields, not in towns and cities, but in the impervious and quaking swamps. When, formerly, I have analyzed my partiality for some farm which I had contemplated purchasing, I have frequently found that I was attracted solely by a few square rods of impermeable and unfathomable bog—a natural sink in one corner of it. That was the jewel which dazzled me. I derive more of my subsistence from the swamps which surround my native town than from the cultivated gardens in the village. There are no richer parterres to my eyes than the dense beds of dwarf andromeda *(Cassandra calyculata)* which cover these tender places on the earth's surface. Botany cannot go farther than tell me the names of the shrubs which grow there—the high blueberry, panicled andromeda, lambkill, azalea, and rhodora—all standing in the quaking sphagnum. I often think that I should like to have my house front on this mass of dull red bushes, omitting other flower plots and borders, transplanted spruce and trim box, even graveled walks—to have this fertile spot under my windows, not a few imported barrowfuls of soil only to cover the sand which was thrown out in digging the cellar. Why not put my house, my parlor, behind this plot, instead of behind that meager assemblage of curiosities, that poor apology for a Nature and Art, which I call my front yard? It is an effort to clear up and make a decent appearance when the carpenter and mason have departed, though done as much for the passer-by as the dweller within. The most tasteful front-yard fence was never an agreeable object of study to me; the most elaborate ornaments, acorn tops, or what not, soon wearied and disgusted me. Bring your sills up to the very edge of the swamp, then (though it may not be the best place for a dry cellar), so that there be no access on that side to citizens. Front yards are not made to walk in, but, at most, through, and you could go in the back way.

Yes, though you may think me perverse, if it were proposed to me to dwell in the neighborhood of the most beautiful garden that ever human art contrived, or else of a Dismal Swamp, I should certainly decide for the swamp. How vain, then, have been all your labors, citizens, for me!

My spirits infallibly rise in proportion to the outward dreariness. Give me the ocean, the desert, or the wilderness! In the desert, pure air and solitude compensate for want of moisture and fertility. The traveler Burton says of it: "Your morale improves; you become frank and cordial, hospitable and single-minded. . . . In the desert, spirituous liquors excite only disgust. There is a keen enjoyment in a mere animal existence." They who have been traveling long on the steppes of Tartary say, "On re-entering cultivated lands, the agitation, perplexity, and turmoil of civilization oppressed and suffocated us; the air seemed to fail us, and we felt every moment as if about to die of asphyxia." When I would recreate myself, I seek the darkest woods, the thickest and most interminable and, to the citizen, most dismal, swamp. I enter a swamp as a sacred place, a sanctum sanctorum. There is the strength, the marrow, of Nature. The wildwood covers the virgin mould, and the same soil is good for men and for trees. A man's health requires as many acres of meadow to his prospect as his farm does loads of muck. There are the strong meats on which he feeds. A town is saved, not more by the righteous men in it than by the woods and swamps that surround it. A township where one primitive forest waves above while another primitive forest rots below—such a town is fitted to raise not only corn and potatoes, but poets and philosophers for the coming ages. In such a soil grew Homer and Confucius and the rest, and out of such a wilderness comes the Reformer eating locusts and wild honey.

To preserve wild animals implies generally the creation of a forest for them to dwell in or resort to. So it is with man. A hundred years ago they sold bark in our streets peeled from our own woods. In the very aspect of those primitive and rugged trees there was, methinks, a tanning principle which hardened and consolidated the fibers of men's thoughts. Ah! already I shudder for these comparatively degenerate days of my native village, when you cannot collect a load of bark of good thickness, and we no longer produce tar and turpentine. . . .

I do not know of any poetry to quote which adequately expresses this yearning for the Wild. Approached from this side, the best poetry is tame. I do not know where to find in any literature, ancient or modern, any account which contents me of that Nature with which even I am acquainted. You will perceive that I demand something which no Augustan nor Elizabethan age, which no culture, in short, can give. Mythology comes nearer to it than anything. How much more fertile a Nature, at least, has Grecian mythology its root in than English literature! Mythology is the crop which the Old World bore before its soil was exhausted, before the fancy and imagination were affected with blight; and which it still bears, wherever its pristine vigor is unabated. All other literatures endure only as the elms which overshadow our houses; but this is like the great dragon-tree of the Western Isles, as old as mankind, and, whether that does or not, will endure as long; for the decay of other literatures makes the soil in which it thrives.

The West is preparing to add its fables to those of the East. The valleys of the Ganges, the Nile, and the Rhine having yielded their crop, it remains to be seen what the valleys of the Amazon, the Plate, the Orinoco, the St. Lawrence, and the Mississippi will produce. Perchance, when, in the course of ages, American liberty has become a fiction of the past—as it is to some extent a fiction of the present—the poets of the world will be inspired by American mythology.

The wildest dreams of wild men, even, are not the less true, though they may not recommend themselves to the sense which is most common among Englishmen and Americans today. It is not every truth that recommends itself to the Common sense. Nature has a place for the wild Clematis as well as for the Cabbage. Some expressions of truth are reminiscent, others merely sensible, as the phrase is, others prophetic. Some forms of disease, even, may prophesy forms of health. The geologist has discovered that the figures of serpents, griffins, flying dragons, and other fanciful embellishments of heraldry, have their prototypes in the forms of fossil species which were extinct before man was created, and hence "indicate a faint and shadowy knowledge of a previous state of organic existence." The Hindus dreamed that the earth rested on an elephant, and the elephant on a tortoise, and the tortoise on a serpent; and though it may be an unimportant coincidence, it will not be out of place here to state, that a fossil tortoise has lately been discovered in Asia large enough to support an elephant. I confess that I am partial to these wild fancies, which transcend the order of time and development. They are the sublimest recreation of the intellect. The partridge loves peas, but not those that go with her into the pot.

In short, all good things are wild and free. There is something in a strain of music, whether produced by an instrument or by the human voice—take the sound of a bugle in a summer night, for instance—which by its wildness, to speak without satire, reminds me of the cries emitted by wild beasts in their native forests. It is so much of their wildness as I can understand. Give me for my friends and neighbors wild men, not tame ones. The wildness of the savage is but a faint symbol of the awful ferity with which good men and lovers meet.

APPENDIX:

Guidelines for Using Web Resources in Your Writing

Guideline 1: Understand the Three Basic Reasons for Acknowledging Your Sources

1. To avoid plagiarism; or, stated positively, to meet your intellectual or professional obligation to acknowledge the sources of distinctive ideas and the words expressing such ideas
2. To allow readers to check your findings in part by checking your sources
3. To establish your standing as someone aware of important and relevant work, individuals, and sources in your field

You can satisfy the first and third reasons for acknowledging sources given here by noting sources directly in your writing: by mentioning authors and the titles of books, articles, and so on directly in the body of your text. The second reason often, though not always, requires the use of a formal apparatus such as a "Works Cited" page.

For important additional information regarding reason 3 for citing material ("To establish your standing . . . ") see the section on The Web as an Information Source (in Getting Started II).

Guideline 2: Always Acknowledge Your Sources, Regardless of How You Cite Them

There is always some flexibility with regard to the exact apparatus used for documenting sources as called for above; but there is little, if any, flexibility with regard to the need for acknowledging sources. Here are three cases, for example, where one would not use any separate apparatus whatsoever for documenting sources: *(a)* certain kinds of popular magazine writing, such as that used in most weekly newsmagazines; *(b)* in-class examinations; *(c)* out-of-class papers where the few sources used can be easily cited in the text itself or where the citations are from the course text.

Guideline 2, restated: There are any number of acceptable ways to refer readers to your sources, some more helpful, more accurate, and less intrusive than others. Erring in the precise style of documentation can be a relatively minor mistake. Acknowledging sources is an entirely different matter: Failure to acknowledge a source, especially if you use direct quotations from that source without quotation marks, is ordinarily *an act of plagiarism* and an extremely serious matter, one that can result in your total loss of academic standing.

Guideline 3: Unless You Are in the Process of Typing a Direct Quotation, Do Not Write Your Paper with Secondary Sources Open in Front of You

Plagiarism in its most serious form results from students' writing their own paper while reading another secondary source on that same topic (for example, with a critical essay or another student essay open in front of them), and then either intentionally or unintentionally using some of the exact wording from that other source in their own paper without proper attribution, often without even quotation marks. Primary sources—for example, the poem, play, or story about which you are writing—are not a problem.

Guideline 3, restated: Writers who compose their own texts while reading other secondary sources create many serious problems for themselves. Students who work with open secondary sources must constantly ask themselves, "Are these my own words or do they come from this other source?" The best advice here is not to write your papers with open sources in front of you unless you have the specific need to quote from them; and then only open the source when you are making that specific reference.

How These Three Guidelines Work with New Electronic Sources

Citing published Web material. The Web is essentially a publishing technology, and thus in some sense anything that appears on the Web can be considered "published" and hence suitable to quote, cite, and document. One might extend this slightly to say that material appearing on people's personal Web pages can be considered as published and hence suitable for attribution, much like any printed source.

A Web page can be cited either in the text or in Works Cited by its URL. In the Works Cited, list the entry alphabetically like others, by author or title; include the original date of print publication, if available, then the URL, followed in parentheses by the date you used this page. This final date is important, because Web page locations change regularly.

If you are submitting your paper electronically and are using a software program that permits direct hotlinking, you can also provide an embedded link directly to the URL, either from the text itself or from the Works Cited citation. In addition to the hotlink, however, also include the full URL somewhere in your paper (preferably the Works Cited section), because the hotlink itself will be useless for people reading a printed form of your paper.

Check out the *Visions* Web page (www.abacon.com/tuman) for links to some more detailed explanations for citing electronic materials.

Citing informal, public online material. Informal remarks are public remarks made by individuals to an electronic group discussion and as such constitute a new class of reference material, especially if the exchange is archived and thus available for direct quotation. Like traditional "published" remarks, such comments are written and distributed by individuals with the intent that they should be widely read; unlike traditional publication, however, they are normally composed quickly, in response to an immediate impulse, and made public without the benefit of revision or rethinking, and often without the chance for careful editing.

As a general rule it is unfair to attribute such informal remarks to specific individuals if your purpose is at all negative or critical (that is, if you are trying to show how wrong or limited some idea is). The only appropriate way to use such informal remarks in a negative way is to paraphrase them and give them a generic attribution ("a classmate," "someone on the Grateful Dead discussion list," etc.). You are free to cite someone's informal comments in an entirely positive fashion, and to attribute those remarks to that person, provided you make the informal occasion of those remarks clear ("as so-and-so said in a message to the UFO-deniers discussion list").

Citing informal, private electronic material (such as email). Messages, electronic or otherwise, sent solely to you or to a well-defined small group (as compared to a public group) should be considered as private communication. You should not quote or otherwise cite such a correspondence without the permission of the original author.

Special concerns for classes sharing papers electronically. In contrast to the situation on the Internet, there is no truly public exchange within a single class: Instead, all exchanges are confined to students registered in the class. Furthermore, all reading and writing of posted work is under the direction of the teacher and is for the sole purpose of meeting specific course purposes. Therefore, it is always appropriate for the teacher in an online class to ask you to read and respond to classmates' papers and/or electronic discussion, either negatively or positively. Unless the teacher says otherwise, a posted paper within any class should be treated as a published source in that class. Hence, students are free to cite a classmate's work in that course, treating that source like any other: That is, they should *(a)* cite as little as possible, so that the citation remains pertinent and well integrated into their own text, and *(b)* use proper documentation.

In general, students in an online class who are citing other students should follow the same procedures that they do for traditionally published materials: That is, *(a)* acknowledge by author's name the source of any distinctive ideas (ideas that they get from another person or ideas that others are likely also to see as distinctive to that source); and *(b)* use quotation marks to indicate exact, word-for-word quotations, as well as to note distinctive individual words or phrases. As stated in Guideline 3, it is best for students to paraphrase classmates' work without having that work open in front of them. If students do follow both *(a)* and *(b)* here—acknowledge sources and use quotation marks—then they need not list their classmate's paper in the Works Cited; this advice applies to citing and documenting the teacher's words as well.

Three Final Warnings for Students in an Online Class

Warning 1. Messages in an online class are private between students, and the rule given above for not citing such messages in public documents applies here as well.

Warning 2. Students have only a highly limited right to copy or print out their classmates' work; namely, *(a)* to cite as part of a course assignment, or *(b)* to complete a specific collaborative assignment given by the instructor (such as reading and commenting on classmates' work at home). Students should destroy copies of classmates' work once they have completed either of these tasks. In addition, they should not copy or print out classmates' work for any reason other than the two indicated here; nor should they retain archive copies of their classmates' work for any reason. Furthermore, under no circumstances do students have the right to circulate or otherwise distribute classmates' work.

Warning 3. Although some copies of student writing have always been available to other students (sometimes hidden in file cabinets or through commercial term-paper services), electronic storage and distribution of papers makes plagiarism much easier to accomplish. Networking makes student essays immediately available to other students within any one class, and there are many student papers available on the Web; and in either place, these other student essays can be electronically copied into one's own paper with only a few keystrokes.

The best advice here is simple: If you are not going to quote from another student's work or otherwise refer to it when writing your own, then there is absolutely no reason for you to have it open before you in the first place. This warning also applies to work that you may have written yourself for another course.

If for whatever reason you do refer to other students' work (or your own previous work) without specific instructions from your teacher, you should restrict yourself to reading through such work for broad, comparative purposes. In general you should not take notes or make a printout of such work. And if you do take notes, make a printout, or otherwise use a student essay in preparing your own work, a standard citation of that work is not enough: In addition to that citation, you need to submit copies of your notes or of the original material to your current teacher along with your final work.

INDEX

Copyright Credits

Sarah Anderson, "Wal-Mart's War on Main Street," from *The Progressive,* November 1994. Reprinted by permission from *The Progressive,* 409 E. Main Street, Madison, WI 53703.

Tim Appelo, "The Future Isn't What It Used to Be," from the *Los Angeles Times,* Home Edition, Jan. 4, 1998. Copyright © The Times Mirror Company. Reprinted by permission.

John Perry Barlow, "Is There a There in Cyberspace?" from the *Utne Reader,* March–April 1995. Reprinted by kind permission of the author.

Reva Basch, "Rapture of the Net." Reprinted from *Equip (Computer Life),* November 1996. Copyright © 1996 ZD, Inc.

Stephen Bates, "The Death of Geography." Reprinted from the March/April 1996 issue of *The American Enterprise,* a Washington-based magazine of politics, business, and culture.

Tracy Baxter, "Into the Outdoors," *Sierra,* Nov.–Dec. 1997. Reprinted with permission of the author.

G. Beato, "Girl Games," *Wired,* April 1997. Copyright © 1998 by The Condé Nast Publications Inc. All rights reserved. Reprinted by permission.

Jack Beatty, "What to Do about Welfare?" An Atlantic Unbound Executive Decision. July 15–July 30, 1996. Reprinted with permission of the author.

Wendell Berry, "Feminism, the Body, and the Machine." Excerpted and abridged from *What Are People For?* Used by permission of Farrar, Straus & Giroux.

David Boaz, "Creating a Framework for Utopia." Originally appeared in the November–December 1989 issue of *The Futurist.* Used with permission from the World Future Society, 7910 Woodmont Avenue, Suite 450, Bethesda, MD 20814. 301/656–8274. http://www.wfs.org.

Paulina Borsook, "Cyberselfish," from *Mother Jones,* July–August 1996. Reprinted with permission from *Mother Jones* magazine, © 1996, Foundation for National Progress.

Eric Brende, "Technology Amish Style," *Technology Review,* Feb/Mar 1996. Reprinted with permission from *Technology Review,* published by the Association of Alumni and Alumnae of MIT, copyright 1998. Reproduced by permission of the publisher via the Copyright Clearance Center, Inc.

John Briggs, "The Promise of Virtual Reality." Originally appeared in the September–October 1996 issue of *The Futurist.* Used with permission from the World Future Society, 7910 Woodmont Avenue, Suite 450, Bethesda, MD 20814. 301/656–8274. http://www.wfs.org.

George Carlin, "A Place for Your Stuff." From *Brain Droppings* by George Carlin. Copyright © 1997, Comedy Concepts, Inc. Reprinted with permission by Hyperion.

Allan Carlson, "Two Cheers for the Suburbs." Reprinted from the November/December 1996 issue of *The American Enterprise,* a Washington-based magazine of politics, business, and culture.

Denise Caruso, "Geeksploitation." Reprinted from *The Site* (ZD Net) November 1996. Copyright © 1996 ZD, Inc.

Arthur C. Clarke, "The Coming 'Cyberclysm.' " From *Asiaweek,* October 4, 1996. Reprinted by permission of the author and the author's agents, Scovil Chichak Galen Literary Agency, Inc.

Douglas Coupland, "Toys That Bind," *New Republic,* June 6, 1994. Reprinted by permission of the publisher.

Kathe Davis, "What about Us Grils?" from *Extrapolation,* Fall 1995. Copyright © 1995 Kent State University Press. Reprinted by permission.

"Dip into the Future, Far as Cyborg Eye Can See: and Wince," *The Economist,* Jan. 3, 1998. Copyright © 1998 *The Economist* Newspaper Group, Inc. Reprinted with permission. Further reproduction prohibited.

Stephen Doheny-Farina, "The Glorious Revolution of 1971," from *CMC Magazine,* October 1, 1995. Copyright © 1995 by Stephen Doheny-Farina. All rights reserved. Reprinted by permission.

Alan Thein Durning, "Seven Sustainable Wonders," *USA Today* (magazine), July 1995. Copyright © 1995 by the Society for the Advancement of Education.

Esther Dyson, "Education and Jobs in the Digital World," from *Communications of ACM,* February 1997. Reprinted by permission of the author.

Gerald Early, "Dreaming of a Black Christmas." Reprinted and abridged from *Harper's Magazine,* January 1997. Reprinted with permission of the author.

Bill Gates, "The Information Age." Excerpted from *The Road Ahead.* Copyright © 1995 by William H. Gates III. Used by permission of Viking Penguin, a division of Penguin Putnam Inc.

Ann Gibbons, Constance Holden, and Jocelyn Kaiser, "Facing the Big Chill in Science," *Science,* March 29, 1996. Copyright, American Association for the Advancement of Science.

Newt Gingrich, "Toward an Opportunity Society." Originally appeared in the July–August 1995 issue of *The Futurist.* Used with permission from the World Future Society, 7910 Woodmont Avenue, Suite 450, Bethesda, MD 20814. 301/656–8274. http://www.wfs.org.

Rube Goldberg, "Keeping Screen Doors Closed," reprinted by permission of Rube Goldberg, Inc. Rube Goldberg ™ and © of Rube Goldberg, Inc. Distributed by United Media.

Amy Harmon, "Virtual Sex, Lies and Cyberspace," from the *Los Angeles Times*, Apr. 10, 1997. Copyright © The Times Mirror Company. Reprinted by permission.

Arlie Russell Hochschild, "Time in the Balance." Reprinted with permission from the May 26, 1997 issue of *The Nation* magazine.

"Home Alone? Home Shopping," *The Economist,* Oct. 12, 1996. Copyright © 1996 *The Economist* Newspaper Group, Inc. Reprinted with permission. Further reproduction prohibited.

bell hooks, "Playing with Dark-Skinned Dolls," from *Bone Black: Memories of Girlhood,* © 1996 by Gloria Watkins. Reprinted by permission of Henry Holt and Company, Inc.

Ann Hulbert, "Bedtime Barbie," *New Republic,* January 9, 1995. Reprinted by permission of the publisher.

David Ignatius, "One World, Ready or Not," *Washington Monthly,* April 1997. Copyright 1997 by The Washington Monthly Company, 1611 Connecticut Avenue, NW, Washington, DC 20009. Reprinted with permission.

Molly Ivins, "Gimme Stuff" from *The Progressive,* July 1, 1994. Copyright © Molly Ivins. Reprinted with permission of the author.

Donald Katz, "The Sports Arena in the Digital Age," originally published in *Sports Illustrated,* July 3, 1995. Reprinted by permission of the author.

Jon Katz, "Birth of a Digital Nation," from *Wired,* April 1997. Reprinted by permission of Sterling Lord Literistic, Inc. © 1997 by Jon Katz.

Alan Kay, "Revealing the Elephant," *Educom Review,* July 1, 1996. Reprinted with permission of the author.

Daniel Kemmis, "Living Next to One Another." Reprinted from *Parabola,* Winter 1993, by permission of the author.

Derrick de Kerckhove, "From Global Village to Global Mind," originally appeared in *UNESCO Courier,* February 1995. Reprinted with permission of the author.

Mark Kingwell, "Not Available in Stores," from *Saturday Night,* July–August 1996. Copyright © 1996 Saturday Night Publishing (Canada). Reprinted by permission.

Jeffrey Kluger, "Rubes in Training," *Discover,* August 1993. Reprinted with permission of *Discover* magazine.

Jeffrey R. Krull, "The Future of Books and Libraries." Paper delivered at the Quest Club, Fort Wayne, Indiana, March 1996. Used with permission of the author.

Richard A. Lanham, "Undergraduate Teaching in the Electronic Age," Coalition for Networked Information. Reprinted by kind permission of the author: Richard A. Lanham, Professor Emeritus of English, UCLA; President, Rhetorica, Inc.

Jackson Lears, "The Simple Life." Reprinted with permission from the March 9, 1985, issue of *The Nation* magazine.

John Leo, "No Books, Please; We're Students," Copyright © *U.S. News & World Report,* September 16, 1996.

Ralph Lombreglia, "In Games Begin Responsibilities." Reprinted from the Atlantic Unbound, December 21, 1996. Reprinted with permission of the author.

Laurie Ann Mazur, "Marketing Madness." Reprinted with permission from *E/The Environmental Magazine;* Subscription Department: Box 2047, Marion OH 43306.

Pamela McCorduck and Karen Coyle, "Women and the Sexed Machine," Brain Tennis, Hotwired, Aug. 26–Sept. 5, 1996. Copyright © 1994–97 Wired Digital, Inc. All rights reserved.

"The Medium and the Message: An Exchange between Sven Birkerts and Wen Stephenson." Originally published in the *Chicago Review,* Summer 1996. Reprinted with permission of the *Chicago Review.*

Susan Mitchell, "Technophiles and Technophobes," *American Demographics,* February 1994. Reprinted with permission.